THE MINERS OF WINDBER

MILDRED ALLEN BEIK

THE MINERS OF WINDBER

THE STRUGGLES OF NEW IMMIGRANTS FOR UNIONIZATION 1890s–1930s

The Pennsylvania State University Press
University Park, Pennsylvania

Library of Congress Cataloging-in-Publication Data

Beik, Mildred A.
The miners of Windber : the struggles of new immigrants for unionization, 1890s–1930s / Mildred Allen Beik.
p. cm.
Includes bibliographical references and index.
ISBN 0-271-01566-7 (cloth)
ISBN 0-271-01567-5 (pbk)
1. Coal miners—Pennsylvania—Windber—History. 2. Immigrants—Employment—Pennsylvania—Windber—History. 3. Trade-unions—Coal miners—Pennsylvania—Windber—History. I. Title.
HD8039.M62U6143 1996
331.88'122334'0974877—dc20 95–47701
CIP

Second printing, 1997

Printed in the United States of America
Published by The Pennsylvania State University Press,
University Park, PA 16802-1003

It is the policy of The Pennsylvania State University Press to use acid-free paper for the first printing of all clothbound books. Publications on uncoated stock satisfy the minimum requirements of American National Standard for Information Sciences—Permanence of Paper for Printed Library Materials, ANSI Z39.48–1992.

In Memory of My Parents
and
Other Windber-Area Mining Families

Contents

List of Illustrations

List of Tables

ACKNOWLEDGMENTS

I used to have a fantasy that I would someday write a novel based on the immigrants and other characters I had known when I was growing up in Mine No. 40, Windber, Pennsylvania. Eventually I replaced the fantasy with the idea of doing a scholarly research project. The change stemmed from an informal conversation I had at Northern Illinois University in the mid-1980s. At a time when I was debating whether to leave graduate school, a friend, Stephen Kern, casually suggested that I write my Ph.D. history dissertation on "that town I was always talking about" and about which I never failed to convey a passionate enthusiasm. Until then, the idea of doing serious research on my hometown and on the coal-mining families I had grown up with had never consciously occurred to me. Nor could I have possibly imagined what rewarding experiences I would have as a result of my decision to take up the suggestion.

Many people and various institutions have contributed to the completion of this work. I am especially grateful to John Bodnar and Irwin Marcus, who read the final manuscript in its entirety and made many valuable suggestions. At an early stage of my dissertation project, Bodnar had generously shared his knowledge of sources and offered encouragement. In recent years, I have benefited greatly from Marcus' advice, friendship, support, and unique scholarly expertise, acquired from his years of research and teaching at Indiana University of Pennsylvania.

I would also like to thank John H. M. Laslett, whom I first met in Bochum, Germany, in 1989 at the International Mining History Congress, for his generous ongoing interest in my work and for his suggestion to Peter Potter that Penn State Press consider my manuscript. As editor, Potter has greatly facilitated the process. He, Peggy Hoover, and others at the press have made the final product better.

I have gained new insights and perspectives on my work in recent years from my participation in a number of interactive community presentations, regional workshops, and statewide seminars on coal, ethnicity, and the labor movement. Among these invaluable experiences were the annual Oral History and Visual Ethnography Summer Institutes and Field Schools, hosted by Indiana University of Pennsylvania

(IUP) from 1992 to 1994; a Pennsylvania History Teaching Institute, organized by IUP's Department of History in 1993; two community presentations in Windber in 1992 and 1993, sponsored by the Folklife Division of the America's Industrial Heritage Project (AIHP); and the "Interpreting Pennsylvania's Industrial Heritage" Seminar, sponsored by the National Endowment for the Humanities and the Pennsylvania Federation of Museums and Historical Organizations (PFMHO) in 1994. I owe a special debt of gratitude to Gary Bailey; Elizabeth Cocke; Eileen Cooper; Jim Dougherty; Jim Harris; John Larner at IUP; Jim Abrams and Kathy Kimiecik, formerly with AIHP; Linda Shopes of the Pennsylvania Historical and Museum Commission; and Jean Cutler of the PFMHO.

Of all the people in the History Department at Northern Illinois University, I am most grateful to J. Carroll Moody, director of my dissertation. At that stage of research and writing, he provided critical support, posed thoughtful questions, and edited the lengthy doctoral treatise. Harvey Smith, Joseph Parot, and Jeffrey Mirel read that version and made many thoughtful suggestions. Otto Olsen and Glen Gildemeister provided additional assistance during the research process.

The Immigration History Research Center of the University of Minnesota provided an Immigration History Research Grant that enabled me to consult its valuable ethnic collections during the spring of 1986. Joel Wurl, Susan Grigg, and other staff members there helped locate sources related to my topic and facilitated scholarly interactions with a number of immigration historians from around the world. I would especially like to thank Director Rudolph Vecoli for his ongoing and enthusiastic support of my project.

Two other financial grants indirectly or directly aided my work. A fellowship from the Council on International Educational Exchange allowed me to pursue my study of the Russian language at Leningrad State University in the summer of 1981. Dissertation Completion Fellowships from the Graduate School of Northern Illinois University for the summer of 1984 and the 1986–87 academic year provided the time and material support needed for completion.

In the Windber area, there are many people to thank. No one has been more helpful over the long run than Thressa Ledney—the catalyst and mainstay of the Windber Museum for many years. From the very beginning on, she took time to call my attention to the town's rare newspapers, photographs, other materials, and she made many useful suggestions. At an early stage, Bruce Williams of the University of Pittsburgh at Johnstown also suggested local bibliographic resources and shared his knowledge of oral history.

A number of parish priests and church officials assisted my effort to obtain oral interviews with elderly people and permitted me to use their parish records. Among these were the Rev. Father Sylvester Bendzella of SS. Cyril and Methodius Church, the Rev. Father Francis E. Luddy of St. Mary's Hungarian Roman Catholic Church, the Rev. Father William M. Wojciechowski of St. Mary Byzantine Rite Catholic Church, and the Rev. Father Stanley Zabrucki of St. John Cantius Church. In addition, the Most Reverend James J. Hogan, D.D., J.C.D., former Bishop of the Altoona-Johnstown Diocese, offered valuable comparative information. Ed Surkosky generously shared his personal collection of local parish histories, religious materials, and photographs.

Other Windber citizens provided access to important records. Joe Elias, a Slovak Club officer, permitted me to use the town's various Slovak fraternal records; George Marcinko, Borough Manager, facilitated use of the town's public records; and Robert Barrett of the Berwind Corporation enabled me to consult Berwind-White's old employment records.

The late Joe Zahurak, union activist and long-term president of Local 6186, United Mine Workers of America, generously made recommendations for the project and provided me with Local 6186's existing records. Unfortunately, Local 5229's records from Scalp Level had been destroyed in a fire. The late Paul Gormish and other officers at the District 2 headquarters of the United Mine Workers at Ebensburg answered my many questions and granted me access to District 2's historical records, which have since been transferred to the Special Collections Department of the University Library at Indiana University of Pennsylvania.

A host of unnamed archivists and librarians at many institutions assisted this project by finding sources and providing services. I am especially grateful to the interlibrary loan staff of Northern Illinois University for their willingness to go the extra mile to locate and obtain rare documents.

There are many others from all walks of life to thank. It is sad to think that so few of those individuals who granted me the privilege of interviewing them are alive today, but I hope that those who are and the relatives and friends of all the interviewees note their invaluable contribution. I would also like to thank all those people who understood that working people are worthy subjects of historical study and who encouraged me to complete this book. Among them are John Brennan, a founder of the Pennsylvania Labor History Society, Rose Czajkowski, the late Elizabeth Dutzman, Janet Greene, Katy Liska, Ted Liska, and the late Jolan Dutzman Steele.

There are several individuals whose contribution to this book stand out in a special way. From the initial decision to undertake this project, to its ultimate completion, they have provided needed moral support, intellectual stimulation, and concrete practical assistance. A good friend, Jitka Hurych, not only helped in moments of discouragement but also shared her remarkable linguistic knowledge with me whenever I ran into problems of translation. My husband, Bill Beik, was a sounding board for many of my ideas and contributed valuable perspectives that stemmed from his own work. He also read the manuscript, made editorial suggestions, and provided welcome computer expertise and emotional support. In addition to furnishing me with a home away from home, my aunt and uncle, Bertha and Pete Gerula, made efforts to locate suitable subjects to interview and otherwise shared their personal experiences and unique knowledge of the Windber community. Judy Gerula Dietz and other family members assisted the project in multiple ways throughout its duration.

I have been fortunate to have had the benefit of so much assistance from so many different men and women in the mining and academic worlds. However, it should be clear that no institution or person other than the author bears responsibility for the views and interpretations presented here.

In a general sense, the experiences of working people and their ongoing quest for justice inspired this study. My grandparents, Andrew and Priscilla Jelinek (later Allen) emigrated from Hungary to Windber early in the century, and in 1914 my own father, John Allen, followed in *his* father's footsteps and became a miner in Eureka Mine 35 at the age of eleven. Family lore has bequeathed a rich personal legacy of the Jelinek-Allen family's experiences of immigrant coal-mining life, company-town oppression, and their own participation in many of the strikes and struggles described in this book.

By contrast, my mother, Mildred Kuhlman, was not from a coal-mining family but became a coal miner's wife by choice during the Great Depression. Her original class and ethnic background differed markedly from my father's. This descendant of an American-born nineteenth-century resident family of Somerset County farmers and schoolteachers, married my father at a time when cultural and class differences were great and when Pennsylvania state law forced her (and all married women) to give up a promising career as a schoolteacher. After their marriage, my parents moved into a company house in Mine 36, and my mother subsequently shared in all the hopes, trials, tribulations, and struggles of a rank-and-file coal miner's wife.

In 1954, after nearly forty years in the mines, my father was permanently laid off. The Berwind mines were closing. With the help of relatives, we moved from Mine 40 to Illinois in 1956, just before I was to enter high school. All members of the family subsequently experienced the trauma of deindustrialization, culture, and class shock. As such, my family's story is typical of that of many immigrant miners and families portrayed in this book. It is my humble hope that I have been able to transmit something of the courageous valuable legacy of these and other working people to my own children, Eric and Carl; to the many other descendants of Windber's mining families; and to present and future generations who have never experienced these particular struggles for social justice or shared in a way of life that has now largely disappeared.

Introduction

The workers who are the subject of this local community study lived, worked, and struggled within the context of a triumphant system of industrial capitalism that transcended national boundaries and increasingly operated on a worldwide scale. The economic imperatives of industrial capitalism—the relentless drive for profits, on the one hand, and the need to earn a livelihood for survival, on the other hand—linked together two unequal, autonomous historical forces that came together by design in the hills and valleys of Central Pennsylvania at the turn of the century. A large modern coal corporation had recruited an ethnically diverse working-class population for its new mines and founded a new company town for them to live in.

From their initial encounters with one another in Windber, Pennsylvania, in the late 1890s, the Berwind-White Coal Mining Company and this diverse working class coexisted in an uneasy relationship characterized by the incessant striving of each to maintain or change its relative position vis-à-vis the other. The company had established an autocratic type of control at the workplace and in the community at the town's founding. Its monopoly of jobs, land, and housing was rapidly supplemented by a compulsory company-store system and control of basic utilities. Outside competition, which would have led to a diversification of industry, was as excluded from the town as were union organizers who sought to provide the masses of miners with a democratic collective voice for their grievances. Berwind-White had vowed to run its mines on a perpetual open-shop basis, and it enforced its policy through discharges, blacklists, an espionage and police system, and control of local organizations and halls. Its political domination of the borough guaranteed its position, and when that position was challenged it could rely on state authorities to send troops or other forms of assistance. The broad struggle of the miners and their families to end this autocratic control and thereby achieve greater control over their lives and work is the raison d'être of this study.

There are many important reasons for this study's focus on the community and mining population of Windber. First, the town itself was of considerable industrial importance in an era when coal was king of the energy world. Indeed, Windber owes its existence to—and is a product of—an era of industrialization in the United States when coal, iron, and steel were the basis of developing industry. Founded by Berwind-White in 1897, Windber rapidly became the largest metropolis in Somerset County, and the most important of a number of towns owned or controlled by a corporation that was later reputed to be the world's largest independent producer of bituminous coal and whose president, E. J. Berwind, was ranked by one author as among the top thirty men who ran America.[1]

At the same time, coal furnished 90 percent of the nation's energy in 1900, when the United States replaced Great Britain for the first time as the world's leading coal producer.[2] Pennsylvania, second in population only to the state of New York, was itself an industrial giant, leading all states in its output of coal, iron, and steel.[3] The newly created borough of Windber was already producing bituminous coal in earnest and continued to do so, while reflecting the fluctuations of the industry and larger economy, throughout the scope of this work and beyond. In its heyday, from 1900 through the 1920s, Windber lay well within the mainstream of American industrial life.

Second, Windber's history has to a large extent paralleled that of industrial America itself. As such, the story of its origins, rise, and decline is not unlike the stories of other industrial towns in important respects. In the past, one-industry mining towns have been notorious for coming into being and passing out of existence in the blink of an eye, or, as the supply of the particular natural resource was successively discovered, utilized, and depleted. In the 1950s and 1960s the Windber-area mines closed down permanently, thereby posing a serious challenge to the town's continued existence and uprooting thousands of former miners and their families. In the twentieth century the people of Windber have had to confront many of the problems inherent in both capitalist industrialization and capitalist deindustrialization.

Today, American industry is itself threatened, not only by foreign competition, multinational corporate ownership, and corporate runaways to other countries, but also by a marked shift in the domestic economy away from the old industrial base to a new postindustrial one dominated by high technology, computers, and service-oriented industries. In this context it is far too easy for the uninformed or noncaring public to forget its own history and to relegate industrial workers and industrial towns to the "dustbin of history."

Third, the population that came to reside within the Windber area was remarkable for its great ethnic diversity. Because of the town's plentiful immigrant groups and industrial importance, the Dillingham Immigration Commission chose Windber as one of two Pennsylvania bituminous towns to use as a case study for its "Immigrants in Industries" series, published in 1911. At least twenty-five different nationalities were represented in a population that totaled about 10,000.[4] Immigrants from southern and eastern Europe, who comprised the bulk of the local working class, were part of the wave of "new" immigrants who came to America in great numbers from 1890 until legislation of the 1920s severely restricted the number of newcomers from these areas of Europe. Windber's rise and prominence as an industrial center coincided with the zenith of immigration from southern and eastern Europe on which it depended for its labor force.

The enormous diversity of Windber's ethnic population makes it a veritable laboratory for analyzing ethnic America during the expansion of industrial capitalism in the early part of this century. Its ethnic composition after 1900 was an almost perfect reflection of the "new" immigration itself, and similar groups of transplanted southern and eastern Europeans simultaneously settled and worked in other industrial towns or in large cities, such as New York, Newark, Cleveland, Pittsburgh. Some transients even migrated back and forth between such places and Windber. The vast majority of these immigrants became industrial workers. By 1920, foreign-born workers and their children were doing most of the nation's work, especially in key industries, such as coal, steel, clothing, construction, and manufacturing.[5]

Fourth, the autocratic structure of this company town demands examination and reminds us that neither political nor economic democracy necessarily accompanied the development of industrial capitalism in the United States. Autocratic company towns were not the only type of industrial town in existence at the turn of the century, but the Pullmans of industrial America were not uncommon either. Company towns—advertised as "model" towns by their owners—were particularly characteristic of the mining industry and far from an anomaly in the steel and textile industries. As late as 1937, when the labor movement / New Deal coalition in the state of Pennsylvania secured passage of an important series of company-town-related laws designed to bring democracy where none had existed before, the Commonwealth of Pennsylvania counted 1,200 such towns within its borders. Windber was one.[6]

Windber was no exception to the general pattern in which unabashed mine owners and operators customarily named their newly founded company towns after themselves or after close relatives.[7] The town's name accurately conveys some sense of the autocratic and paternalistic nature of its social structure. Transposed, the two syllables of the originating company's family surname, "Berwind," simply became "Windber." As a name, "Berwind" had been preempted, so to speak, in 1897, when Windber was founded. Berwind, Colorado, immediately adjacent to Ludlow, which subsequently became one of the most significant and infamous sites in American labor history, already existed. The western town had derived its name from that of the Berwind-White president in an earlier era, when he had been associated with a predecessor company of Colorado Fuel and Iron.[8]

Even older, medium-size cities with preindustrial histories sometimes became subject to a company-town type of control when one-industry companies monopolized resources, jobs, and municipal authority. In a recent monograph on one such city, for example, sociologist Ewa Morawska has described Johnstown, Pennsylvania, a neighbor of Windber, as "essentially autocratic."[9] In such settings as these, local working-class populations faced the perennial and largely undisguised problems of paternalism and control from the very start and had to devise strategies for contending with the fundamentally undemocratic forces that impinged on every sphere of their lives.

Fifth, Windber-area miners and their families were active participants in the historic and sometimes bitter labor struggles which took place in industrial America. While the Ludlows, Matewans, and Lattimers are better known instances of class war, Windber has a history of militancy and struggle that deserves to be better known. Ethnically-diverse miners and their families overcame many obstacles to challenge the numerous autocratic structures they encountered in their daily lives and, on several occasions, fought fierce battles for their right to join the United Mine Workers of America—the one democratic class-based alternative available to them.

Scholars would be hard put to find any firmer, more loyal, more popular base of support for the New Deal than that which existed among Windber-area miners. Until the 1930s Windber had had a history of "lost" struggles and strikes, along with some victories. Three strikes—in 1906, 1922, and 1927—were technically lost; two of these, in 1906 and 1922, were especially protracted, bitter, and violent, leading to the occupation of the town by the state police. Despite this militancy, union recognition and the right of collective bargaining were not formally won until 1933, when Windber miners, like American workers elsewhere, eagerly seized on section 7(a) of the National Industrial Recovery Act. Heart and soul, area miners backed Franklin Delano Roosevelt and the New Deal from the start, and enthusiastically continued to do so in subsequent years.

How and why unionization eventually came to Windber is important to understand because labor historians have sometimes neglected the foreign-born, especially southern and eastern Europeans, and at other times unfairly characterized them as inherently poor union material.[10] Historians making such judgments have tended to rely extensively, if not exclusively, on corporate sources or English-language, even nativist, materials that assumed this characterization as verity. This mistaken view is also bolstered by an uncritical and superficial but accurate observation that unionization was not successful in many such places until the New Deal.[11] A history that considers ultimate victors and victories to be the only subjects worthy of study is distorted history at best. It has the effect of dismissing as unimportant the lives of the vast majority of working people and the alternatives realistically available to them in given moments of time.

Windber calls such characterizations and judgments into question. Its history suggests that peoples and industries and geographical areas that have traditionally been dismissed as nonunion need much more scrutiny and critical thought than they have received in the past. Any such reexamination must be based on methodologies that focus on the workers themselves, their circumstances, and their particular cultural and linguistic sources. Ethnic groups and communities are not the monolithic entities that outsiders have frequently envisioned them to be. New questions, not old irrelevant ones, such as why was there no socialism in America, need to be raised to understand the complexity and diversity that did exist within the American labor movement.

The significance of Windber's labor struggles lies beyond the local community level in other important ways. The most populous town of the nonunion stronghold of Somerset County in 1906 and 1922, Windber was key in the struggle of previously unorganized or nonunion miners to win union (United Mine Workers of America) recognition. What happened there during these strikes and the events and personalities surrounding them became a source of considerable controversy within the labor movement itself. Union policies in the strike of 1922 were in part responsible for the development of a radical rank-and-file miners' movement that aimed at reform, first, from within the union and, then, having failed in that, from without. Moreover, if historian David Montgomery is even remotely correct in his assessment that the strikes of 1922 marked the end of the nation's most important strike wave, which lasted from 1916 to 1922, and the defeat of a broad-based, working-class postwar drive over wage issues and "worker control"—a turning point in American history—then Windber's importance and historical role may be even more impressive.[12] Certainly its history deserves our attention.

This study spans the entire nonunion period of Windber's history, from its inception in the 1890s to the New Deal era of the 1930s. The fundamental analytical concept employed is "class." Capitalism is usually defined as an economic system characterized by private property, the profit motive, and wage labor. Capitalistic social relations and classes are therefore determined by who owns the means of production and who does not, who decides how wealth—and power—are distributed and who does not. As such, "class" is an objective category that is predetermined by the economic system itself and that exists regardless of subjective opinions or whether a class is class-conscious or not.

This definition does not imply that class-consciousness is unimportant or that working people are not active participants in the making of their own lives or history. Quite the contrary. Class-conscious workers have historically behaved differently from other workers. Nor is the stress on the objective economic underpinnings of class formation meant to suggest that other categories of analysis—gender, race, ethnicity—should be ignored or minimized. The real experiences of diverse working people in industrial capitalist societies can be understood only by viewing class through broad lenses and by focusing on the historical intersections of economics, power, politics, culture, human agency, and consciousness in its many forms.

Thus, this work attempts to combine the best insights from the "old" and the "new" labor histories. Power comes in many forms. To interpret industrial history, we need to look at both the dominant power structure and the much less powerful classes and groups in society. We cannot afford to dismiss economics, culture, or politics. The old labor history is correct in that working-class institutions, organized labor, the role of the state, and strikes *are* important. These were arenas in which the major contests between capital and labor were fought out on a large scale. But power also existed and was contested on other terrains—the realms of everyday life and culture, for example. The not-so-new "new" labor histories have centered attention on the very groups—women, African-Americans and other minorities, the unskilled, and immigrants—who once received little or no attention. By exploring new categories of analysis—race, gender, and ethnicity—and their intersections with class, social historians have greatly expanded our knowledge and our conceptions of labor history. They have given us additional perspectives from which to examine the world of the worker, and of power relationships, on various levels.[13]

Unequal power relationships, exploitation, and the development or muting of class-consciousness are therefore important underlying problems that are examined throughout this study. If, on the local level, a coal corporation could succeed in establishing the company, ethnic leaders, the church, or the town as the primary claimant for worker loyalty, instead of the working class, class-consciousness was inevitably muted and the behavior and actions of workers took a correspondingly pacific course. If, on the other hand, a coal company failed in this endeavor, if workers understood at least something of their class position, they could then perhaps overcome internal divisions and use their diverse cultural backgrounds as a resource for either reformist or radical goals and common action.

The residents of Windber and elsewhere in the United States in the early 1900s would not have been surprised by this use of the terms "class" and "capitalism," terms that were in widespread use at the time. Journalists, teachers, economists and others publicly touted the virtues—or criticized the failings—of capitalists and capitalism. At the same time, ever since Haymarket at least, the "labor" or "working-class" *problem* had been widely discussed in print in the United States.[14] Certainly intellectuals, captains of industry, the middle classes who formed the base for within-the-system progressive reforms, and socialist and radical critics who were seeking a fundamental change in the existing economic order, defined the labor "problem" differently and thus proposed differing and contradictory solutions. What all of these disparate interests and classes had in common, however, was their mutual preoccupation with an ongoing issue of the day: the position and power of the working class in an industrial capitalist society.

Because Windber's working class was predominantly of foreign-born origin, questions pertaining to immigration and ethnicity inevitably arise. Historians of immigration have rightly called attention to the premigration roots of immigrants to the United States as a means of understanding their subsequent history. Scholars on both sides of the Atlantic have debated the extent of the penetration of the market, of railroads, and of industrialization into the places of origin of southern and eastern Europeans. They have similarly disagreed about the degree of cultural persistence and change among the various populations throughout disparate regions on the eve of the twentieth century. While the exact rate of social and economic change in their native lands, and the precise character of their various premigration cultures and traditions, is debatable, scholars have increasingly agreed that change in the direction of industrial capitalism was occurring unevenly in the Old World as well as in the New. In a recent synthetic study, *The Transplanted,* John Bodnar argues persuasively that immigrants had already encountered the imperatives of capitalism in Europe and were facing the problem of finding survival strategies. Their initial decision to adapt, which did not necessarily imply acquiescence, originated in Europe and not in the United States.[15]

While the premigration history of Windber's immigrants is not unimportant, the focus of this work is nevertheless on their lives in the United States. What the vast majority of the many ethnic groups in Windber from the 1890s to the 1930s had in common was the experience of being industrial workers—usually miners—or members of miners' families. To be sure, disparate workers and the two sexes underwent the class experience in different ways, but to what extent did inequality and exploitation shape the character of their respective lives and struggles? What role did work, family, culture, the church, fraternal societies, and political and civic institutions play in the formation and subsequent development of this class? How and in what specific ways did such diverse ethnic workers overcome fragmentation as a class to challenge the dominant social structure at significant historical moments?

The central aim of this study is to explain how and why class-based unionization came to Windber. Parts One and Two are designed to be complementary contributions to the achievement of this goal. Part One, a social history of work and community, describes the many structural constraints within which working people lived, worked, and struggled. Among the broad issues examined are the company's labor, ethnic, and governing policies; the nature and composition of the social structure; the functioning of immigrant communities and institutions within the larger American community; the importance of the family economy in the mining town; and the class, religious, and occupational divisions that existed within the stratified ethnic communities themselves. Part One seeks to demystify the autocratic structures of a company town and highlight the difficulties that diverse working people encountered in achieving working-class integration.

If Part One focuses on structural constraints to protest and unionization, Part Two describes the many ways in which both foreign-born and American-born workers acted to try and remove objectionable constraints and establish democratic class-based alternatives in their place. The emphasis here is on the various efforts of class-conscious miners of disparate nationalities to unionize, from the 1890s to the 1930s. Worker aspirations, wages, working conditions, grievances, political conditions, and the lack of fundamental rights, which generated ongoing resentment, protests, and major strikes in 1906 and 1922, are documented. Considerable attention is given to the many external obstacles that they encountered—the company and all its powerful resources; hostile local, state, and federal governments; nativist prejudices and legislation; internal union politics—but internal local obstacles are not overlooked. American labor historians have too often treated immigrant communities as homogeneous units, but Windber illustrates that the class struggle was fought out internally in this arena as well.

Perhaps no local community study can be truly comprehensive. Practically speaking, even if it were desirable it would be impossible for any single researcher to thoroughly study every person, every ethnic group, every institution, and every organization in a community of any size or diversity, or to devote equal attention to every happening there. What is possible, and in fact demanded, of historical analysis is intelligent selectivity, which is always partially dictated by existing available sources but also partially the result of conscious choice.

Thus, five groups of southern and eastern European immigrants—Slovaks, Hungarians, Poles, Italians, and Carpatho-Russians—have received most of the attention in this work. These five ethnic groups constituted the vast majority of the town's working-class population. Furthermore, the availability of historical sources justified a greater concentration on these groups than on other ethnic nationalities. Probably the most readily usable information was that pertaining to the American-born or English-speaking sectors of the community, which, although a minority, were neither unimportant nor neglected in the course of research.

This study is based on a wide variety of primary and secondary sources. In addition to printed documents, written records used include church records, censuses, union files, Slovak fraternal society papers, borough council records, the papers of individual organizers, company employment records, and rare newspaper collections. Oral histories proved invaluable in the effort to ascertain how working people felt, thought, and acted. Among those interviewed for this project were thirty-five of the oldest living immigrants in the community.

What happened in this autocratic company town—that is, how and why class-based unionism came about—is an important, if somewhat neglected, chapter in American history. The pluralistic roots and traditions of Windber's working class are undeniably part of the story of diverse working people struggling to democratize the undemocratic America they knew. Their history suggests something of the possibilities and limitations, strengths and weaknesses, of worker protest in the early twentieth century in the United States.

PART ONE

Structure and Society

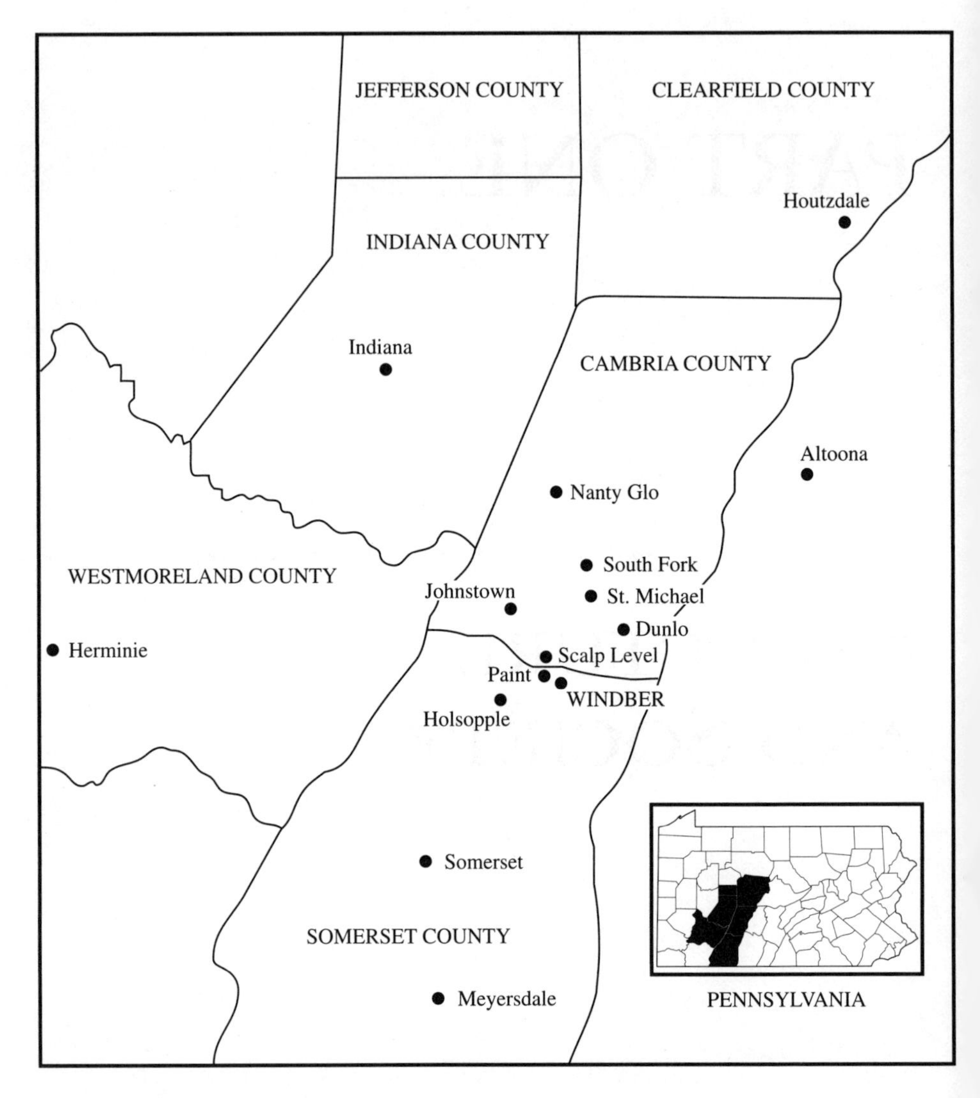

The broader setting of Windber, with towns mentioned in the text.

1

From Berwind to Windber

Neither the Berwind-White Coal Mining Company nor an immigrant population from southern and eastern Europe was present in Pennsylvania's Somerset County at the time of the 1890 census. In the decade between 1890 and 1900, however, the county experienced industrial development, urban growth, and changes in the social and ethnic composition of its population. This transformation of a predominantly rural region into an important industrial center marked its incorporation into a maturing system of international industrial capitalism. Windber, founded in 1897, was in large part responsible for this local transformation. But why was Windber founded, and what kind of a town was it?

The Setting

Somerset County was a picturesque setting for a new industrial town. Bounded by the Allegheny Mountains on its eastern edge and by the

Laurel Mountains on the western edge, it had often been noted for its beautiful hills and scenery. Artists from Pittsburgh had come regularly every summer since 1878 to a small country village on the northern border of the county in order to sketch the region's lush forests and pastoral landscapes. In an address marking the centennial celebration of its history in 1895, the Honorable William H. Koontz recalled that early pioneers, impressed by the abundance of natural meadows situated near the headwaters of many of its streams, had frequently and affectionately called Somerset County the "Glades."[1]

By 1890, industrial development had made few inroads into this rustic setting. It is significant that state historian Sylvester K. Stevens barely mentioned Somerset County in his scholarly account documenting the rise of Pennsylvania industry. Stressing the uneven nature of nineteenth-century economic development, he argued that rural sections of the state, as late as 1870 or beyond, often resembled the colonial era more than the emerging industrial one. In any event, the first railroad to enter county limits did not do so until 1873, and in 1890 most residents continued to make their livings as farmers, much as they and the nation had done throughout the century. As a town, Windber did not yet exist. Rolling hills, dense forests, and the extensive farmlands of David Shaffer of Paint Township would become the site of the future metropolitan coal center.[2]

The Berwind-White Coal Mining Company and immigrant workers ultimately met one another in a county that was inhabited by a predominantly native-born population. In fact, the percentage of native-born in Somerset County was much higher in 1890 than it was in the nation as a whole. Nearly 97 percent of its inhabitants had been born in the United States; the corresponding national figure was slightly more than 85 percent. While the county's population had increased from 28,226 in 1870 to 37,313 in 1890, the number of foreign-born had actually decreased during that period. Thus, by 1890, the county's foreign-born constituted only a little more than 3 percent of its total population, in the nation they constituted nearly 15 percent.[3]

If the relative balance between the native-born and foreign-born population in Somerset County differed significantly from that of the nation, the ethnic distribution of its foreign-born more closely mirrored it. In both cases, northern Europeans, headed by Germans, constituted the vast majority. On the national level, Germans alone accounted for more than half of the total 9,249,547 foreign-born. In Somerset County, where they had been settling throughout the nineteenth century, they made up 64.3 percent of the total. Nationally, the Irish were second in importance, followed by a third group of English, Scots, and Welsh. In Somerset County the position of these two groups was simply reversed, but immigrants from the British Isles, Scandinavia, and northern Europe nevertheless comprised 94.24 percent of its foreign-born.[4]

In neither the national nor the local instance did those who were considered part of the new immigration from southern and eastern Europe constitute more than a mere fraction of the total foreign-born population. In the nation as a whole, no southern or eastern European nationality group numbered more than 100,000 or 200,000 in 1890, and the sum total of all such groups in Somerset County was only 68, or 5.59 percent of the total foreign-born. Italians, who comprised 4.7 percent of the county's immigrants, formed the only sizable cluster in the scattering of these nationalities.[5]

Thus, Somerset County's population was predominantly white and native-born. It was also, not surprisingly, Protestant. Even as late as 1906, when the influx of new immigrants into the area was sizable, a historian counted only six Roman Catholic congregations but a multitude of older, well-established Protestant ones representing many different denominations. Lutheran bodies alone numbered fifty-five, and there were thirty-eight Reformed congregations.[6]

In politics the county traditionally favored the Republican party, which had easily carried its presidential and gubernatorial candidates to victory in every election since 1860. For example, in the critical realignment contest in 1896, William McKinley outpolled William Jennings Bryan, 15,861 votes to 2,295. Political contests in the late nineteenth and early twentieth centuries were more likely to occur between rival local factions of the Republican party than between Republicans and Democrats, and at the turn of the century the Prohibition party was the most powerful of a number of minority parties. Temperance and Sabbatarianism were issues that many Protestants took seriously. Anti-Saloon League and Women's Christian Temperance Union organizations were growing in numbers, size, and influence, reflecting a nationwide movement in the direction of prohibition.[7]

It was in this setting, on July 4, 1895, that residents gathered in Somerset, Pennsylvania, to proudly celebrate the county's centennial amid predictions that the next one hundred years of its history would greatly outshine the prosperity of the previous century. The basis for such optimism was the value of the vast deposits of bituminous coal that geologists had reported underlay much of the Somerset and Cambria county region. Although Somerset County had, in terms of development, historically lagged behind other areas endowed with similar riches, change was imminent. The editor of the *Somerset County Democrat* proclaimed: "Her [Somerset County's] vast fields of coal are only commencing to be developed; in the next ten years the development will be in full blast and that means prosperity." He then lauded the belated arrival of railroads into the mountainous area: "Her desirable territory will be crossed and recrossed by bands of steel on which the products of her soil, the wealth of her mountains, and the results of the industry of her people can be carried to a market, and this means greater prosperity." The editor concluded on this bright note: "In fact Somerset County in the next decade is destined to advance to the proud position of one of the leading counties in the state, a position which belongs to her by right of her natural resources and by reason of her hardy, progressive, intelligent and capable class of citizens."[8]

These optimistic predictions of imminent industrial development of the county turned out to be substantially correct. Somerset County came into its own just as the state of Pennsylvania and the nation were concluding a period of historical development that has often been dubbed "the golden age of industry." In 1900, Pennsylvania ranked second only to New York in terms of its industry and its population. By any standard—such as the percentage of wage earners in manufacturing or the value of manufactured products in the state—Pennsylvania attained its greatest relative importance in industrial production vis-à-vis the nation during the period 1870–1900. While still important for decades afterward, the share of its contribution to national industrial productivity declined after 1900, as did that of most other older, established states in the eastern United States.[9]

The great industrial growth of Pennsylvania and the United States in the late nineteenth century was in large part based on coal. By 1900 the two different types—anthracite and bituminous—were supplying approximately 90 percent of the nation's energy, with soft coal furnishing more than 70 percent of that amount and hard coal about 30 percent. Post–Civil War railroad development, combined with technological developments in the iron and steel industries, such as the open-hearth furnace, had produced a great demand for soft coal. Approximately half of Pennsylvania's bituminous production from 1880 to 1900 went into the making of Pennsylvania coke, which was subsequently used in the new blast furnaces of the Pennsylvania steel industry, which in turn came into historical importance in its own right.[10]

The Relative Significance of Windber-Area Production

Throughout most of the twentieth century, Somerset County has been among the top six or seven bituminous-producing counties in Pennsylvania, which led all other states in such production until 1927, when West Virginia surpassed it. According to a report published by the Pennsylvania Geological Survey in 1928, Somerset County ranked fifth in original bituminous deposits in the state, and Cambria County ranked sixth. Moreover, the amount of coal actually mined bore a close relationship to these rankings. Only the counties of Fayette, Greene, Washington, Westmoreland, and Allegheny, all of which are located in southwestern Pennsylvania, historically produced more.[11]

Coal mining did not become an industry of any importance in Somerset County until the 1880s, but after 1897, when Berwind-White opened the first of its thirteen mines, its growth was spectacular. By 1903, local newspapers were citing geological statistics, commenting on the previous five years of growth of Somerset's coal industry, and attributing the recent rise in county production solely to the new Berwind-White mines.[12]

Windber-area mines continued to be the mainstay of the county's coal industry in subsequent years. Peak production was reached during the years 1910 to 1913, when nearly 4 million tons were produced annually. The zenith of all Somerset County production was attained somewhat later, in 1920, while bituminous production in the state reached its high mark in 1918. Nonetheless, throughout the 1920s and beyond, Berwind-White and an affiliate, the Reitz Coal Company, produced approximately half of all the coal produced in the state's important 24th bituminous inspection district, which encompassed Windber and the adjacent Johnstown region.[13]

Only mines owned or controlled by major steel companies regularly outproduced Berwind-White mines in Pennsylvania. Table 1 illustrates the company's relative importance in all state production. From 1901 on,

Table 1. Berwind-White's state production

Year	State Ranking	State Tonnage	Windber Mines Tonnage	Windber's Share of Company's State Tonnage
1901	5th	4,608,981	2,471,286	54%
1902	4th	4,354,006	3,126,460	72
1903	4th	4,042,968	2,908,432	72
1904	4th	3,870,920	2,911,833	75
1905	4th	4,209,464	3,205,867	76
1906	5th	3,457,721	2,511,896	73
1907	4th	4,388,103	3,394,990	77
1908	4th	3,774,951	3,223,426	85
1909	4th	3,998,040	3,432,373	86
1910	4th	4,356,886	3,887,051	89
1911	4th	4,285,521	3,877,617	90
1912	4th	4,337,508	3,920,457	90
1913	4th	4,289,406	3,841,715	90
1914	4th	3,743,333	3,351,406	90
1915	5th	3,204,585	2,823,900	88
1916	4th	3,466,270	3,005,123	87
1917	6th	3,124,953	2,961,567	95
1918	6th	2,952,701	2,680,986	91

SOURCES: Pennsylvania Department of Internal Affairs, Bureau of Mines, *Report* (Harrisburg, Pa., 1902–1903), annual reports, 1901–1902; Pennsylvania Department of Mines, *Report* (Harrisburg, Pa., 1904–1920), annual reports, 1903–1918.

the annual reports of the state mining bureau or department contained a separate table listing Pennsylvania's largest bituminous producers along with their yearly tonnage. In the first year when such data was compiled, Berwind-White ranked fifth, but for thirteen of eighteen years from 1901 to 1918 it placed fourth. Its lowest ranking was sixth, which occurred twice, during the war in 1917 and 1918. Throughout this period, the only producers who regularly surpassed it in total output were the Pittsburgh Coal Company, the H. C. Frick Coal and Coke Company, and the Monongahela River Consolidated Coal and Coke Company, companies owned by or affiliated with steel corporations. Table 1 also indicates that Windber-area mines were primarily responsible for the company's prominent and prosperous standing. They contributed 72 percent or more of the company's state output from 1902 to 1918, and as much as 90 percent in the peak years of 1911 to 1914.

History of the Berwind-White Coal Mining Company

By the time the Berwind-White Coal Mining Company came to Somerset County, it exemplified the new type of impersonal, large-scale corporations that were coming into being in the United States in the period following the Civil War. Four of five Berwind brothers, the children of a Prussian-born cabinetmaker who had settled and prospered in Philadelphia in the 1840s, did eventually take active part in the coal business through the Berwind-White Coal Mining Company. The company had grown from the Berwind brothers' earlier, small-scale partnerships in the 1860s and 1870s until it reached modern form in January 1886, when it was reorganized and incorporated as the Berwind-White Coal Mining Company.[14]

Throughout the 1860s and 1870s the Berwinds initiated and expanded their business interests. They leased mines, bought up mineral and surface rights, invested in related enterprises, and entered into contracts that guaranteed markets for their coal. Their first mine, Eureka No. 1, was opened in the early 1870s in Houtzdale, Pennsylvania, where the Berwinds also established their first company store, known as the Eureka Supply Company. Through direct ownership, leases, interlocking directorates, rebates, links to leading financial houses, pools, and other cooperative arrangements, the company rapidly achieved considerable success in the bituminous coal industry. Like other large corporations, the company's success in the highly competitive industry of bituminous coal had been ensured by its growing vertical integration, which included control not only over the coal resources themselves, but also over transportation, docks, finance, marketing, and other enterprises related to mining.[15]

Berwind-White's development was greatly aided by its important and controversial long-term association with the Pennsylvania Railroad. In 1875, Edward J. Berwind had gone to New York to obtain contracts from steamship companies, who agreed to buy coal for the transatlantic trade from Berwind mines and the Pennsylvania Railroad. Until then, the Pennsylvania Railroad reportedly had no share of this important soft-coal market. Berwind was highly successful in this endeavor, in a manner that was repeated over and over again in the future with other companies and public utilities. With or without an intermediary, he sold coal from his own mines to other enterprises in which he had become an important investor or director. Such questionable conflict-of-interest practices led to a number of scandals and governmental investigations. One of the most important occurred during the strike of 1922, when New York City investigated the company's financial and labor practices because Windber coal fueled the city's subway system, on whose board Berwind sat.[16]

The vast holdings and wealth that Edward J. Berwind acquired in his lifetime led one contemporary competitor to dub him "the 'biggest' name in bituminous coal mining."[17] Upon his death in 1936 at the age of 88, the *New York Times* reported that Berwind was "reputed to be the largest individual owner of coal properties in the United States" and the last survivor of a circle of J. Pierpont Morgan's followers and associates. During his lifetime, he was president of six coal companies, including Berwind-White, and director of four others. Yet, the obituary noted, Berwind was also noted for his business activities outside the coal industry. He had served as a director of approximately fifty of the nation's other leading corporations, including railroad, banking, insurance, steamship, and communications companies.[18]

Nonetheless, the Berwind-White Coal Mining Company played a key role in the Berwinds' accumulation of wealth and power, and evidence suggests that Windber-area mines were the most productive and profitable of all its historical operations. In the 1870s and 1880s the Clearfield region of Pennsylvania had been the chief focus of its activities. Seven years after Windber's founding, the company established new towns and mines in West Virginia. Yet Windber was chiefly responsible for Berwind-White's claim for coal empire status early in the twentieth-century, and throughout the period 1897–1940 Windber occupied the central place in the corporation's extensive operations. It was the showpiece of the company's 50th-year anniversary advertisement in 1936 and was important enough for a woman who married into the Berwind family to discuss her introduction to—and growing consciousness of—the town in her memoirs.[19]

By Windber's founding, E. J. Berwind's contracts with the Navy, railroads, and steamship companies had led to what coal industry journals have described as a monopoly on the bunkering and steamship trade. Long before Windber-area coal began to supply naval ships or the transatlantic trade, Berwind-White's Eureka coal from the Clearfield region had been doing so. But by the early 1890s the expanding bituminous coal market and the growing depletion of coal reserves in the Clearfield region were causing company officials concern. They began to look elsewhere for future profitable resources and locations. Windber was the direct result of their search.[20]

Defining "Windber"

Windber was at the center of thousands of acres of valuable coal lands that the company came to own in Central Pennsylvania. A Philadelphia journalist who had toured the Windber region in 1899 described the coal underlying Somerset and Cambria counties as "one of the richest steam-making coal deposits in the country."[21] The Wilmore basin, in which the desirable coal deposits were located, contained the famous B-seam coal, considered ideal for steaming purposes. A leading coal journal described this particular B seam as "perhaps the most uniform and reliable coal of the Allegheny formation, running high in fixed carbon, low in volatile and ash—a truly 'smokeless' coal."[22]

These coal reserves, however, did not neatly follow man-made boundaries, and the Wilmore basin actually ran along county boundaries. In order to develop mining operations there, the company eventually had to cross portions of two counties—Somerset and Cambria, several townships, and three boroughs—Windber, Scalp Level, and Paint.[23]

Thus, when Windber was founded in 1897, there was considerable confusion over names and boundaries. Until the late 1890s the only settlement in the area had been Scalp Level, a small village that straddled the Cambria-Somerset county line approximately eight miles southeast of the city of Johnstown. The new town of Windber was located on forested land only a mile away from historic Scalp Level, and the names of the two places were often used interchangeably in regional papers. More confusion resulted in 1900, when Scalp Level itself divided into two separate boroughs. Residents in the Cambria County portion retained the name Scalp Level, while those in the Somerset County portion chose the new name of Paint Borough.[24]

Windber had been consciously located at the center of the company's various mine holdings. The only sensible way, therefore, to define "Windber" in a community study is to include more than the territorial limits of Windber itself. Although the new town became the focal point of business and other activities, the Windber area nevertheless included the three boroughs and three townships in which the surrounding mines and coal camps were located. The Dillingham Immigration Commission logically concluded that it needed to incorporate this adjacent larger region into its community study of "Windber."

Buying Up Lands

The company's first order of business was the tremendous task of acquiring title to properties owned by many different people in Somerset and Cambria counties. To coordinate its diverse operations, Berwind-White created the Wilmore Coal Company, which functioned as its holding company. Different enterprises of the company were carried on under different names, with the mineral rights held by the Wilmore Coal Company and surface rights by the Wilmore Real Estate Company.[25]

In acquiring its valuable deposits, the company took advantage of the economic situation and acquired a substantial portion of its 30,000 acres in the Scalp Level (Windber) district during the depression of 1893. From 1892 on, a number of company agents busily canvased the region, offering hard-hit and more prosperous farmers cash for coal rights, surface rights, or both. One of the most important of these emissaries was Berwind-White's advance agent, James S. Cunningham, who rode horseback from one farm to another and personally bought up more than 100 square miles along Paint Creek. In 1893 he successfully convinced David Shaffer to sell the extensive farm holdings on which Windber was subsequently built.[26]

Other farmers also sold rights. Many were pleased, at least at first, provided the price was right. David and Rachel Shaffer, contented with the $12,706.11 they had received for the 356 acres of property they sold on December 29, 1893, retired to what soon became Paint Borough to live out their lives. Another wealthy property owner, Peter Hoffman, was one of the earliest to sell his coal lands; he then became a director in the Windber National Bank and an investor in other commercial enterprises.[27]

Sometimes agents bought up surface and mineral rights directly in the name of Edward J. Berwind, who in turn transferred accumulated properties to the Wilmore Coal Company. On December 1, 1898, for example, Berwind and his wife conveyed fourteen separate deeds to more than 1,000 acres in a single transaction to the company for the sum of $1.00.[28] At other times, agents bought up land rights and titles in their own names and then conveyed them to the coal corporation. The son of one prominent farmer who never sold his land or mineral rights noted that this method was used by Berwind-White in order to conceal the identity of the real purchaser. Independent farmers were more likely to sell to individuals than to a coal corporation. Those who did sell would have demanded a higher price had they known who the real buyer was.[29]

Robert H. Sayre of South Bethlehem, Pennsylvania, chief officer of the Wilmore Coal Company, was an agent who bought up properties in his own name instead of the company he represented. He concentrated on buying up land rights and titles to ninety different properties located mainly in Cambria County. On May 4, 1893, he conveyed title to these deeds, along with all accompanying coal, surface, minerals, and mineral rights, to the Wilmore Company for the nominal sum of $1.00. While some deeds like Isaac Lehman's two acres were small additions, others, like Abram Weaver's 265 acres, were more substantial. In the end, more than 8,000 acres were involved in this single transaction.[30]

By December 1900 the *Windber Era* was reporting that Berwind-White owned 70,000 acres in Somerset and Cambria counties alone and that the company expected to expand its holdings in future years.[31] Only a few highly independent farmers in the Wilmore basin had refused to sell their coal rights.

Even at this stage, however, the company's relations with farmers were not always serene, as the experience of E. H. Butterbaugh shows. Butterbaugh was another Berwind-White agent who had begun securing coal lands from Paint Township farmers in Somerset County in 1892. Through some misunderstanding, a number of farmers had not received payments for the rights they had sold to him. Consequently, they sold them a second time—and for a higher price—to a competitor. In July and August 1893, Butterbaugh took these cases to court for arbitration, and the farmers eventually settled with Berwind-White for the original payments.[32]

For the first time, farmers had expressed some dissatisfaction with their dealings with the company. In future years, other farmers would legally challenge the company for mining coal beyond prescribed limits or for damaging the land surface or water sources on which their livestock and farms depended. A number of independent farmers who had never sold away their rights would later allow the miners to hold union meetings on their property when other sites were denied them.[33]

Railroad, Coal, and Timber Interests

Berwind-White could not have developed its new coal holdings without an adequate transportation system. Throughout the early 1890s, Somerset County papers had been speculating about potential railroad competition and development. In the end, the Pennsylvania Railroad Company built or controlled the lines necessary for development.

By the 1890s the Berwind-White Coal Mining Company was widely reputed to be a Pennsylvania Railroad concern, and the widespread perception of their interlocking interests in the region was such that at the turn of the century newspaper reports sometimes used the names Berwind-White and Pennsylvania Railroad interchangeably. The railroad's president, Alexander Cassatt, occasionally came to Windber to inspect railroad lines to the mines. More important, in 1906, Interstate Commerce Commission investigations into charges of car and rate discrimination by the Pennsylvania Railroad on behalf of Berwind-White substantiated such charges, uncovered other favors and cooperative arrangements, and revealed that railroad magnates and lesser officials had received gifts of stock in the coal company.[34]

Railroad and coal development in this portion of Cambria County and in northern Somerset County did not begin until the Johnstown flood of 1889 had made new routes into the area possible. In 1891 the South Fork Railroad Company completed construction of one line, which facilitated access to the new coal fields. Built directly through the break in the breast of the dam, this railroad, linked to the Pennsylvania system, connected South Fork and Dunlo, where Berwind-White had opened a mine in Cambria County adjacent to the Windber-area mines.

In March 1897 a state charter was issued to the Scalp Level Railroad Company, which the Pennsylvania Railroad and the Berwind-White Coal Mining Company controlled, to build a line connecting South Fork and Scalp Level (Windber). Since surveys had already been made, contracts were quickly let and construction rushed in hopes that mines might be opened by July 1897.

A test run of the first passenger train, with dignitaries aboard, took place early in August 1897; the pioneer shipment of coal from Berwind-White's first mine in the area, Mine 30, occurred in September 1897. The South Fork Railroad Company and the Scalp Level Railroad Company merged in 1902. Built to develop Windber-area coal lands, they were absorbed directly into the Pennsylvania system in 1903.[35]

Rumors of railroad and coal development in the Central Pennsylvania region attracted the attention of a third interest, which usually accompanied the other two in the opening of new territories. Timber was a commodity of great commercial value, and lumber industrialists sought to exploit the virgin hemlocks and other forests in the area. While a number of firms were involved in this industry, the largest by far was the Babcock Lumber Company of Pittsburgh. In October 1897 its president, Edward V. Babcock, acquired an initial tract of 6,800 acres of forested land in the Ashtola district of Somerset County. Berwind-White, however, retained ownership and rights to the underlying minerals.[36]

By 1900 the Babcock Lumber Company was employing 400 men, and by June 1902 its payroll had increased to 1,244. In subsequent years, it purchased many additional acres of forests in the region and bought out most competitors. Babcock closed the last of its local operations in 1913, after the timber had been depleted, but during its active operations in the early years of the century its various lumber camps were located on the outskirts of the new town of Windber. Although it established its own housing and stores, newspaper reports suggest that Windber's hotels, clubs, and speakeasies attracted many lumberjacks, as well as miners, on Saturday nights.[37]

As must often have been the case with timber and coal interests in a given area, the Babcock and Berwind companies were complementary and not in competition with one another. Their respective owners and officers entertained each other socially, especially in the rustic setting of the Babcocks' newly constructed luxurious lodge, and if newspaper reports are accurate, they typically supported the same political candidates.[38] Certainly they shared in policing the region in the frontier era of Windber's history. Although the first burgess of Windber was a superintendent of Babcock Lumber Company, Berwind-White, not Babcock, had originated regional development, and coal, not timber, was the reason Windber had come into existence.[39]

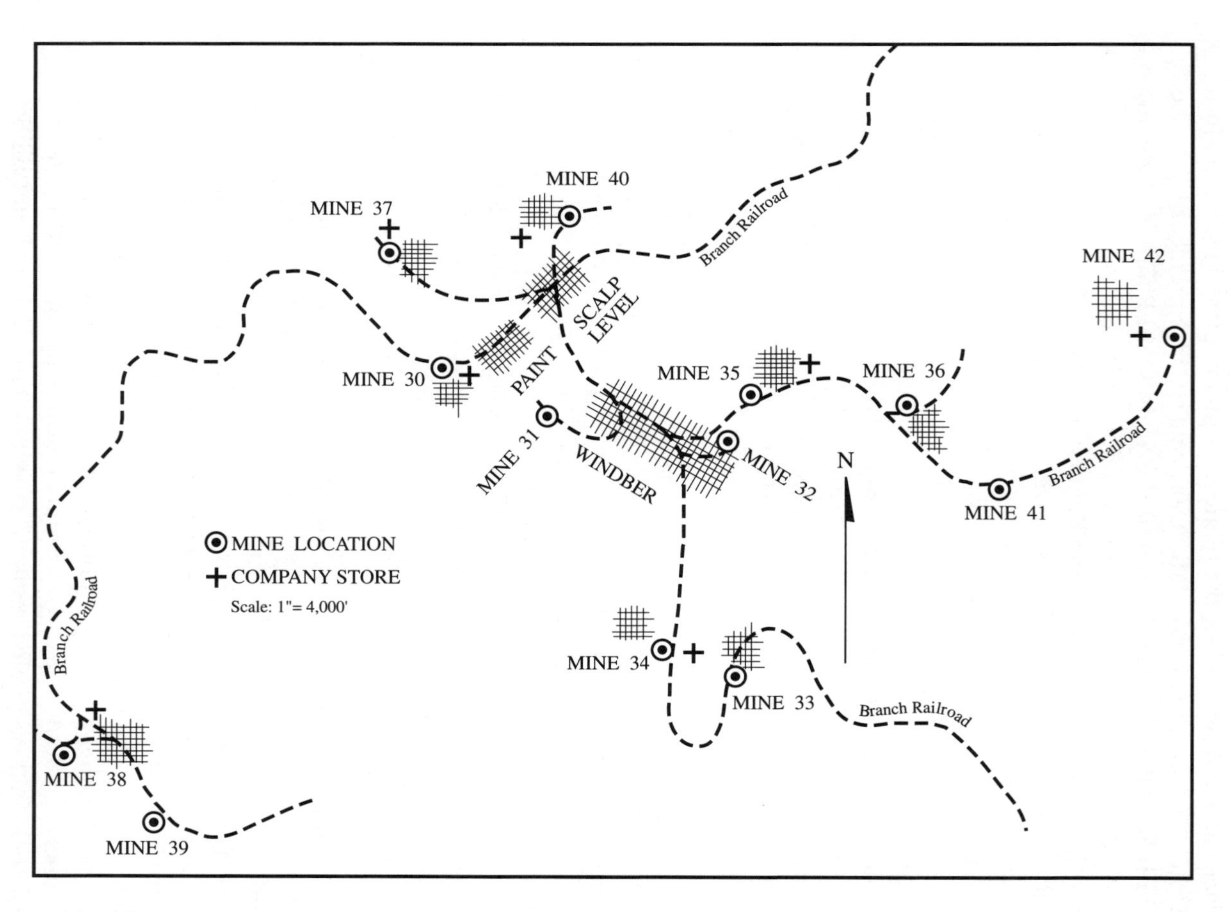

Map of the Windber area, c. 1911.

Mine Development and Construction

The location and development of Windber mines can be seen in the Dillingham Commission map. In 1897, Berwind-White officers and engineers proceeded to gradually open thirteen new and highly productive mines in the area. Numbered sequentially from 30 to 42, the mines rapidly expanded. The first mines, Nos. 30 and 31, started operation in 1897; four more mines, Nos. 32–35, opened in 1898; No. 37 followed in 1899, and No. 36 in 1900. When the *Windber Era* published its special "Illustrated Industrial Edition" in December 1900, eight mines were operational. From 1902 to 1908, five additional mines, Nos. 38–42, were added, completing this development.[40]

As the number of mines gradually grew, the number of employees did also, and production mounted. In March 1898, only 550 were reported on the company's payroll, and only fifty carloads of coal were shipped out daily. By October of that same year, the number of employees had increased to 1,600 and the daily number of coal-laden cars to 160. In December 1900 the *Windber Era* reported 3,000 people employed and 450 cars shipped daily from town. The Dillingham Commission estimated that 4,000–4,500 workers were on the company's local payroll later in the decade.[41]

In opening these mines, Berwind-White enjoyed all the advantages that came with belated industrial development. The company could select the most modern techniques and equipment for its new mines. There were no preexisting structures, mines, or investments to prohibit modernization. Furthermore, the company had the capital necessary for investment in power plants, electric haulage, Baldwin locomotives, boiler houses, steel fans, Ingersoll-Sergeant compressors, and Sullivan pick machines. The modern nature of Windber's mines received favorable attention in coal journals and mine inspection reports.[42]

Shortly before the first shipment of coal left Mine 30 in September 1897, Berwind-White announced the founding of its new town. The *Somerset County Democrat* reported: "J. S. Cunningham is the agent for the new town and all the preliminaries have been arranged, the property having been plotted and plan perfected for putting the lots upon the market."[43]

Construction, under the auspices of Berwind-White, began to boom in Windber in 1897 and continued to do so for several years. The first company houses, double houses, were completed in September 1897. In October 1897, C. J. Duncan moved from Dunlo to build one of Windber's first hotels, and the company awarded contracts for the waterworks system of the town. Carpenters, contractors, plasterers, masons, and painters took part in the boom.

By March 1899, an imposing new building, the Berwind-White clubhouse, stood perched on a hill overlooking central Windber, where it was visible to all. Located near it were the company's headquarters and the main company store. Unlike the outlying mining camps where most miners resided, the urban center and hill district where prominent company officials and business people lived would have modern conveniences. Superintendent James S. Cunningham told reporters: "After we get everything else in good shape . . . we will organize a borough or city and then provide for first class streets."[44]

Town Structure and Borough Status

Berwind-White had the resources and derived benefits from belated development that enabled it to set up its mines as it wanted to, but it also enjoyed the same advantages when it came to establishing a town of its own making. Not only did it enjoy monopolistic ownership or control over land and resources, but its mining operations were the raison d'être of a new town whose population would be directly or indirectly dependent on the company for jobs, housing, and other necessities.

A new town was a tabula rasa. There were no preexisting powers, no independent authorities, no business or political competition, and no organized labor movement with which to contend. Moreover, Berwind-White's managers had gained considerable experience through its operations in other towns. In the 1890s, for example, it had successfully smashed a chapter of the Knights of Labor in Horatio, Pennsylvania.[45] Thus, Windber did not spring into being spontaneously, but was the conscious creation of company officials who knew precisely what type of town they wanted.

The town's name, "Windber," an anagram of "Berwind," accurately reflected the type of autocratic town structure that the Berwind-White Coal Mining Company ultimately created in Windber through its agent, Superintendent Cunningham. The company's monopoly of land ownership, housing, and jobs, along with its dominance of the borough's political offices, guaranteed its control, which from the town's inception was

Paymasters leaving Berwind-White headquarters on Somerset Avenue en route to the mines, c. early 1900s.

overt and undisguised. However, as subsequent chapters show, company control was exercised in many different ways and extended well beyond its enormous economic and political power into cultural, religious, and other arenas as well.

Table 2 conveys some notion of the pervasiveness of the company's economic influence in the town and of how interlocking directorates were one device used to ensure continued dominance. Six key industries and organizations, including the company store, the first bank, two utilities, one newspaper, and one recreational association, had as officers or directors one or more of seven top absentee owners or resident Berwind-White officials. By 1900 all six enterprises had been incorporated and were part of the economic structure or cultural foundation on which the community was built.

With few exceptions, notably a brewery, a brickyard, and a farmers' bank, this pattern of interlocking directorates and stockholder owner-

Table 2. Interlocking directorates: Berwind-White officials and Windber's key industries, 1897–1900

Business	Edward J. Berwind, Pres., New York	Harry A. Berwind, Sec., Philadelphia	William A. Crist, Gen. Mgr., Windber	Thomas Fisher, Gen. Supt., Windber	Frederick McOwen, Treas., Philadelphia	James S. Cunningham, Supt., Windber	Edward L. Myers, Philadelphia
Eureka Supply Co.		X	X	X	X		X
Windber National Bank	X		X			X	
Windber Electric Co.		X	X	X		X	X
Paint Township Water and Power Co.			X			X	X
Windber Publishing Co.			X			X	
Windber Park Assoc.						X	

SOURCES: Somerset County Courthouse, *Books of Deeds*, 90:371–372, 97:333–335, 99:152–154, 106:565–567; J. L. Fehr, comp., *Windber–Scalp Level and Vicinity: Eureka Mines, Berwind-White Coal Mining Co., Operators* (Windber, Pa., December 1900), 4, 7.

ship persisted as major new enterprises were founded in subsequent years. Between 1900 and 1910, Berwind-White officials were instrumental in establishing the Windber Trust Company, the Windber Hospital, the Windber Building and Loan Association, the Windber Heating Company, and various new branches of the Eureka Stores, all of which were under the company's control.[46]

Berwind-White enjoyed an indisputable monopoly of the commanding heights of town industry. Consequently, no new manufacturing industries or other possible industrial competitors of Berwind-White were able to secure a foothold in the new town. On various occasions throughout the nonunion era, union-oriented miners, concerned small businessmen, and other progressive elements in the community sought unsuccessfully to bring in other industries so that diversification would break the coal company's stranglehold on independent unionism, competitive business, and community autonomy. Berwind-White bitterly fought off such efforts.

Yet Windber did initially attract a number of aspiring small businessmen, in part, because the new settlement had been widely advertised. These smaller businesses were never monolithic entities. Big fortunes were occasionally made, especially when company patronage was involved. Mitchell McNeal, for instance, arrived early in town and grew wealthy in the lumber industry by supplying timber for Berwind-White mines. Smaller merchants like Joseph Levant, who owned a jewelry store; A. M. Bloom, who bought a prospering hotel; and George Rudolph, who employed five journeymen in his tailor shop, had all come to Windber in 1898 or 1899 and achieved much more modest success. Scattered newspaper accounts of business failures, changes in ownership, and businessmen leaving Windber for other places convey the impression that many aspiring entrepreneurs found Windber's constricted business environment to be less than profitable or, perhaps, not to their liking.[47]

From the outset, American- and foreign-born business people occupied an ambivalent or dependent position vis-à-vis Berwind-White and the working population. All had to operate within the constraints set by the controlling corporation in a company-town setting. For some, especially those who derived considerable material benefits and became part of the local power structure, the constraints were no burden. The businessmen who sat on the borough council throughout the nonunion era were among the town's most prosperous people, and they willingly agreed to anything the company wanted rather than take an independent stance. Yet other owners of small businesses had deep resentments against the company's constraints, including the company-store system, and sympathized with union-minded miners. During the major strikes and struggles that occurred, some business people consequently threw in their lot with the company, while others sided with the miners.

By the end of 1899, James S. Cunningham, among others, believed Windber was ready for borough status. On December 22, 1899, a mass meeting of citizens was held in town to organize a borough, and on February 16, 1900, a petition for incorporation was filed in court in Somerset. Because numerous farmers protested being included in the new municipality, action was delayed until July 3, 1900, when Judge J. H. Longenecker resolved the conflict by granting incorporation rights to two separate boroughs: Windber and Paint. Company officials, who preferred to have only one borough authority, accepted the result.[48]

A total of 361 individuals and 9 businesses or organizations signed the document. Leading the businesses was the Berwind-White Coal Mining Company, followed by five of its subsidiaries or affiliates. The organizations included the Windber Park Association and the Presbyterian church. Property-holding must have been deemed essential for the petition's success, as petitioners claimed to have 314 freeholders and 56 renters among the signers.[49]

Notably absent from the Windber petition were the names of the various Berwinds. As absentee owners and indisputably nonresidents of Paint Township, they did not qualify as signatories. Instead, their local agents signed on behalf of the company and its associated enterprises. For example, Superintendent Cunningham endorsed the petition eight different times as a result of the different positions he held.

A cross-check of the petitioners' signatures with the manuscript schedule for Paint Township for the census of 1900, taken in June, revealed that the largest number of identifiable signers were miners, mining supervisors and officials, merchants, or skilled contractors and tradespeople. While most were born in the United States, those who were not came primarily from northern European countries. At least ten prominent people, including Superintendent J. S. Cunningham, were nonresidents of the township and thus not enumerated in the local census. Census figures also indicate that a smaller percentage owned homes than the petition claimed, although the majority of those identified were owners. In any event, the petitioners for a borough included a greater number of wealthier, more skilled, and native-born or northern European naturalized citizens and property owners than the larger transient population beginning to flood the Windber area by 1900.[50]

Ordinary working people were without political representation as a result of Windber's first borough election, held on July 14, 1900, when 423 votes were cast. In the main contest, the superintendent of the Babcock Lumber Company defeated a prominent druggist to become the town's first burgess. Seven council members were also elected: three were coal company or railroad officials, and four were prominent businessmen. Berwind-White's corporate influence in the town's government was further increased when the Windber council immediately appointed the Assistant Chief Engineer of Mining as its clerk.[51]

Republicans dominated the majority of higher and lower offices, although three council members were Democrats. In this era, on the municipal level and in town affairs in general, Democrats seemed to differ little from their Republican counterparts. Neither party was critical of Berwind-White in any regard. Consequently, borough contests were less partisan than other elections, and personality often counted more than party labels. On other electoral levels, however, Windber voters, like the county's voters, normally endorsed Republican candidates by a large plurality.[52]

Even before the borough's first election, Windber had already earned a dubious political reputation in certain quarters outside the borough limits. For at least some of those in the county's minority party, Windber was already the example par excellence of fraudulent, unqualified voting and undue corporate political influence.

In his report on the primary elections of April 1900, the outraged editor of the *Somerset County Democrat* had charged: "Windber as usual did herself proud; they polled 729 votes there; nobody believes that there are more than half that many qualified Republican electors at that place." He then referred to a fiercely fought factional contest currently taking place within the Republican party and charged: "A circular issued by J. S. Cunningham, Supt. of the B. W. Coal Co. placed the company and its influence on the side of the Stalwarts." Saloonkeepers, he maintained, supported the insurgents. The *Democrat*'s editor concluded that the intra-Republican fight was a contest "between boodle, on one hand, and whiskey, on the other."[53]

In subsequent years, numerous miners and other residents within the town frequently charged that, at election times, Berwind-White officials ordered its employees to vote for the company's candidates or risk loss of their jobs and homes. Many succumbed to such coercion and pressures. Moreover, a majority of the population—foreign-born miners and their families—did not vote at all throughout the nonunion era because they were not American citizens and were therefore ineligible to vote.

Neither American-born miners nor naturalized citizens experienced authentic political democracy in Windber, because none existed there. Working people had to go through major struggles to get the right to freely exercise the franchise in the autocratic company-town setting. It was only after successful unionization in the 1930s, and after the United Mine Workers established citizenship schools, that the franchise and American citizenship become linked to the practical possibility of independent democratic voting in Windber. Until then, ordinary working people born in the United States or abroad—the vast majority of the population—remained effectively without political representation.

The newly elected borough council met frequently during its first months. Legal fees, taxation rates, dog licenses, appointment of officials, establishment of a police force, setting of salaries, business licenses, and building-permit fees were concerns that occupied their immediate attention. Considerable care was devoted to issues related to preservation of order. On July 24, 1900, for example, a committee was appointed to supervise erection of a jail; on August 3, 1900 the burgess was granted authority to hire extra police as needed; and on August 10 the council approved the purchase of the first police uniforms along with "one dozen maces, one dozen shields, one half dozen pairs hand cuffs, one half dozen pairs of nippers and one half dozen belts for use of police force."[54]

From the start, the Windber council and the Berwind-White Coal Mining Company, or its associated enterprises, worked hand in hand to develop the town, as the company wanted and often at the public's expense. The landowning coal corporation donated and in fact chose the site for the lockup; the borough took over company-built sewers on terms that the council left unspecified in its books. Meanwhile, Berwind-White escaped paying taxes comparable to its corporate wealth. On the county level, it sought and often won reduced county assessments. In Windber itself, it paid virtually no taxes.

Windber's tax assessors seriously undervalued the company's vast property holdings. In 1900, for instance, the company's entire local tax assessment was only $495.92. Moreover, instead of paying these assessed taxes in cash, Berwind-White had made special arrangements with the council to work out borough taxes through road, electric, construction, or other work. Competitive bidding was therefore bypassed in the process. From August to December, the coal company performed for the borough street work that totaled $1,190.10 and included a 5 percent interest charge. In December the council found that the borough owed the corporation money, not vice versa, and had to authorize a payment of $694.18 to Berwind-White![55]

Water, sewer, and electric franchises and street contracts were likewise routinely granted to company enterprises. The council neither actively encouraged nor seriously entertained bids from outside competitors. Instead Berwind-White habitually got what it wanted on the terms that it wanted. The coal company even occasionally intervened on behalf of other corporations, as it did in petitioning the council to grant the Atlantic Refining Company the right to erect oil tanks and to lay pipes. This interlocking pattern of town and business relations, along with the exclusion or rejection of serious outside competitive bids, was established from the outset.[56]

In December 1900 the elite of Windber society hosted a banquet to honor the judge who had officially granted Windber borough status.[57] During that same month, the *Windber Era* published a special edition that lauded the town's achievements and presented short biographies of its leaders.[58] It is significant, once again, that the most wealthy and powerful town leaders—absentee owners like the Berwinds—did not appear in that edition. Instead, it contained the biographies of 106 resident leaders in Windber, Scalp Level, and Paint Borough. This source is invaluable, however, in that, even more than the petition or census, it illustrates the social composition of the local upper classes.

The 106 biographies indicate that the vast majority of Windber's resident municipal, business, and professional leaders were white, male, young (in their twenties or thirties), born in the United States (most often in Pennsylvania and quite often in Somerset, Cambria, or nearby counties), and Protestant. Only one female, employed in printing, was included in the selection, and of fourteen people who were born in foreign countries, only two were clearly identified as having come from southern or eastern Europe. One was a Roman Catholic priest, the other was an Italian ethnic leader and Berwind-White contractor.[59]

The social composition of the town's leadership and local upper classes—white, male, American-born, and Protestant—was firmly entrenched. It would remain so for many years, even though it soon proved to be highly unrepresentative of the masses of the population. With the exception of a few notable individuals, the absentee owners of Berwind-White were able to count on this local elite for the company's continued political, as well as economic, dominance of the borough.

In three short years, then, "Berwind" had in fact become "Windber." A modern, nationally prominent coal corporation with absentee owners had set up a company town in a sparsely populated rural Pennsylvania setting in order to enable it to get the valuable coal deposits that underlay it. By 1900 the mines, the municipality, and the local social structure desired by Berwind-White were in place. Nonetheless, even as the company was advertising the existence of what it called a model mining town, the masses of immigrant and American working people who would ultimately challenge autocratic company rule in the community as well as at the workplace were beginning to arrive on the scene.

2

From Europe to Windber

From the town's inception in 1897, immigrant workers were of crucial importance to the development of Windber and the area's mining industry. The controversial but influential Dillingham Immigration Commission, which studied Windber in 1909, reported that "operations in the mines of Community A [Windber] were begun with immigrant labor, and the general expansion of the mining industry and the development of the locality have been due principally to immigration from Europe." It concluded: "The mining company has been successful principally through the fact that it could secure immigrant labor, there being no supply of native labor available."[1]

From the standpoint of Berwind-White, from 1897 until the 1920s, the company faced a perennial struggle to secure and retain a labor force of sufficient size to meet the needs of its new mines in the Windber area. In this era of industrial expansion, scarcity of labor was a problem that coal operators frequently cited in coal journals. E. J. Berwind himself stressed recruitment problems in testimony before the U.S. Industrial Relations Commission on January 21, 1915: "I know one thing—that there has never been a day in this 10 years, until this war broke out, that we could get the men to man our works."[2]

Berwind went further. He insisted to the commission that this labor shortage was chronic, and therefore a reason to reject Congressional proposals that would legally restrict immigration. At the time, the Berwind-White Coal Mining Company was a large financial contributor to the National Liberal Immigration League, an important lobbying organization that sought to convince the public that immigration should remain open and unrestricted. When a commissioner suggested that labor might be too abundant and that large employers sometimes took advantage of this overabundance to suppress wages, Berwind admitted: "The scarcity of labor raises wages more than anything else." Yet, he persisted, he needed more, not fewer, workers for his mines.[3]

Berwind's continual search for miners resulted in a polyglot company work force drawn from diverse sources and numbering approximately 4,000 at any given time throughout these early years. The Dillingham Commission reported that, as representatives of twenty-five or more nationalities found jobs in area mines during the decade from 1900 to 1910, movement into Windber was constant and movement out frequent.[4] But what do we know about who these people were, and what had prompted them to come to Windber?

Initial Operations

Opening up new mines required expertise and skill, so in 1897, when Berwind-White began to develop Windber-area mines, it turned first to its own supervisory and engineering staff in such towns as Houtzdale. By transferring a number of its own officials as well as experienced engineers and miners, it laid out the new mines and prepared operations so that less-skilled miners and machines could then take over the rigorous task of extracting coal and loading it for shipment to market.

Many of the top company officials who served in the Windber area in the town's early years had worked for the company or associated enterprises for a decade or more. Among these were W. A. Crist, Harry D.

Edelblute, and John Lochrie. Having been general manager for Berwind-White in Houtzdale for years, Crist became the first general manager in the Windber region in 1897. Edelblute, the first head of the company-store system in the new town, had risen to a supervisory position in the company's Eureka store in Horatio, Pennsylvania, while the assistant superintendent of mines at Windber, Scottish immigrant John Lochrie, had begun his family's long-term working relationship with the Berwinds in Houtzdale in 1878 shortly after his arrival in the United States. On the other hand, the initial ties that certain ethnic leaders, such as Frank Lowry, had to Berwind-White are less clear. Yet Lowry, an important labor contractor who supplied and supervised work gangs composed of fellow Italians for the company's many Windber-area projects, had built a career as a labor contractor for railroads even before 1890.[5]

Local management was easily transplanted, but Berwind-White still needed skilled operatives to carry out its initial operations. It therefore asked a few individuals who possessed special skills to leave its operations in Houtzdale and move to Windber. John Adam Novak, a recent Slovak immigrant, was one of those who moved to the new town to help open mine headings there. Yet the numbers involved were not sufficient for the tasks at hand. Because the local rural population was sparse and there were no immigrants in the immediate area, except a few Italians employed in railroad construction, the company turned to outside employment agencies to obtain about 1,500 workers from other locations. The Dillingham Commission reported that 80 percent of this initial work force of 1,500 was English in origin and had had previous experience in mining. The remaining 20 percent was a mixture of many different ethnic nationalities.[6]

As mine development progressed in 1899 and 1900, the company continued to increase the size of its labor force, and as less-skilled labor was needed for its ongoing operations, southern and eastern Europeans began to come into the community in great numbers. The Dillingham Commission asserted, but did not prove, that the English miners hired earlier had begun to leave the area en masse before 1900, presumably as a result of this influx.[7] The commission's hostility to new immigrants was such that it habitually assumed and then concluded—without evidence—that new immigrants would push out or displace Americans and northern Europeans, thereby lowering wages and working conditions.[8] In fact, however, as the company's president later said, Berwind-White's new operations required many workers and then more workers, and the company was able to absorb all the skilled and unskilled labor it could recruit and retain in these early years. Where that labor force came from was of secondary importance, when important at all, to the primary need for workers.

The Size and Foreign Character of the Population

In 1909 the Dillingham Commission estimated that the Windber-area population, which included Scalp Level, Paint Borough, and outlying mining settlements, totaled approximately 10,000 and that Windber residents alone numbered 7,500. One year later, federal census-takers counted a slightly larger population. According to the official census, Windber then numbered 8,013 inhabitants and the two other towns 2,423. Outlying camps brought the mining area's total to nearly 12,000 people.[9]

The region's growth in the preceding decade is evident when these figures are compared with those in the 1900 census. Then Scalp Level had only 450 people, compared with 1,424 in 1910. When the census was taken in early 1900, neither Windber nor Paint Borough had yet been granted formal borough status. As a result, the populations of the two emerging boroughs were included with and not distinguished from other residents of Paint Township. The entire township, however, totaled only 6,835 in 1900.[10]

After 1910, the area continued to grow. Windber's urban population peaked in 1920, but Scalp Level and Paint Borough's did not do so until 1940.[11] When the populations of the three towns are combined, it is evident that, overall, the regional pattern showed growth until 1920, stagnation or a tiny decline by 1930, small growth again by 1940, followed by a relatively sharp decline and the beginning of a trend that reflected increased exodus from the region due to the onset of permanent mine-closings there.

If the Dillingham report and the 1910 census were in relative agreement about the area's population size at the end of the century's first decade, they present somewhat different pictures of the community's ethnic composition, and especially of the balance between Windber's foreign-born and native-born populations. The Dillingham Commission referred repeatedly to "the preponderatingly foreign character of the population" and had in fact selected Windber as one of two bituminous coal towns in Pennsylvania to study precisely because of its influx of new immigrants. According to commission calculations, three-fourths of the 10,000 residents, or 7,500 people, were foreign-born, but that included all children whose fathers were of foreign origin. This method of calculation—classifying native-born children according to their father's birthplace rather than their own—automatically diminished the size of the native-born population and exaggerated the numbers of foreign-born. If, through ignorance or error, English-speaking census enumerators often underrepresented southern and eastern Europeans, the commission did exactly the opposite. Its focus on immigrants and its sympathy for immigration restriction may have contributed to its use of a methodology that systematically distorted and underrepresented American-born and English-speaking people. In any event, its investigators estimated that only 2,500 Americans resided in the entire area.[12]

When another standard of judgment is used, a different conception of Windber's ethnic composition emerges. For example, when all those born in the United States are considered native-born and U.S. census figures are used, Americans constituted a majority of 55 percent of the area's total population in 1910, instead of the minority of 25 percent cited by the Dillingham Commission. By these calculations, 57 percent of Windber borough residents had been born in the United States, and usually in Pennsylvania.[13]

Despite these reservations, the Dillingham Commission was absolutely right to select Windber to study because of its preponderance of immigrants, most of whom came from southern and eastern Europe. After all, 45 percent of the area's population, including 43 percent in Windber proper, were born in foreign countries. At the same time, in 1910 only 17 percent of the inhabitants of Somerset County or 19 percent of the residents of the state of Pennsylvania could claim a foreign heritage.[14]

The zenith of foreign-born representation in the overall population of Windber and surrounding areas was reached in 1910. By 1920 the outright foreign-born comprised only 29 percent of Windber's population. Immigration restriction in the 1920s contributed to a further decline of this portion of the population. By 1930 the foreign-born constituted 22 percent of the total population; by 1940, 18 percent; and by 1950, 15 percent.

The Breakdown of the Foreign-Born Population

Four nationalities—Slovaks, Magyars, Poles, and Italians—unquestionably constituted the bulk of Windber's immigrant influx. These four leading ethnic groups comprised 84 percent of the area's entire foreign-born population and 38 percent of its general population in 1910. According to the manuscript schedules of the census taken that year, the Slovaks were overall the most numerous of the four, but only slightly ahead of the Magyars; Poles were in third place, followed by Italians.[15]

The numerical preponderance of these four dominant foreign nationalities varied somewhat according to specific place of residence. Each of the three towns or outlying mining settlements had a slightly different ethnic makeup. Italians were highly concentrated in Windber proper and considerably outnumbered the Poles there. However, there were so few Italians in the two other towns and outlying mining areas that they did not rank in the top four ethnic clusters in any of them, and other, generally less numerous, nationalities surpassed them in these localities. Also, Magyars, not Slovaks, led the roster of nationalities living in Scalp Level, Paint Borough, and Mine 37, while the reverse was true in Windber proper.

The four leading nationalities were the basis on which the various immigrant communities were founded. They pioneered in establishing their own churches and fraternal societies. Their size itself gave them a certain degree of power and influence, especially among other less numerous ethnic groups from southern and eastern Europe. No attempt to integrate disparate nationalities into a working-class movement, such as the United Mine Workers union, could afford to ignore any of the big four nationalities.

It is more difficult to be precise about the numbers of the less numerous nationalities living in the area. There were many inherent problems in identifying and classifying people according to nationality, which was not always self-evident to either enumerators or immigrants. If the Dillingham estimates erred by underestimating the American-born and English-speaking population, the census erred in the opposite direction by underrepresenting people from southern and eastern Europe. Nevertheless, it is possible to draw some basic conclusions about the size and composition of the less numerous resident ethnic groups.

First, many different nationality groups had a presence of some sort in the Windber area. Outside of the big four, twenty-five different nationalities were easily represented in the general population. Among these were people from northern and western Europe, central and eastern Europe, southern and Mediterranean Europe, and northern Africa. Ethnic diversity was a distinguished feature of the local mining communities.

Second, additional evidence suggests that the Dillingham figures about the numbers of southern and eastern European nationalities in the area are closer to the truth than was the census. If this is so, then new immigrant groups, not the old ones, constituted a majority of the less numerous foreign-born populations. Three groups in particular—Romanians, Jews, and Carpatho-Russians—were almost certainly larger than the modest numbers counted in 1910. Each of these examples is worth examining.

Evidence that the Romanians once had a local population of some size and importance derives from religious sources and newspapers. An article in the *Catholic Encyclopedia*, published in 1909, cited both Windber and Scalp Level as chief areas of settlement for Romanian Greek Catholic immigrants in the United States. Also, Scalp Level's Romanian Greek Catholic Church, founded in 1908, was only the second such church to be established in the entire country, and it served as the center for all Romanian Greek Catholic missionary activities in Pennsylvania.[16] The Dillingham report claimed that this church had a membership of seventy-five families, or 400 people, and a regular Sunday attendance of 150.[17] While the census itself showed only one resident of this nationality in Windber proper, it found a concentration of Romanians in Scalp Level and Mine 37; the church was located close to this cluster in what was described in local newspapers at the time as "Roumanian hill."

Jewish people in Windber certainly numbered more than the twenty-two listed as Yiddish in the census of 1910. Individuals known to be Jewish were enumerated as Germans, Magyars, Poles, or Russians, reflecting their place of origin or degree of previous assimilation. While no known Jewish people were involved in mining, a number were active in Windber's early, rapidly changing businesses. A historian of Johnstown's Jewish community claimed that frontier areas like Somerset County attracted itinerant Jewish peddlers and businessmen early in the twentieth century because the opportunities there seemed great. But not all who came remained. For example, one of Johnstown's most prominent businessmen, Moses Glosser, had stopped off and operated a store briefly in Windber before moving on to the larger city, where he opened a highly successful department store in 1907.[18]

Windber's Jewish community was not large enough to support a synagogue or temple of its own, so it relied on the Johnstown population for an active social and religious life. The Reform rabbi from the larger city traveled to Windber frequently in the town's early years, to perform traditional ceremonies. In turn, according to local newspapers, Jews from Windber supported and attended social and religious festivities en masse in Johnstown. However, Windber Jews were organized enough on their own to have established a short-lived B'nai B'rith chapter, which conducted a fund-raising drive for the relief of persecuted Russian Jews in 1905. In subsequent years, they formed fraternal insurance societies and social clubs or joined Johnstown chapters and groups. But the internal divisions that split Johnstown Judaism into Reform and Orthodox factions in 1902 also affected Windber. Windber's Orthodox wing, whose membership came primarily from Russia or Poland, was numerous enough to have had its own kosher butcher and rabbi, Isaac Slesinger. Although a synagogue was never erected, quarters in various town buildings were rented for weekly services. During the decade from 1910 to 1920, this community was of sufficient size and importance to attract Jewish people from more isolated or adjacent towns to come to Windber to celebrate important religious holidays there.[19]

Carpatho-Russians—the third example—were vastly underrepresented in both the census and Dillingham report because they were almost always enumerated as other nationalities, especially as ethnic Russians or other groups who had originated in the old czarist empire. One important reason for this confusion is that the masses of Carpatho-Russian immigrants had no sense of a distinct national identity until after 1914. The term "Rusyn" had been used in early centuries to denote all Eastern Slavs, otherwise undifferentiated, and was derived from the word *Rus,* the geographical region where they had all once lived. Gradually, through the years, Russians, Belorussians, and Ukrainians became differentiated nationalities, so that by the twentieth century only Eastern Slavs in the Carpathian Mountains were still known as Rusyns. The name Carpatho-Russian today suggests this "Rusyn" people's geographical origins in the Carpathians, their Slavic culture and linguistic origins, and their religious connections to Eastern Christianity. In the past, Carpatho-Russians were frequently called Ruthenians, but they have also been known as Rusnaks, Uhro-Rusyns, Carpatho-Rusyns, or Carpatho-Ukrainians.[20]

Parish histories and local newspaper accounts reported that Ruthenians were instrumental in 1900 in founding one of Windber's first two Catholic churches, the one that adhered to the Byzantine, or Greek Catholic, rite. This new Windber church, the first Greek Catholic church to be established in Somerset County, attracted an original ethnically mixed membership of 1,500, many of whom were Ruthenians. Because of its Ruthenian, Slovak, and Hungarian Greek Catholic populations, in 1924 the Vatican placed it under the jurisdiction of the Byzantine Ruthenian Rite Catholic Church in the Pittsburgh diocese, not its Ukrainian counterpart in Philadelphia, when ethnic and religious conflicts on the national level led to the creation of two separate Byzantine Rite dioceses in the United States. Despite a bitter national and local schism that split Greek Catholics into two churches in 1936, it is significant that the two local churches retained the use of the Ruthenian language in their Sunday services for years afterward.[21]

Provincial Origins and Their Significance

That Windber-area immigrants came from Austria-Hungary, Italy, or the Russian empire in 1900 might tell us something about them, but such information is clearly insufficient or even misleading for many purposes. Immigration is not a random occurrence, and not all countries, provinces, towns, or villages sent immigrants in equal numbers to the United States. Sometimes the pioneers influenced relatives, friends, and neighbors from the same place to leave in what is usually described as a chain migration. In other villages and towns, few people, or perhaps none, left their birthplaces. Nor did all nationalities emigrate equally or even in numbers proportional to the general population.[22]

Knowing the specific provinces or regions from which Windber-area immigrants left in great numbers can provide us with important information about the areas they left behind and their motivations for leaving. Moreover, at the turn of the century, immigrants frequently defined themselves in terms of their province or region of origin, rather than in terms of a nation-state. Contemporary nationalism had barely affected the peasant masses of a number of southern and eastern European societies, even though, by then, middle-class intellectuals and religious leaders in middle Europe were espousing nationalistic causes. Carpatho-Russians have already been cited, but there are other examples. Thus, Italians settling in American towns and cities usually identified themselves not as Italians but as people from northern Italy or southern Italy, from Abruzzi or Sicily, while Slovak newcomers to the United States typically referred to themselves as Šarišania or Zemplínčania, or some other term that reflected their county of origin. Although there were some exceptions, the masses of new immigrants from many of these societies developed a national consciousness and interest in nationalistic causes in Europe only after they were living in America.[23]

Because immigration was a selective process and migrating people often thought in regional terms, it is useful to break down ethnic populations into smaller geographical units of origin, whenever possible. Doing so requires going beyond censuses and most governmental reports, to church and fraternal society records, which often contained the name of the province or town of a person's origins. Unfortunately, the plethora of foreign nationalities, churches, and fraternal groups in Windber makes a comprehensive breakdown of the area's population into smaller units virtually impossible, however desirable. But it is possible to convey an analysis of certain sectors of that population in this and subsequent chapters.

The vast majority of all southern and eastern European immigrants who came to Windber in the early twentieth century were Roman or Greek Catholics in terms of religious affiliation, although discernible numbers of Jews, and Protestants, especially Slovak Lutherans and Hungarian Reformed Church members, had also migrated. Nonetheless, by 1900 this immigrant majority, adherents of Catholicism, had established two large Catholic parishes in the new town. St. John Cantius Church was originally designed to serve all Roman Catholics regardless of nationality, while St. Stephen's Byzantine Catholic Church (later renamed St. Mary's Greek Catholic Church) served a similar purpose for all Greek Catholics. In time, four additional Roman Catholic churches and one Byzantine Orthodox church eventually emerged or seceded from these two original bodies. Consequently, the membership of these two churches provides us with a useful overview of the origins of many of the town's new immigrants early in the century. The records of the 363 Greek Catholic and

430 Roman Catholic marriages that took place in these churches from 1899 to 1912 are the most useful gauge for assessing church membership and origins.

These marriage records indicate that each church drew constituencies from particular regions of Europe. Greek Catholics came exclusively from counties in northern, especially northeastern, Hungary and the province of Galicia, otherwise known as Austrian Poland. Two-thirds of all 363 brides and 363 grooms married at St. Mary's during this era had emigrated from three provinces, Zemplén and Sáros in Hungary and Galicia in Austria. Except for a small number of the betrothed whose origins were unknown, and a few who had either been born in the United States or whose records had incorrectly cited current residences rather than place of origin, all other Greek Catholics had come from Hungarian counties adjacent to these three provinces. Thus, Windber's Greek Catholic community originally came from a limited and concentrated region that encompassed the northern and southern slopes of the Carpathian Mountains and traversed the territory along the Hungarian-Galician border. At the present time, this area is located in the Lemkian region of southern Poland and the Prešov region of Slovakia.[24] Table 3 shows the detailed breakdown of the provincial origins of St. Mary's brides and grooms.

The ethnic diversity that characterized Windber, then, was not new to St. Mary's parishioners, who had originally come from ethnically mixed provinces in a multiethnic empire. All three of the major nationalities that made up the church's membership—Carpatho-Russians, Slovaks, and Magyars—had coexisted in this region. According to the Hungarian census of 1910, Zemplén's population at the time consisted of 56.5 percent Magyars, 27.1 percent Slovaks, and 11.4 percent Ruthenians. Sáros had a majority of Slovaks (58.3 percent), more Ruthenians than Zemplén (22 percent), relatively few Magyars (10.4 percent), and a smattering of Germans (5.4 percent). As far as Galicia was concerned, the Austrian census of 1900 indicated that Poles made up 46 percent of the province's population, Ruthenians 42 percent, and Jews 11 percent.[25]

In contrast to Byzantine Rite Catholics, Roman Catholics in Windber came from a much more dispersed geographical area, and one that went far beyond northern Hungary and Galicia, into other parts of Europe and across the Atlantic to the United States. A considerable percentage, 16 or 17 percent of the 430 brides and 430 grooms married in St. John's, listed towns in Pennsylvania or other states as their place of

Table 3. Provincial origins of brides and grooms, St. Mary's Greek Catholic Church, Windber, Pennsylvania, 1900–1912

Province	Brides	Grooms	Total Number
Galicia, Austria	23.4%	27.5%	185
Zemplén, Hungary	25.9	20.4	168
Sáros, Hungary	18.7	19.8	140
Ung, Hungary	5.0	5.2	37
Szepes, Hungary	3.9	5.0	32
Szatmár, Hungary	2.8	4.1	25
Borsod, Hungary	3.0	3.6	24
Abaúj-Torna, Hungary	1.9	2.2	15
Bereg, Hungary	1.1	1.7	10
Szabolcs, Hungary	1.1	0.3	5
Szilágy, Hungary	0.8	0.6	5
Gömör, Hungary	0.6	0.6	4
Ugocsa, Hungary	0.3	0.3	2
Máramaros, Hungary	0.3	0.8	4
Nyitra, Hungary	0.0	0.8	3
Komárom, Hungary	0.0	0.3	1
Hungary (province unspecified)	0.0	0.3	1
Lomza, Russia	0.0	0.3	1
United States	3.9	0.6	16
Unknown	7.5	5.8	48
Total	100.0	100.0	726

SOURCE: Marriage Records, St. Mary's Greek Catholic Church, Windber, Pa., 1900–1912.

origin. This figure is misleading, though, because in half of these cases respondents or visiting priests had erred by citing Windber or other current residences, not birthplaces, as the place of origin. Nevertheless, more Roman than Greek Catholics had either resided in the United States for a long time or been born here.

Ireland, which had historically supplied so many Roman Catholic adherents in large cities, never furnished more than a handful of church members in the Windber parish, but Italy provided a far greater number than the marriage data indicated. Linguistic and ethnic differences in the parish had led priests, as early as 1903, to keep separate records, not included in these totals, for Italians, who split off from the church in 1906. Meanwhile, Slovaks and others from northern Hungary also formed a separate parish derived from St. John's in 1905. In the earliest years of St. John Cantius, then, the percentage of church members from Italy and Hungary was higher than in subsequent years and higher than these marriage totals suggest. Despite all these qualifications, these marriage records do convey a good sense of the original geographical and ethnic diversity represented within the church and in Windber. Table 4 summarizes these findings.

Although the parish population was diverse, the core constituency of St. John's membership from its inception had been the Roman Catholic immigrants from the sections of historic Poland partitioned and annexed by Austria and Russia in the eighteenth and nineteenth centuries. Austrian Poland or Galicia alone supplied 35 percent of all brides and grooms married in the church during this period. Another 10 percent came from provinces in Russian Poland, in particular Lomza, Plock, Warsaw, and Lublin, while 6 percent had migrated from Russia's Lithuanian provinces of Vilna, Kovno, and Suval, where numerous Poles also lived. Except for a small number of Lithuanians, then, these people were Polish in nationality.

Northern Hungary supplied the next largest number of brides and grooms married at St. John's in these early years. Zemplén and Sáros, which had furnished so many Greek Catholics, led a list of nine counties in northern Hungary that were represented in the Roman Catholic marriage data. Hungarian and Slovak Roman Catholics originating in these counties therefore constituted an impressive 24 percent of all those married. Smaller percentages of the brides and grooms came from Prussia, Belorussia, or Austria-Hungary's Slavonia and Carniola provinces, which suggests that the church also had members of the German, Belorussian, and Slovenian nationalities.

Several striking observations result from this breakdown of the membership of the two churches into respective provinces of origin. First, St. John's broader geographical and ethnic inclusion should not obscure the fact that, regardless of political boundaries and ethnic differences, virtually all Greek Catholics and about 60 percent of the Roman Catholics in the Windber area came from the same contiguous region in central Europe—the Carpathian Mountains of Galicia and northern Hungary. As Emily Balch, an economist, activist, and pioneer of immigration studies, pointed out long ago in her classic study of Slavic immigration, Slavic and other nationalities converged in this area. Once immigration had begun, it was a contagious phenomenon, not

Table 4. Provincial origins of brides and grooms, St. John Cantius Church, Windber, Pennsylvania, 1899–1912

Province	Brides	Grooms	Total Number
Austria-Hungary			
Galicia, Austria	35.1%	35.3%	303
Slavonia, Austria	0.0	0.2	1
Carniola, Austria	0.7	0.7	6
Austria, province unspecified	0.0	0.2	1
Zemplén, Hungary	7.7	6.0	59
Sáros, Hungary	10.9	11.9	98
Szepes, Hungary	3.7	2.8	28
Trencsén, Hungary	0.5	0.5	4
Liptó, Hungary	0.2	0.9	5
Ung, Hungary	0.0	0.5	2
Abaúj-Torna, Hungary	0.2	0.2	2
Bereg, Hungary	0.0	0.2	1
Gömör, Hungary	0.2	0.0	1
Hungary, province unspecified	0.7	0.9	7
Russian Empire			
Lomza, Russia	5.6	8.8	62
Plock, Russia	2.8	1.9	20
Lublin, Russia	0.5	0.7	5
Warsaw, Russia	0.7	0.2	4
Russian Poland, province unspecified	0.0	0.2	1
Suval, Russia	3.5	3.3	29
Kovno, Russia	2.6	2.3	21
Vilna, Russia	0.2	0.5	3
Belorussia, Russia	0.5	0.2	3
Russia, province unspecified	0.2	2.1	10
Others			
United States	17.2	16.3	144
Italy	1.2	1.2	10
Prussia	0.7	0.9	7
Ireland	0.2	0.2	2
England	0.5	0.0	2
Europe, not specified	0.2	0.0	1
Unknown	3.5	0.7	18
Total	100.0	100.0	860

SOURCE: Marriage Records, St. John Cantius Church, Windber, Pa., 1899–1912.

limited to any one nationality or any one nation-state but one that gradually spread eastward throughout the Carpathian region.[26] More recently, historian Julianna Puskás and others have traced some of the specific geographical, ethnic, and occupational currents involved in this wave.[27]

Second, the great diversity of peoples, cultures, and languages coexisting in this area meant that their juxtaposition in a new place like Windber was not a new phenomenon. Old World conflicts were sometimes transplanted, but, as in Europe, the peasants who formed the bulk of the immigrant influx were least likely to be hostile to peasants of other nationalities or religions. The emerging nationalism that affected upper-class elements in these societies, and the Hungarian government's oppressive discriminations against non-Magyars, had not injured relations at the village level. Balch herself stressed the nonexistence in Europe of ill-feeling between the masses of Slovaks and Magyars and concluded: "The peasants of both races are profoundly unconscious of any reason for hating one another, regard one another as friends, and intermarry freely."[28] In many cases, it was the experience of having become a foreigner in the United States that brought a new consciousness of ethnic identity and nationalism. Whether ethnic awareness ultimately led to interethnic cooperation or to conflict in Windber would depend on many variables, and not on the mere juxtaposition of populations that spoke in different tongues. As Victor Greene demonstrated in his work on Slavic miners in the anthracite region, miners of disparate nationalities were capable of transcending ethnic boundaries and uniting as workers for common class action.[29]

Third, that Windber's new immigrants generally mirrored the larger wave of new immigration, in terms of provinces of origin and multiple ethnic composition, can be seen in the specific case of Hungary. Between 1880 and 1914 the most important regional and mass migration from Hungary as a whole occurred in eight northeastern counties located on the right bank of the Tisza, the same area that had furnished so much of Windber's foreign-born population. Slovaks, Ruthenians, and Germans left this ethnically mixed geographical region of Hungary in unusually large numbers early in the century. But so did the resident Magyars. The general heterogeneous nature of migration from concentrated areas suggests strongly that its root causes lay in particular geographical, social, and economic conditions rather than in national or political oppression.[30]

Regional Social and Economic Conditions

Despite territorial boundaries, similar social and economic conditions prevailed in the areas of central and eastern Europe that sent so many immigrants to Windber at the turn of the century. The entire region was in the midst of an uneven social and economic transformation. Austria-Hungary and Russia, which had ended serfdom in their empires only in 1848 and the 1860s, generally lagged behind western Europe in terms of the capitalistic development of agriculture and industry.

Ending serfdom was a precondition for such development to take place. In no cases were the terms of emancipation generous to the peasants, who won personal freedom and onerous taxes or redemption payments but little or no land. By the turn of the century, landholding, which was already in the hands of a small minority, often became even more concentrated. For example, peasant holdings in Hungary declined from 502,346 units in 1869 to 428,111 by 1896, while the holdings of the nobility increased by 11 percent during this time. Consequently, by 1900 only 30 percent of the population of Hungary had enough land to support an independent existence. Large estates and their owners coexisted with a landless proletariat, who then constituted 25 percent of the population.[31]

Nor was the situation different in Austrian Poland. The combination of concentrated ownership of large estates, backward agricultural methods, and excessive subdivision of small holdings after emancipation gave special meaning to the contemporary term "Galician misery." By 1900 some 80 percent of all plots in the province were small (less than 12.5 acres), and 50 percent consisted of 5 acres or less, when 5 acres were considered the minimum necessary for survival. The trend to commercial agriculture had contributed to greater economic inequality and a stratified social system that had pauperized many of its inhabitants.[32]

High birth rates sometimes, but not always, aggravated the rural crisis in districts with high emigration rates. For instance, the population of Russian Poland doubled from 1860 to 1900 from 5 million to 10 million, and that of Austrian Poland also greatly increased in the late nineteenth and early twentieth centuries, despite the simultaneous exodus of 10 percent of its people. By 1930, Galicia had the highest population density, 162 people per square kilometer, in Europe. On the other hand, demographic growth is a relatively unimportant factor in explanations for the exodus from northern Hungary. Hungary's population growth was moderate, compared with western standards, and although the regional birth rate was somewhat higher than the national average, the mountainous region was never densely populated.[33]

With the development of capitalistic agriculture and concomitant social inequality, peasants in northern Hungary and Galicia found themselves increasingly unable to support themselves and their families on small parcels of land. Earning wages through other means had become essential to survival, but in neither case did provincial industry fill the need for wages and employment. Meanwhile, railroad penetration and the arrival of manufactured goods made elsewhere were destroying local crafts and displacing artisans. Industry as a whole was not well developed in Hungary, in part because the Austrian government consciously followed a policy that favored Austrian manufactures, on one hand, and maintained Hungary as a source for raw materials and foodstuffs, on the other. That the food industry was one of Hungary's most advanced industries indicates its second-class status in the Dual Monarchy.[34]

In a similar manner, the Hungarian government discriminated against Slovak businesses as part of its broader discriminatory nationalistic policies. Economically, Galicia fared no better, and of the three Polish partitions, it was noted for being the most backward, because no industry of any importance existed there. Its economic development, or lack thereof, contrasted sharply with that in Russian Poland, where the czarist government had consciously fostered industrial growth in the late nineteenth century. By 1900, Warsaw and Lodz had become important industrial centers, while coal, steel, textiles, and sugar-refining simultaneously emerged as leading industries in the Russian Polish sector.[35]

Migration for work was a well-established tradition for many of the peoples who inhabited central and eastern Europe. Julianna Puskás and Ewa Morawska, among others, have argued that the regions with high emigration rates were usually the regions in which this tradition existed. In many cases, geographical mobility had gradually extended to more-distant locations. For example, Slovaks, who had been migrating to southern parts of Hungary from the eighteenth century on, traveled to neighboring countries in the early nineteenth century and to the United States by the 1870s and 1880s. The unevenness of economic development, the fragmentation of landholding, and the increased dependence on wages eventually led migrants to seek work beyond the European continent even as migration was becoming an integral part of an international world capitalistic economy.[36]

Under certain circumstances, the general economic and social conditions existing in the Carpathian region became significant "push" factors that prompted mass emigration from the area. Emigration was never automatic, and historians have traditionally cited a variety of "push" and "pull" factors, rather than a single reason, to explain large-scale migration from one particular province or country to another. In an insightful study of emigration from Hungary, Julianna Puskás has integrated push and pull factors in an original way while stressing that emigration was highest in regions where the two forces interacted and reinforced each other.

Generally speaking, Puskás argued, the peripheral areas that supplied so many of Hungary's immigrants were more mountainous and had poorer natural resources than other parts of the country. They also lacked industrial opportunities. Moreover, the numerous nationalities in these regions had traditions involving geographical and seasonal mobility for work and personal contacts with successful overseas emigration as well. Emigration rates varied from year to year, but the overall rate of departure from Hungary seems to have borne a rough correlation to economic conditions in the United States. At least it was lowest during years of depression, when remigration rates also rose. In the opinion of Puskás (and many other historians), the primary pull of America consisted of the steady jobs and relatively high wages paid in industry. She calculated that American industrial wages were five to six times higher than could be earned during this period in agricultural work in Hungary. Because money was needed to purchase land and otherwise provide for the economic well-being of oneself and one's family, migration was a possible strategy or solution designed to cope with existing realities, to achieve basic security, and to maintain or elevate one's social status at home.[37]

A Profile of the Immigrant Populations

Windber-area immigrants reflected the general wave of new immigrants. Many are known to have come from rural districts where they farmed their own land or that of other people. Although displaced artisans and small merchants often preceded the rural peasant exodus, 70 to 80 percent of all immigrants to the United States from central and eastern Europe after 1890 came from rural areas where the lack of industrial development and parcelization of land offered them few alternatives to migration. Small and medium-size landholders came first, with the landless coming in greater numbers after 1900. The proportions of small landowners and the landless varied considerably, reflecting national and regional differences. Thus, approximately 50 percent of all Magyars, Carpatho-Russians, and Poles who arrived in the United States from 1900 to 1914 were landless, perhaps young sons of small landholders, while only 10 percent of the Magyars, 19 percent of the Slovaks, 24 percent of the Carpatho-Russians, and 29 percent of the Poles were small independent farmers.[38]

Moreover, on the national level, 80 percent or more of certain groups, such as Slovaks, came to the country as unskilled laborers, most of whom entered industrial occupations afterward. The Dillingham Commission reported that Slovaks alone comprised 13.1 percent of the nation's steel workers and 12.8 percent of all bituminous coal miners. The rural background of the new immigrants and their lack of previous mining experience were also cited. Only 8.2 percent of all the southern and eastern immigrants employed in coal mines in the United States had been employed in that occupation in their native lands. By contrast, 87.6 percent of the Welsh miners and 82.6 percent of the English miners who worked in American mines had worked in European mines before their emigration.[39]

The initial influx of new immigrants into Windber mirrored national trends in that it was predominantly male. Mining was "men's work," and thirteen years after the town's founding, males still comprised a far greater portion, or 59 percent, of the local population than did the females, who made up only 41 percent of it. Census officials noted that a greater percentage of males than females had emigrated to the United States during the decade from 1900 to 1910, and that the ratio of males to females among the foreign-born was higher (131.1 males to 100 females) than it had been in previous decades. Certainly the immigrant influx of particular ethnic groups was predominantly male. Paul Magocsi has pointed out that single or recently married young men made up a full 75 percent of the Carpatho-Russian immigrant population before 1914. Moreover, men who came without their families in search of work predominated during the peak years of immigration from places such as Hungary.[40]

It is not surprising that immigrant males who expected to work in difficult, strenuous, industrial jobs emigrated when they were in the prime of life. Puskás has shown that nearly 60 percent of those who left Hungary during the peak emigration years of 1905–1907 were under 30 years of age and that most of these were in their most productive working years. About 7 percent were under the age of 14; only 3 percent were 45 years of age or more. Miners interviewed often said that mining required strong (presumably youthful) backs. The Dillingham Commission indicated that Berwind-White had found just such men during Windber's early years. Of 2,833 foreign-born company employees in Windber, the largest segment, or 52 percent, were between the ages of 20 and 29; those in their 20s and 30s made up 90 percent of this work force.[41]

National census figures suggest that female immigrants were similarly youthful or even younger. The bulk of females who entered the country from all immigrant groups during the era were clustered in what was considered the primary marriageable age range, 14 to 21; only small percentages of new immigrant women were over 30. These figures varied with each nationality, but 73.7 percent of arriving Magyar females and 84.6 percent of Slovak females arriving fell within the age range of 14 to 21.[42]

A majority of the men who came to Windber to find work were married. Of the 2,821 foreign-born miners for whom Dillingham investigators had information, 55.5 percent were married while 44.1 percent were not. The comparable rate for a much smaller number of native-born miners varied only slightly from this figure. Moreover, the 1910 census survey of the large boarding population living in the Windber area indicates that approximately one-third of all boarders were married men whose wives lived elsewhere.[43]

The immigrants who came to Windber during its early years were predominantly recent arrivals in America. At least 75 percent of all foreign-born residents in the Windber area in 1910 had come to the United States sometime between 1900 and 1910. More than 42 percent of them had come within the previous five years, and it seems reasonable to conclude that many of these either migrated directly to Windber or moved there within a short time of their arrival. The groups with the highest proportions of their local populations to arrive in America after 1900 included the four most numerous nationalities from southern and eastern Europe. By contrast, the vast majority of northern Europeans had originally arrived twenty or more years earlier.[44]

The time Windber's various ethnic groups arrived in America closely parallels national immigration trends. Locally and nationally, the bulk of northern European immigrants historically preceded the bulk of southern and eastern nationalities by at least a decade. Slovaks were relative pioneers of the new immigration wave. They had begun to migrate to the anthracite mining region of Pennsylvania in the 1870s and 1880s, when coal companies recruited new immigrants under contract labor conditions in the aftermath of Molly Maguire labor troubles there. A relatively large proportion, 30 percent of the local Slovak population, had arrived in America before 1900.[45]

Circumstantial evidence derived from oral histories, contemporary newspapers, parish records, fraternal society correspondence, naturalization documents, the census, and the Dillingham report indicate that many Windber-area immigrants initially expected to be in America only temporarily. Historians of immigration generally concur that, until World War I, most immigrants from Hungary and other places in central and eastern Europe left there with the intention of finding temporary work, accumulating savings, and eventually returning to their homelands. Estimates of actual return rates have ranged anywhere from 25 percent to 60 percent, depending on particular nationalities, regions, and historians. It is significant that northeastern Hungary—the primary locus of origin for many of Windber's immigrants—had one of the highest return rates in Europe. John Bodnar cited estimates that the Magyar return rate was 64 percent and that the Slovak rate was 59 percent.[46]

Decisions to remain permanently in the new country were usually made after immigrants had lived here for some time and were often prompted by some compelling national or familial circumstance. Nor was it unusual for laboring men to travel back and forth across the ocean several times before a final decision was reached. In its national findings, the Dillingham Commission itself concluded that 40 percent of the new immigrants eventually returned to Europe and that two-thirds of these stayed there permanently.[47]

Throughout the early twentieth century, steamship agents advertised widely in local newspapers that routinely carried items about immigrant workers who were returning to Europe. In one such announcement, in December 1902, the *Windber Era* reported that agent Andrew Zemany had sent eleven people to Slavonia via Bremen on the *Kaiser Wilhelm.* Other articles stated that work in the Windber area was migratory and seasonal in nature for a number of the foreign-born who owned land in Europe. On March 31, 1903, for example, the *Windber Journal* reported that sixteen people were leaving to resume farming in Europe and that more were expected to follow them in the next few days. Other workers seem to have left Windber in the winter but were expected to return in the spring.[48]

According to the papers, economic recession also influenced return rates. For example, late in 1903 the *Windber Era* noted a great exodus of foreigners from the coke regions of western Pennsylvania during an economic lull, but also commented on the departure of many local Italians and Poles. On occasion, contemporary journalists acknowledged that some immigrants who had accumulated savings were returning home permanently. In one such instance, in July 1903, the *Windber Journal* reported that sixteen foreigners, including several women, were about to leave for Europe. Apparently they did not expect to return to the United States, because the steamship agent who had arranged their passage told the reporter that all "were well supplied with American coin, and they can now live luxuriously in their fatherland."[49]

One standard way of measuring the expected temporary or more permanent nature of an immigrant group's residence in America is to see where the wives of miners lived. Caution is needed here in drawing conclusions. In general, it is true that ethnic-group males who emigrated with wives and children intended to stay permanently from the onset. On the other hand, married men who came alone may have simply lacked the resources necessary to bring their families over immediately.

The Dillingham report, the only source for information on the location of wives, found that of the 1,507 whose whereabouts were known, 54.1 percent were living elsewhere in the United States and 45.9 percent were still abroad. Among the recently arrived ethnic groups, Italians, Magyars, and Russians had more than 50 percent of their wives in Europe. More northern Italian wives, 58.2 percent, than southern Italian wives, 50.9 percent, were abroad, as were 54.5 percent of the Magyar spouses. Of the four most numerous nationalities in Windber, two groups, Slovaks and Poles, had a majority, approximately 60 percent, of their wives in America. Because Slovaks had been a pioneer ethnic group in immigration to the United States, the higher resident rate of Slovak wives is perhaps understandable, but this rationale does not explain the Polish finding.[50]

Also, although northern European immigrants had a greater percentage of their wives located in America, a significant portion also had wives abroad. A surprising 20 percent of the German miners and 19 percent of the English miners had wives who still lived overseas.[51] That a sizable segment of these immigrant groups had not yet reunited their families, even though their ethnic groups generally had been in America longer, casts doubt on artificial distinctions sometimes made between the nature of the old versus new immigration and strongly suggests that the immigrants' class position and length of residence, and the life cycle and stage of immigration itself, explain many differences better than national origin does.

References to wives and children residing abroad appeared frequently in local newspapers in routine reports on fatalities resulting from mining accidents. Typical was the short obituary that followed a longer description of the accident that resulted in the death of Vidor Roseleski in August 1904: "Roseleski was thirty years old and had a wife and two children in the old country." In addition, fraternal societies kept detailed information for insurance purposes. A perusal of national Slovak fraternal records showed that it was still common in the 1920s and 1930s for wives or other relatives living abroad to write letters of inquiry about benefits owed the beneficiaries of a deceased husband, son, or brother who had once worked in Windber.[52]

Another traditional barometer of the transience of various ethnic groups is the amount of money sent abroad. One major goal of most immigrants was to earn high wages and save as much money as possible. Money was essential for supporting the family unit and for purchasing land in the home country, if returning was contemplated, or for a house in the United States, if a decision to stay permanently had been made. The amount of money sent to Europe also paid the passage for other relatives to emigrate. Government officials in Austria-Hungary sometimes first became aware of mass immigration from a particular region because of the amount of dollars and money orders sent there from America, and the Dillingham Commission cited statistics to show that large amounts of money were involved. More than $500 million had been sent abroad via the postal service alone from 1900 to 1909.[53] Windber's immigrants took part in this characteristic immigrant activity. Again, the only source for concrete information on the local level comes from the Dillingham study.

Commission investigators collected related but disparate data for a two-year period from the town's postal service, two national banks, and a Slovak enterprise, all of which transmitted money orders domestically and internationally. One surprising finding that emerges from these sources is that the largest portion of money transmitted by the post office and the Slovak bank went to domestic rather than foreign destinations. The postal service had sent $85,000 to other places in the United States during the two years, and only $20,000 abroad. Even more surprising, the Slovak bank, which sent a total of about $21,000 to other locations, had transmitted only 1 percent of this or approximately $210, to foreign nations. About 82 percent of its money orders were destined for residences in Pennsylvania; the remainder went to other states.[54] Virtually all nationalities did send some money abroad, but the large amounts sent to domestic sites suggests that the native-born also transferred funds and that many foreign-born had spouses, parents, or other relatives already in residence in the country. Slovaks readily come to mind in this regard because numerous members of this ethnic group had spent time in the anthracite region before emigrating to Windber.

Nationalities also sent dollars abroad in amounts that were disproportionate to their numbers in the area's population, and findings for Windber confirm a general national pattern. By all calculations, the unquestioned leader in accumulating and transmitting savings to Europe was the Italian contingent. This one ethnic group single-handedly purchased more than $17,000 of the $20,000 mailed abroad in two years via the postal service. An additional $90,000 was sent to Italy during the same period by the two national banks, which transmitted foreign money orders only to Italy. The Italians even surpassed the Slovaks, Magyars, and Poles in the percentage of their money orders sent abroad but purchased in the Slovak bank![55]

Another striking fact that emerges from the data on dollars sent abroad from the postal service in Windber is that, next to Italy, the countries receiving the greatest amount of money were England, Wales, and Scotland, in that order. Nearly $2,000 of the $20,000 sent in two years went to these places. Given their small representation in the overall foreign-born population, these northern European nationalities seem to have greatly outdistanced most southern and eastern ethnic groups in financial transmissions. Perhaps this result should not be surprising. These nationalities generally occupied the more skilled and higher-paying mining positions. Smaller numbers of these groups could afford to send large sums of money. By contrast, large numbers of Italians sent many small individual remittances. But other sources and information are needed for a complete picture. New immigrants used many other mechanisms to transfer modest amounts of currency to their homelands. Sometimes individuals personally carried funds for themselves or relatives. Moreover, private groups and ethnic societies had financial links of their own, and other immigrant and American banks were located in nearby cities, such as Johnstown.

No written sources convey a sense of the poignancy of the original intention of many immigrants to return as well as oral histories do. One elderly couple interviewed spoke movingly of the lot of the man's Magyar parents, immigrants who had come to the United States early in the twentieth century. From the time of their arrival in Windber before 1910 until their deaths decades later, the parents had dreamed of accumulating a nest egg to use to return to their native village, where they hoped to buy a modest house and open a small store. They even sent their eldest American-born son to religiously oriented summer schools with the conscious intention that he learn to read and write Hungarian better than they themselves could. He had to be educated to run the store in Europe and keep its books. Once or twice in their lifetimes the unfortunate couple did manage to save enough money to make the return, but some event or other intervened to deplete their funds or otherwise cause them to change their plans. In one case, a world war prevented their return. In a second instance, they were bilked of their entire savings by an unscrupulous salesman involved in a fraudulent land scheme. Never fully at home in their new environment, the man labored in the mines for the rest of his working life. Lacking the means necessary even for a visit, the foreign-born couple were never able to realize their dreams or see their native village again. In the late 1960s, the eldest son and his wife were able to visit the cherished birthplace.[56]

Berwind-White needed men to mine coal in its new operations in Windber. Slovaks, Magyars, Poles, Italians, and other nationalities supplied the work force. They made the journey from Europe to Windber, either directly or indirectly, because an international system existed that brought working people from the Carpathians and elsewhere together with industrial employers in the Alleghenies. The diverse populations of Europe had by that time become an integral component of an emerging worldwide labor market, and the formative stages of American industrial capitalism required plentiful labor.

Contrary to popular myth, until about World War I most new immigrants who came to the United States expected to return to their native countries, where they hoped to buy land or other property from the wages they had earned during a work stint in the New World. The primary reason for the general exodus from selected areas at the turn of the century was economic, although other reasons, such as czarist persecution of Jews, were influential in particular instances. The regions that immigrants were leaving behind were undergoing uneven social and economic transformations of their own, and migration for work—neither a new nor necessarily permanent phenomenon—became one strategy for confronting the imperatives of the new economic system. Once they had settled in Windber, they continued their struggle for a better life in an autocratic setting that presented them with many new problems and challenges.

3

THE WORK OF MINING

Berwind-White's dependence on immigrant labor from the outset forced the coal corporation to develop recruitment practices that were consciously designed to appeal to the foreign-born. These practices, however, were only the first step in a broader corporate policy that linked the company to its multiethnic work force. The company's success in obtaining a diverse population to mine its coal enabled it to engage in deliberate, if commonplace, employer strategies and techniques that aimed at dividing and dominating its workers. The company-town setting that Berwind-White had created ensured that its efforts to maintain authoritarian control, maximize profits, and avert unionization would not be confined to the workplace but would extend to the entire community. Nevertheless, the workplace rapidly became a key arena of contention between capital and labor.

This chapter examines the organization of work in the mines and in the community, with a view to exposing the underlying pattern of occupational ethnic clustering and concomitant social stratification that characterized Windber throughout the nonunion period. That pattern, in turn, should reveal many of the ongoing work-related grievances of the miners, as well as many of the obstacles that workers had to overcome to unionize and improve their situation.

Ethnic and Racial Hierarchies in Town Occupations

The new immigrants who came to work in Windber-area mines found themselves inhabitants of a town that was predominantly working class. The vast majority, 78 percent, of the area's working population, enumerated in the 1910 census, held unskilled, manual laboring jobs. Another 8 percent were employed in the various trades, while 4 percent worked in white-collar clerical, bookkeeping, or sales positions. Only a minority, 10 percent, were engaged in business, the professions, or management, which were traditional middle-class occupations. Berwind-White mine owners, it is important to remember, were absentee owners.

The new immigrants were also entering a society that was highly stratified. The town's prevailing class structure embodied a particular hierarchical arrangement of ethnic groups that left the Americans and certain northern European nationalities at the top of the social order and clearly distinguishable from the new immigrant groups in terms of occupation, political clout, economic power, residence, and status. As a result, the major dividing line within the community was simultaneously a class and ethnic one.

Cross-tabulations of nationalities and jobs in the 1910 census reveal the privileged occupational positions of Americans and northern Europeans. Americans dominated the professions in Windber. All dentists, two-thirds of the local physicians, all druggists, and 88 percent of area teachers were native-born. Moreover, lawyers, law enforcement officials, and justices of the peace were exclusively products of Pennsylvania. Reflecting the importance of the ethnic populations, the clergy—priests and ministers—were ethnically more diverse, but Americans still made up 50 percent of the total number enumerated. All bookkeepers whose birthplaces were listed were American-born, and 75 percent of all clerks were of American or English descent. Merchants were a mixed lot, but the native-born still comprised more than 40 percent of the businessmen and owned the bulk of the most important enterprises. With a couple of exceptions, Americans, Scots, Welsh, Germans, or Scandinavians filled such occupational categories as coal operators, railroad and mining bosses, and foremen. These same nationalities also dominated skilled jobs, such as engineering, and trades, such as carpentry. The brewmasters at the local brewery were exclusively German or Bohemian, and the concern employed thirty German workers.[1]

Southern and eastern Europeans were more likely to be in certain trades and businesses. For example, Italians made up 38 percent of the town's stonemasons, 39 percent of the barbers, 9 percent of the blacksmiths, 100 percent of the shoemakers, 25 percent of the bartenders, and 10 percent of the tailors. In turn, Slovaks made up about 29 percent of the insurance agents, 3 percent of the blacksmiths, 20 percent of the butchers, 7 percent of the clerks, 8 percent of the merchants, 80 percent of the musicians, 20 percent of the tailors, and 5 percent of the teachers. The new nationalities also monopolized a few of the least desirable jobs. Thus, Magyars, Jews, and Russians were the sole junk dealers and peddlers in town. On the other hand, a surprising finding that emerged from cross-tabulation was that two-thirds of all servants were American-born and that Magyars and Slovaks were only 25 percent of the total. But, except for these female servants, the jobs listed above occupied only a small minority of the employed ethnic populations. The vast majority of the new nationalities worked for Berwind-White and filled the area's numerous unskilled and semi-skilled laboring and mining jobs.[2]

When we look at specific nationalities, we find that certain ethnic groups tended to specialize in particular occupations in Windber. For example, census tabulations indicate that 25 percent of all Syrian residents, including unemployed women and children, were merchants and that approximately 45 percent of all enumerated Jewish people were merchants, peddlers, or salesmen. At the same time, the infinitesimally small numbers of the lowest groups in the social hierarchy, Orientals and blacks, were confined to certain manual laboring jobs. All three Chinese residents were laundrymen, and of the thirteen blacks in the area, nine were employed, six as laborers, primarily hod carriers, one as a teamster, one as a hotel porter, and one as a combination housekeeper and servant.[3]

As a group, southern and eastern Europeans were not the bottom rung but close to it, of the ethnic and racial hierarchy that pertained to jobs in the community at large. Aside from the small minority of their fellow compatriots who were engaged in other jobs in town, new immigrants were clustered in the many unskilled and semi-skilled laboring and mining jobs.

Labor Recruitment and Ethnic Policies

Over the years, Berwind-White employed various mechanisms to recruit miners for its Windber-area operations. After its initial use of employment agencies, transfers of company personnel from other places, and publicity about the opening of new mines, the company turned to overt advertising in American newspapers. In 1899, for example, in the midst of its first serious battle against unionization, it sought to lure men away from neighboring mines by placing a want ad for 500 miners in a Johnstown paper. In another instance, a want ad placed in a Pittsburgh paper asking for a store clerk who could speak several languages brought a future company-store manager to Windber in 1907. In later years the company advertised widely in foreign-language newspapers in languages and in terms thought to appeal to ethnic populations.[4]

But steamship agents, immigrant bankers, and ethnic labor contractors were even more important as suppliers of labor early in the century. Particular ethnic leaders had railroad, steamship, and financial connections that enabled them to serve as intermediaries for new arrivals in America and as de facto or authorized agents of the company. In addition, the coal corporation periodically sent a representative who was noted for his diverse linguistic talents to New York and other ports to meet incoming ships to try to persuade men to come to Windber. This individual was also sent to other mining towns for the same purpose. Once foreign-born miners had settled in the Windber area, they themselves became active, if sometimes unintentional, agents of recruitment through letters and dollars sent to friends and relatives abroad.[5]

Recruitment agents had to sell Berwind-White and Windber to prospective immigrant miners in a competitive labor market. In a certain sense, company-town employers enjoyed advantages over competitors who had only jobs to offer, and the company's monopolistic ownership and control over area jobs, housing, and community institutions initially attracted some immigrants to Windber. One elderly miner interviewed noted that, for immigrants who had arrived in the country with sparse resources, an immediate job was a necessity and that, unlike some other enterprises, the local mining corporation could simultaneously promise prospective workers jobs, housing, and credit from the company store for food, mining clothes, tools, and powder. Moreover, the Windber area's diverse foreign-born populations offered many newcomers ethnic and religious kinship that would help in the transition to a new environment.[6]

The Dillingham Commission ignored these advantages and cited only one inducement—steady employment—for men to come to Windber. Berwind-White's markets, especially the transatlantic steamship trade, made the company relatively immune from the periodic recessions experienced by competitors. State mining reports confirm that the company's annual total of workdays regularly surpassed the average for bituminous coal. During these early years, lulls in work were rare, and usually resulted from car shortages or strikes by company dockworkers in New York rather than from general industrial slowdowns. Furthermore, the company's extensive use of modern mining machinery meant that inexperienced and unskilled people could find abundant work in the mines. That the mines were free of large accumulations of gas was a welcome bonus.[7]

Retaining a labor force was ultimately as important as recruiting one. As early as 1875 a steel-industry employer in Johnstown had advised Carnegie Steel Company officials that steady work was crucial in keeping a labor force intact, even as it simultaneously enabled the company to get away with paying low wages.[8] Berwind-White's ability to provide steady work did help obscure that wages paid were often not the rate paid by competitors in unionized mines. In addition, oral histories have revealed that the company considered the method of payment to be an important recruitment device that brought people to Windber instead of to South Fork or other unionized places. When miners received cash after all company deductions were made, their low wages were paid in gold and silver rather than in paper currency, because immigrants were believed to trust and prefer the hard metallic variety to its alternative.[9]

But these policies were not always successful and the immigrants did have minds of their own. This is evidenced by the Dillingham report, which stressed outmigration from the community as an ongoing significant reality and the "constant shifting of the population" as a major hindrance to Americanization. In its view, the company's relative low wages, general lack of unionization, and specific grievances, such as the requirement that miners push cars themselves to prescribed locations, caused workers to leave area mines during prosperous times when other jobs were plentiful. During times of general depression, however, miners poured into the area.[10]

From the onset of its Windber operations, Berwind-White was noted for its virulent antiunion policies, and it indulged in a time-honored employer practice designed to keep workers apart. In a manner advocated by contemporary steel companies, it "judiciously mixed" the nationalities by hiring people on the basis of an ethnic quota system.[11] Unskilled work at the mines was similarly and purposely mixed, according to the Dillingham Commission. Voluntary ethnic work segregation was not permitted. Instead work was given out to gangs apportioned equally among the nationalities. Commission investigators abruptly rejected the company's claim that this system maintained efficiency that would be lost if any one nationality dominated, and singled out a very different motivation for the corporation's hiring and gang policies:

> The existing situation . . . seems to have arisen from the efforts of the operators to drive out organized labor in the past and to prevent any organization of the miners in the future. Members of the same race working together would be united in sympathy and language and would offer a good field for the efforts of the labor organizer. At the same time unanimity could be easily secured in presenting grievances to the employer. On the other hand, by mixing the races employed, the barrier of language, together with traditional enmities and prejudices, prevents concerted action among the miners and renders it easy for the operator in the event of grievances to play one race against another.[12]

The Work of Mining: Job Classifications

The laborers who took part in the activity of bituminous mining performed a variety of skilled, semi-skilled, and unskilled jobs. Traditionally, mine work was organized on the basis of the numerous occupations entailed in the operations of the mines.

Miners at Eureka Mine No. 35, 1905.

Mining jobs can be classified in a number of ways, but a first important division, one incorporated in state mining reports, was between the occupations that were carried on inside the mines and those that were performed on the surface or outside the mines. The work underground included the basic tasks necessary to the extraction and transportation of the mineral to the surface, while the outside jobs involved subsidiary tasks, such as the weighing and shipping of coal to market, bookkeeping, carpentry, and blacksmithing. Underground employees were necessarily preponderant, and statistics show that work below the surface was vastly more dangerous. Although the number of Windber's work force varied considerably from 1897 to 1918, the ratio between those employed inside and outside of the mines remained more constant. On the average, 89 percent of the entire work force held underground jobs, while 11 percent had jobs outside the mines.[13]

Another traditional way of classifying the bulk of mine employees is to distinguish between those who were termed "miners" and those who were termed "company men." By the twentieth century, the word "miners" generically included everyone who shot, cut, or loaded coal at the face of the working place or room in the mine. Consequently, "miners" might be pick miners or machine miners, loaders, cutters, machine runners, or scrapers. By contrast, company men included the various other employees who worked in or around the mines and who were not directly engaged in the shooting, cutting, or loading of coal. Company men employed on the surface included tipple gangs, who dumped coal from the mine cars to railroad cars, carpenters, blacksmiths, and machinists. Inside company men included motormen, maintenance men, tracklayers, timbermen, bratticemen, and wiremen.[14]

Miners and company men differed substantially from one another in terms of skill, method of payment, and degree of independence at work. While some company men had unskilled common labor jobs, such as road cleaning, approximately half of this category of the work force held skilled jobs in maintenance or transportation work. By contrast, especially after the widespread introduction of mining machines in the late nineteenth and early twentieth centuries, mining and loading, with few exceptions, were considered unskilled work. The disintegrating craft of the miner, who was once an independent contractor, has been described in a number of works devoted to this subject, but for our purposes it is sufficient to note that by 1900 the increased use of mining machines in the industry had made it possible and profitable for operators to hire masses of unskilled "greenhorns" from southern and eastern Europe. "Strong backs" were so frequently cited by operators and miners alike as the primary requirement to do the work that the saying "A mind that's weak and a back that's strong" eventually became embodied in mining folklore and song.[15]

Miners were pieceworkers. Unlike the company men, who were often called "day men" because they were paid a set daily rate of wages, miners were paid by the amount or tonnage of coal they produced. This distinction is crucial because the method of payment affected ways of thinking and the work process itself. For example, because nonunion miners typically received no pay for deadwork—the removal of rock, slate, or other obstructions in order to get to the coal itself—they might tend to be somewhat careless in blasting coal if they thought they could escape hours of hard, tedious, and financially unrewarding work. On the other hand, although motormen, trackmen, electricians, and other company men undoubtedly had schedules to keep, their earnings remained constant for each day worked, regardless of schedule changes or variable tonnage outputs.[16]

The two methods of payment and two types of categories of occupations reflected a major division within the ranks of a mine's labor force, a division that had important implications for the success of unionization efforts in general and the outcomes of strikes in particular. As a rule, the United Mine Workers of America, an industrial rather than craft-based union, successfully incorporated both groups of workers into its organization, and during the strike of 1906 in Windber the UMWA rejoiced that it had won a great victory when the inside day men joined the tonnage men on strike.[17]

Because miners worked largely unsupervised, alone or with a buddy, in rather isolated rooms of the mine, they had also traditionally enjoyed a large degree of freedom at the workplace. Company men did not have the same freedom. Never having been independent contractors, they had always been hired directly by the operator or owner and had also been more closely supervised. Operators generally considered them adjuncts of management, but miners distinguished those holding elite positions, such as weighmasters, from inside company men who held lesser jobs and shared the miner's tradition of independence and comparative freedom from supervision. Transportation men, for example, were always on the move and difficult to locate. It is probably not accidental that the massive walkout in the Windber mines in 1922 was led by these men. Outside men were more easily supervised than inside men. In general, the traditional autonomy of the miners was contagious, spreading outward from the face of the miner's working place. Carter Goodrich reported that the advent of new technology and a concomitant reorganization of work in the mid-1920s threatened to destroy the miner's freedom and quality of working life even as the changes presented new opportunities for new collective forms of self-direction.[18]

Unskilled miners working at the face of the coal seam unquestionably constituted the bulk of the mining labor force, but exactly how numerous were those men dubbed "miners" in the Windber mines? Carter Goodrich estimated that 75 percent of a given mine's workers were miners. Windber's figures vary only a few percentage points from his estimate. According to calculations based on occupational figures contained in state annual mining reports from 1897 to 1918, miners totaled 71 percent of the entire number of employees in Windber-area mines. Nearly 66 percent of all Berwind-White mining employees in the area during this period were classified by the state as either miners or loaders. Another 5 percent were machine runners and scrapers, thereby bringing the overall percentage of miners, as usually defined, in the total labor force to 71 percent. Miners constituted an even higher average or 79 percent of all local underground workers during this period.[19]

Table 5 lists the changing number of miners throughout the years 1897 to 1918 and indicates what percentage of the inside employees and total labor force they constituted. Especially striking is the 40 percent figure for 1906—one indication of the widespread participation in the strike by men at the working face. The smaller numbers and changed percentages from 1916 to 1918 reflect, in part, the company's increased difficulties in securing and retaining labor recruits during the war years.

Superintendents, Foremen, and Assistant Foremen

Clearly distinguishable from the masses of miners and company men were the managers of the mines. While a general superintendent and an assistant superintendent were responsible for supervising the overall operation of Windber-area mines, the foremen and assistant foremen had the major daily managerial responsibility at the various mines. Pennsylvania state law required that operators hire competent state-certified foremen "in order to secure efficient management and proper ventilation of the mines, to promote the health and safety of the persons employed therein, and to protect and preserve the property connected therewith." These men were a tiny but powerful minority, ranging from less than 1 percent to 2 percent, of the entire work force during the period 1897–1918.[20]

Ethnic Clusters and Hierarchies

Southern and eastern Europeans dominated the unskilled mining and loading jobs of the Windber-area mines. Figures taken from the manuscript schedules of the United States census of 1910 indicate that 90

Table 5. Miners and their representation in the labor force, Berwind-White mines, Windber, 1897–1918

Year	Number of Miners	Percent of Inside Employees	Percent of All Employees
1897	175	74%	55%
1898	1,173	87	82
1899	1,613	80	75
1900	2,261	82	75
1901	2,286	81	74
1902	3,065	87	78
1903	2,965	76	71
1904	3,259	87	82
1905	3,304	88	80
1906	1,522	46	40
1907	3,114	74	65
1908	3,698	83	74
1909	3,599	83	74
1910	4,044	86	81
1911	3,331	78	73
1912	3,103	78	75
1913	2,841	81	77
1914	3,393	85	79
1915	2,544	80	70
1916	2,113	77	66
1917	2,285	79	67
1918	1,506	76	59
Average	2,600	79%	71%

SOURCES: Pennsylvania Department of Internal Affairs, Bureau of Mines, *Report* (Harrisburg, Pa., 1898–1903), annual reports, 1897–1902; Pennsylvania Department of Mines, *Report* (Harrisburg, Pa., 1904–1920), annual reports, 1903–1918.

percent of those listed as inside miners, loaders, or coal cutters were new immigrants. The big four nationalities—Magyars, Slovaks, Poles, and Italians—alone constituted 85 percent of such workers. Magyars made up 28 percent, Slovaks 26 percent, Poles 20 percent, and Italians 11 percent of the local Berwind-White miners.[21]

At the same time, southern and eastern Europeans held other unskilled jobs in the mines. They were the majority of tracklayers and trackmen and a sizable portion of the company's laborers. Only a handful of these ethnic workers held semi-skilled jobs, such as motormen and spraggers, or more highly skilled jobs, such as electricians or coal carpenters. Blacksmithing was one exceptional arena in which new immigrants were more highly represented. Nearly one-third of all company blacksmiths were of new immigrant origin.[22]

By contrast, Americans and northern Europeans held the vast majority of skilled company jobs. Without a single exception in the census of 1910, they monopolized Berwind-White's entire force of engineers, weighmasters, surveyors, car inspectors, and bookkeepers. Moreover, although Americans held only a small minority of all the jobs inside the mines, they greatly predominated in outside jobs.[23]

A certain amount of specific ethnic clustering occurred even within the skilled and unskilled dichotomy. Two instances stand out. The first case involved the Italians who exercised ethnic dominion over work at the tipple. Those employed as outside miners, tipple miners, and tipple laborers were exclusively Italian in national origin. State mining reports covering 1914 to 1918 tend to corroborate this finding, since Italians were second only to the Americans in the number of outside employees in the district. The car shop was a second instance of specific ethnic clustering. Here the skilled work force was divided rather equally between Americans and Scandinavians. Norwegians and Swedes did much of the car repair work.[24]

In sharp contrast to their numerical preponderance in the less-skilled mining occupations and in the overall labor force, southern and eastern Europeans occupied only a small number of the managerial jobs. In fact, their proportional share of these positions was exactly the reverse of their share in the unskilled jobs. While they made up 90 percent of the miners in 1910, they were only 10 percent of those listed as mining foremen and assistant foremen. Table 6 contrasts the ethnic composition of the men employed in the two types of occupations.

That Americans comprised the core of absentee owners, superintendents, and foremen who ran Windber-area mines is hardly a surprising finding. However, it would be wrong to draw the conclusion that the few southern and eastern European individuals who occupied managerial positions were unimportant. In fact, the opposite case might well be argued as it has been in another chapter. These strategic elevated positions gave a handful or two of key ethnic men a stature, prominence, and power which extended far beyond their small numbers. Along with

Table 6. Miners and foremen, Berwind-White mines, Windber, 1910

	Miners	Foremen
Americans	7%	56%
Northern Europeans	3	34
Southern and Eastern Europeans	90	10

SOURCES: Pennsylvania Department of Mines, *Report*, part 2, *Bituminous 1910* (Harrisburg, Pa., 1911), 310, 902; Bureau of Census, Manuscript Schedules, 13th Census 1910: Pennsylvania (1910; Washington, D.C.: National Archives, 1978), microfilm rolls, 1324, 1420, 1421.

ethnic contractors and other ethnic leaders, they formed an "ethnic aristocracy" which played a crucial role in maintaining the Berwind system at the workplace and in Windber's various ethnic communities.

Despite the interruption of the flow of new immigrants to the United States during the war years and the outflow of local labor to more lucrative jobs in war industries located in American cities, all evidence suggests that southern and eastern Europeans continued to constitute the bulk of Windber's mining force. Annual figures on the nationalities of mining employees in each of the state's inspection districts from 1914 to 1918 reveal the continued preponderance of these particular ethnic groups in District 24, where Windber-area mines were located. However, their overall share of the larger work force did decline as the war progressed. In 1914, southern and eastern Europeans constituted 79 percent of District 24's mining employees, and in 1918 they made up 66 percent of the district's mine workers.[25]

Previous Experience, Discipline, and Steady Work

Few Windber-area miners had been employed as miners in Europe. The Dillingham Commission found a major difference in previous mining experience between northern Europeans who had entered American mines en masse in an earlier decade and southern and eastern Europeans who did so in the early twentieth century. Only 7.7 percent of southern Italians, 13.7 percent of northern Italians, 9.8% of Poles, 10.9% of Magyars, and 10.7% of the Slovaks who came to the United

States and mined coal had been miners in their native lands. Immigrants who had had previous jobs in mining before coming to Windber were more likely to have held such jobs in the United States. As already indicated, Berwind-White miners sometimes migrated from company enterprises in Houtzdale or other Pennsylvania towns, and numerous Slovaks are known to have emigrated to the area from the anthracite region.[26]

Because mining was a new occupation for the many Europeans who had farmed land as small landowners or agricultural day laborers in Europe, prospective miners often faced a necessary and difficult transition from a rural to an industrial mining work-life. Even the relative absence of a factory-like discipline still made mine work unbearable to many, and the new and unfamiliar working conditions did not appeal to everyone who was looking for work. One elderly miner described the lure of America for people who remained in Europe after family and friends had emigrated: "When they [the immigrants] come here, they would write letters and tell them how nice it was here and everything. The ones that were left back there . . . thought that the streets in America was paved with gold, opportunity. But," he emphasized, "they never told them what had to be done."[27] This same miner had a family member who was enticed to emigrate but who rejected mine work as an occupation after a short stint in the mines:

> Myself, I had a cousin—my mother's brother's son. He always wanted to come here from Europe. He was left back on the family farm over there in Europe. . . . And he came over here. He wanted a job in the mines. His father was here so he took him in the mines. When he went, he loaded for about a week. He worked in the mines, and he said he's not going to mine no more. He said, "There's no windows or anything in there in the mine." He went in there in the dark, underground. Then he worked one week and quit.[28]

Turnover was a continual problem for Berwind-White, although there are no exact statistics on the subject. The Dillingham Commission referred repeatedly to the "constant shifting of the population," the exodus of a younger generation from the community, and the lure of higher wages and union conditions elsewhere, especially when times were prosperous. However, Berwind-White also enjoyed the advantage or relative immunity from industrial depressions and even attracted workers in harsh times because of the company's transatlantic steamship and bunkering trade and contracts. World War I seriously hurt company recruitment efforts, more than earlier depressions had, because it cut off the supply of new immigrants from Europe just as the demand for coal grew to new heights. Nevertheless, immigrants and others found steady work in Windber-area mines from 1897 until 1921, when the recession that followed the war and preceded the strike of 1922 occurred.[29]

Irregularity of work, along with periods of boom and bust, were characteristic of the mining industry as a whole. Although Berwind-White was not totally immune from cyclical swings and intervening developments, such as labor strikes or car shortages, its privileged position in the industry enabled the company to work its mines for more days during a year than the operators of most other companies. The years of 1897 to 1921 were a golden era in this regard for the company. James Roderick, chief of the Pennsylvania Department of Mines, considered a work year of 280 days to be one in which mines were working at full capacity or close to it. Although Windber mines did not always work 280 days or more a year, they enjoyed relatively regular work, compared with the important Pennsylvania industry as a whole.[30]

Never again would work be so abundant for so long a period in area mines. In the 1920s, following the strike of 1922–1923, the number of workdays diminished, not attaining previous levels and reflecting the onset of depression in the coal industry. But the nadir was reached in the period of the Great Depression. Berwind-White mines in the Windber area had worked an average of 270 days a year from 1900 to 1921 but only an average of 171 days during the years 1928 to 1940.[31]

Hours and Working Conditions

The eight-hour day had been a major goal of the American labor movement since the nineteenth century. The United Mine Workers believed that they had achieved an important national victory in 1898 when the union won operator recognition of the shorter day in the central competitive field, a region that set the standard for settlements in other mining districts. In 1903, union officials in the UMWA's District 2, where the Windber-area mines were located, won the eight-hour day for unionized

Miners outside Mine No. 36 in Windber, c. 1907, with the work clothes, hats, and lunch buckets of ordinary immigrant and American miners of the day.

mines. However, Windber miners did not share in this victory, because Berwind-White mines were nonunion and did not honor many of the working conditions specified by union contracts. It is not surprising, then, that one of the demands of local miners in the 1906 strike had been the eight-hour day.[32]

The eight-hour day came to Windber-area mines in 1917, at a time when the war had created a huge demand for coal, when labor was scarce or leaving for higher-paid jobs in factories and war industries, and when the UMWA was taking new initiatives on the regional and national levels. Although local mines were not unionized, Berwind-White could not ignore the growing strength of the labor movement or the ongoing need to be competitive vis-à-vis other operators in order to keep its labor force. In January 1917, for the first time in Windber's history, the company reduced the length of its workday—from ten hours to nine hours. Five months later, it cut the hours again from nine to eight. Modest wage increases accompanied each of the hourly reductions.[33]

From 1897 to 1917, then, the nominal workday for Windber-area miners had been ten hours, and two shifts of workers had been employed throughout much of this period. But the hourly standard had meaning mostly for the minority of workers—day men or company men—whose income was not based on tonnage output. For the masses of the labor force—the miners—nominal hours, whether ten, nine, or eight, had much less meaning because their wages continued to be based on the amount of coal they produced. Moreover, because deadwork was not remunerated, the struggle to make a living led many individuals to work exceedingly long days. For this reason, people interviewed often described work in the nonunion era this way: "There was no quitting time or starting time for those men," or, "In the wintertime they [the miners] didn't see no daylight. They went to work early and come home at 5:30, [when] it was dark. Actually they only seen daylight on Sunday." As one man put it, they went in "to get their place ready—to have coal ready because they pushed coal cars in from the place where they were. . . . All the men went in to have this coal ready so they could shovel it out whenever the cars was pushed into their place of working."[34]

Individual miners responded in different ways to the piecework mentality and to the pressure to produce coal to earn money. One man commented: "Some [miners] went to work at two o'clock in the morning, three o'clock in the morning. . . . My Dad never went that early though. He says it's foolish. . . . Before the day's up, you wear yourself out."[35] Other families reported that they rarely saw their husbands or fathers and that some men worked in split-shift fashion, working a full eight or nine hours then coming home for dinner and a short nap before returning to work for an additional time.[36]

From its initial victory in achieving a shorter working day, the UMWA had difficulty enforcing the eight-hour standard even in unionized mines, but there were no such limits (or none beyond physical endurance) in nonunion mines. Old habits of going to work early or staying late persisted into the union era in Windber-area mines. Once the Berwind-White mines were organized in 1933, Windber UMWA Local 6186 had to enforce a uniform starting time that was set by contract, and deal with the men who continued to enter the mines hours before the official starting time.[37]

The length of the working day was closely connected to the issue of pay for deadwork. Throughout the entire nonunion era, more than one Windber miner became a strong union advocate because of company policies regarding deadwork. One union leader of the 1930s described the crystallization of his support for the UMWA shortly after an experience he had in the 1920s soon after he had begun to work in the mines. He and his father had spent an entire workday clearing away rock and slate from their working place, and when they finished he naively asked his father how much money they had made that day. The father answered: "Son, we made nothing today. You didn't load no coal. You only get paid when you load coal." The son replied bitterly: "Hey, Dad, I don't like this. . . . From now on I'm a strong union man. If a guy says we're on strike, I'm dumping my water and away I go. I'm on strike, Dad. I don't care what you do." He then summed up the reason for his outrage: "I do not intend to work for nothing!"[38]

Other specific grievances regarding working conditions have been discussed elsewhere, but the importance of honest weights needs to be emphasized. On each car a miner loaded he routinely placed a metal mark or check that bore his individual number so that the weighmaster would properly credit him for his tonnage. Throughout the nonunion era, however, miners generally believed that they were regularly cheated out of full credit for the amount of coal they had loaded. An elderly businessman, whose father and brothers had worked in the mines described Berwind-White's weighing system, which he had observed as a youthful paper boy:

> They'd bring the coal out in the car, and they'd run it on the scale and then dump it. Well, you know, that was the big problem. . . . They [the miners] weren't being paid by the coal at the mine by weight. I actually watched this happen in my life. When I delivered the paper to the weighmaster there, he would start reading the paper and boom-boom-boom-boom-boom-boom-boom-boom-boom-boom-boom [the sound of coal passing, unnoticed and unweighed, over the scale and into railroad cars]. All the time he was reading. Then suddenly he'd start to write something.[39]

In union mines it was standard procedure for checkweighmen to verify and keep company weights honest, and, since 1883, Pennsylvania law had formally permitted, but not required, checkweighmen on each tipple. In Berwind-White mines in Windber, the weights issue was not resolved until the summer of 1933, when the UMWA won formal recognition and the miners elected the mines' first checkweighmen.[40]

Accidents, Safety, and Worker Compensation

The golden era of Berwind-White's productivity and steady work in Windber-area mines encompassed the years 1897 to 1921, but it also witnessed the most numerous number of fatal mining accidents in the company's history. In other words, the life of a Windber miner was considerably cheaper in the first decades of the twentieth century than it was in later decades. What was true for local mines was generally true for mines all over the nation as well. Fatalities in American mines peaked during the five-year period from 1906 to 1910, when an average of 2,658 deaths a year occurred, and critics began to point out that American mines were more lethal than their counterparts in leading European countries.[41]

Table 7 shows the fatalities that occurred in Berwind-White mines in the Windber area from 1898 to 1940. The greatest number of both local and state mining deaths occurred in the years from 1906 to 1910, but the periods 1901–1905, 1911–1915, and 1916–1920 also took high tolls on human life. For Windber, the most lethal year was 1907, when twenty-three men met their deaths in area mines. By contrast, one man was killed in 1930 and one in 1939. Despite continued high productivity on the local level, accident rates in the first two decades of the century were greater than in the next two decades. In order to appreciate this difference, we can compare the early decades, 1901–1920, with the next two decades, 1921–1940. In the first twenty-year period, 242 men lost their lives, compared with 93 in the second period. The number of employees per fatality and the amount of tonnage produced per fatality also provide further comparisons.

Because the annual reports of the Pennsylvania Department of Mines contained rather thorough statistics and information from 1897 to 1918, it is possible to draw a partial profile of the victims of fatal accidents in Windber-area mines during this period. Most of the victims were young or in the prime of life. A majority, 51 percent, were under the age of 30, and only 6 percent were age 50 or over. The detailed age breakdown is: 9 percent were 19 years or younger; 42 percent were in their 20s; 29 percent in their 30s; 14 percent in their 40s; 4 percent in their 50s; and 2 percent in their 70s.[42]

We also have information on the ethnicity of those killed in mining accidents from 1900 to 1918, but unfortunately we do not have figures on ethnic representation in the general Windber-area work force

Table 7. Fatalities, Berwind-White mines, Windber area, 1898–1940

Year	Number of Fatalities	Average per Year	Employees per Fatality	Tonnage per Fatality
1898–1900	21	7.0	308	220,282
Total or average	21	7.0	308	220,282
1901–1905	59	11.8	330	260,856
1906–1910	81	16.2	308	214,648
1911–1915	54	10.8	467	392,442
1916–1920	48	9.6	310	294,919
Total or average	242	12.1	354	290,716
1921–1925	33	6.6	558	397,828
1926–1930	22	4.4	933	393,284
1931–1935	20	4.0	722	489,144
1936–1940	18	3.6	1,282	836,325
Total or average	93	4.65	874	529,145

SOURCES: Pennsylvania Department of Internal Affairs, Bureau of Mines, *Report* (Harrisburg, Pa., 1899–1903), annual reports, 1898–1902; Pennsylvania Department of Mines, *Report* (Harrisburg, Pa., 1904–1959), annual reports, 1904–1940.

throughout the period. However, because we do know that a substantial portion of the employees, including 90 percent of the miners in 1910, were new immigrants, it is not surprising to discover that they made up the bulk of the accident victims. Of the 232 men killed in Windber mines from 1900 to 1918, some 83 percent were southern and eastern Europeans, 12 percent were Americans, and 5 percent were northern Europeans. Magyars made up 27 percent of the victims; Slovaks 24 percent; Poles 18 percent; and Italians 8 percent.[43]

The lack of previous mining experience on the part of many Windber-area miners has already been noted, and there is no question that miner carelessness contributed to the overall rate of accidents. Some falls of roof—the major cause of fatalities—would undoubtedly have been preventable if miners had taken better care in timbering or blasting, if greater supervision by foremen had been exercised, if incentives for safety and a system of payment based on something other than direct piecework had been in effect, and if a rigorous training program had existed.

Nevertheless, a few examples gleaned from newspapers and oral histories strongly suggest that the first days, weeks, or months spent in a mine were particularly dangerous, for a number of the mining casualties are known to have worked in Windber mines only a short time before their deaths. Thomas Washko, brother of a town druggist, was killed on his first or second day in the mines. William Polony, one of a number of victims reported as a mining fatality in local papers, but not in state records, "had never worked in a mine until the day before the fatal accident." John Hunter, listed as 13 years old in official records but as 11 years old in newspapers, was killed at the end of a workday when he was leaving the mines. Neither the 9 percent of the casualties age 19 or under, nor the "greenhorns" who were older, could have been very experienced. Joe Arnoski, a Polish miner, was one of several whose obituary noted that he "came to America only a month ago."[44]

In any event, coal-related accidents were not limited to the mines, and consciousness of elementary safety was so undeveloped on the part of the company, the miners, and the public early in the century that Windber miners were expected to buy and store blasting powder in their congested boardinghouses, where a forgetful smoker or a random spark from the stove could endanger an entire household and even neighboring homes. Only after several spectacular explosions of stored powder had occurred in company houses, injuring and fatally wounding miners and residents, were measures taken by the borough council to enforce restrictions on the amount of powder that could legally be kept in homes and stores. It is somewhat ironic that, for safety purposes, the state mining code of 1893 had set strict limits on the quantity of powder miners could take *into a mine* each working day.[45]

Given a certain degree of naiveté and carelessness on the part of "greenhorns," were new immigrants any more careless than miners of other nationalities? After all, it was the fashion in the early twentieth century for leading experts and mining inspectors, as well as nativists, to blame the high number of coal-mining fatalities on the inexperience and ignorance of the southern and eastern Europeans who increasingly mined the nation's coal. In 1905, James Roderick, chief of the Pennsylvania Department of Mines, attributed the recent increase in accidents to "the introduction of coal-cutting machinery and the great influx of foreign-labor." While miners traditionally knew how to be careful, he concluded, the new immigrants did not:

> In many cases, however, these men [new immigrants] are not miners at all in the true sense of the word; they are ordinary foreigners, often without experience, and invariably ignorant of the language used by the foreman or fire boss. This lack of experience and the inability to understand the instructions given them cause frequent accidents. The only qualification deemed necessary seems to be, in many cases, a pair of strong and willing arms.[46]

Roderick's views were challenged even in his own day. His predecessor in office, Robert Brownlee, chief of the Pennsylvania Bureau of Mines, argued strongly that new immigrants were not especially accident-prone:

> Some people attribute the cause of so many accidents to the large foreign element employed in and about the mines. I have my doubts as to that being the cause. My experience and observation have been that this class are as careful of danger, if not more so, than many of the experienced miners. I know I will be criticized for those expressions, but they stand good until proven otherwise. In many sections the foreign element is largely in the majority, and produces the greater portion of the coal. It is not, therefore, strange that we find a great number of accidents charged to their account.[47]

In his *Coal-Mining Safety in the Progressive Period,* William Graebner challenged the assumption that new immigrant miners were more careless than native-born or northern Europeans, and he used the statistical evidence for the state of Pennsylvania from 1899 to 1912 to make his case. Overall, Americans had more fatalities than their proportion of the bituminous mining population, while Slavs, Italians, and Poles had victims in rough proportion to their share in the mining work force.[48]

The occupational profile of the 245 accident victims in Windber-area mines supports Graebner's findings. Unskilled work inside the mines was vastly more dangerous than other types of work, and there is a close correlation between occupational representation in the overall labor force and occupational representation in the fatality statistics. Miners, who constituted 71 percent of the general work force of the mines, made up 72 percent of all the fatalities. Laborers, otherwise undesignated, constituted another 6 percent; spraggers made up 4 percent; rockmen 2 percent; motormen 2 percent; tracklayers, track cleaners, and trackmen

2 percent; road cleaners and roadmen 2 percent; and the remainder 6 percent. The last group included one managerial person—a night foreman—and one outside worker, a tipple man.[49]

Beyond miner inexperience, many other factors—such as conditions at the mines, the number of openings or tracks, the placement of electrical wires, and the degree of supervision and training—contributed to high accident rates. As Graebner pointed out, the mentality of blaming the victim was tantamount to acceptance of the fatalities as a necessary price and a rationalization for not taking measures to prevent accidents. Significant safety measures and training cost operators money, and rigorous state inspections and enforcement of safety measures were unacceptable costs for many taxpayers. Graebner also argued rather convincingly that the competitive nature of the coal-industry economy and the politics of federalism, which left basic regulation and control of the industry in the hands of the individual states, prevented the adoption of more effective safety measures during the Progressive Era, vitiated federal efforts, and ultimately doomed more-modest efforts to make state laws uniform.[50]

In the first decade of the century, Berwind-White had no training program and no safety program and provided no compensation of any kind for accident victims or their families. Unlike H. C. Frick, an even larger coal producer, or many other nonunion coal operations, the Berwinds did not institute voluntary programs or welfare work. The one instance of what might be considered welfare work—the company hospital—had been conceived and pushed to fruition in 1906 by progressive doctors, newspaper editors, and others in the community. The company did build and control it, but it was the miners who ultimately paid for it out of regular deductions from their paychecks.[51] One early Windber resident sarcastically described how he viewed the company's attitude toward accident victims: "Another Hunky will come over and take his place. That's how serious they [Berwind-White officials] took anybody getting killed or getting hurt."[52]

From 1911 to 1915, however, Berwind-White was forced to adopt certain safety measures and programs in order to be competitive with other major companies and to fulfill the mandate of new state laws. In 1911 it belatedly created a first-aid program to train employees to help the injured and to compete against first-aid teams from other mines in contests sponsored by the YMCA or other organizations. First-aid work had first been organized in the United States in the anthracite region in

1899, and James Roderick had recommended it to bituminous operators as early as 1906. In 1911 many large bituminous operators, including Berwind-White, began to participate in first-aid programs in accordance with a new state law. The new bituminous mining code of 1911 also mandated additional safety measures, including closer supervision of workers, the posting of uniform danger signals in designated places, making safety regulations available in various languages, and raising the minimum age for work inside the mines. Until 1915, Berwind-White did not pay any death or accident benefits, but passage of a worker compensation law in Pennsylvania that year led Berwind-White to organize its own insurance company, the Eureka Casualty Company, to oversee claims of accident victims or their families.[53]

Training and Child Labor

Even after passage of these new laws and an increased consciousness of safety reform as a conservative business measure designed to create economic efficiency and social stability, Berwind-White still had no formal training program for new employees. In the 1920s, as in the earlier years, it was miners who trained miners how to work. Men took their own sons, relatives, and other "greenhorns" from Europe into the mines as their buddies and taught them what they knew. One retired miner, who had himself gone into the mines with his father, described the system:

> Well, somebody had to train them [new employees]. Somebody had to show them . . . how to . . . mine coal and so on. Even when my Dad came here [to Windber in the 1890s]. Well, . . . other men start coming from the old country. So my Dad used to take them in as buddies. . . . He'd teach them how to work. And he'd, you know, well, after they learned how to work, well, he'd go with somebody else. Oh, I know my Dad always worked with younger men than he was. All the time. I know this one guy told me, . . . "I can't keep up with your Dad." And he [the father] knew how to work it, to make money, to get the coal out. He knew the tricks right. I'm telling you. I learned a lot from him.[54]

Contrary to official state records, miners have suggested through interviews that child labor was a common occurrence in Berwind-White mines in the first quarter of a century. In fact, violations of the law seem to have occurred with the complicity of the company, the miners, their families, and school officials. Moreover, newspaper obituaries of young boys killed in local mines sometimes gave a younger age for the victims than official records did.

Miners and their families routinely related that a boy got illegal working papers by going to certain school officials who, knowingly or unknowingly, falsely certified the person's age. The certificate was then presented to the company and the boy was given a job. Most often, a boy went with his father to work, but fatherless boys either went with another relative or were given company jobs so that they could work alone. In Pennsylvania, the legal minimum age for boys to work inside bituminous mines was 12—until 1909, when the age was raised to 14. The 1911 bituminous code kept the age limit at 14 but added qualifications: "No boy under the age of eighteen years shall be permitted to mine or load coal in any room, entry, or other working place, unless in company with an experienced person over eighteen years of age." Boys under 16 also had to have on file in company offices the requisite age certificate endorsed by a school official.[55]

The nature of the family economy and the piecework wage payment system gave miners and their families a vested economic interest in having sons work with their fathers as buddies. The tonnage of the pair would then be credited to members of the same family, as would wages. One miner explained that another important reason fathers took their underage sons to work with them was that an additional person of any age at the working place entitled the pair to an additional car. Because an experienced miner could sometimes load coal faster than he could get cars to put it in, the mere presence of the child—even if he did little or no work—added to the family's biweekly paycheck.[56]

One instance in which boys systematically went into the mines at the earliest possible time, legally or illegally, occurred when a mother was widowed and young children were left fatherless because of a mining accident or some other misfortune. In 1906, James Roderick argued that the issue of raising the age limit for boys' employment in the mines was related to the issue of worker compensation. Widows and young children needed financial support when the miner—the male breadwinner—died. If boys were realistically expected to stay in school until a later age, he warned, some compensation had to be given the families of fathers killed in the mines.[57]

Miners have also suggested in interviews that nativist prejudices reflected in the local public school system played some part in decisions to leave school at an early age. In their eyes, the dominant society composed of Americans and Johnny Bulls pushed them into fulfilling the occupational role expected of "hunkies."[58]

It was in a father's interest to train his sons as well as he could, and presumably he had greater concern for the novice's safety than the company or others would have. Still, the family and kinship ties of the immigrants, reinforced by the family economy so common in the mining industry, spared the company the expense of training "greenhorns" or young boys. Because the company paid miners on the basis of tonnage, it lost nothing by employing youths, and it did not have any monetary interest in their safety—or anyone else's—until the passage of compensation laws.

Despite what seems to be a rather widespread effort at evading the law, there were miners and families who hoped that their sons could avoid going into the mines. The philosophy of one such miner was that although mining was good enough for the father it was not good enough for the sons. Miners had always valued practical knowledge and knowledge of life, but formal education became an increasingly important value in subsequent generations. Nonetheless, in 1940 some 69 percent of all Windber residents, male and female, over the age of 25, had completed only eight grades of school or less. Work at an early stage of life remained the fundamental reality, and in Windber, for males, work usually meant the mines.[59]

Machinery and Reorganization of Work

Technological developments in mining eventually produced reorganizations of the work process, but no such major changes occurred during the nonunion era in Windber-area mines. By the time these mines were opened, mining machines had begun to replace the traditional handpick in the nation's more advanced mines. Windber's belated development had enabled Berwind-White to begin operations with state-of-the-art equipment, including the Sullivan pick machine and the

Ingersoll-Sergeant coal-cutting machine. Over the years, the company introduced, and miners used, a wide variety of coal-cutting machines, including bottom cutters and top cutters, pickhammers, jackhammers, and machine loaders. Yet the new mining machines, a new form of hand tools, were not the thorough mechanization of the mining process which economists later described as machinery and conveyors that linked "every operation, from face to tipple, into a single continuous process." Nor did new machines change the basic organization of work at the face of the coal seams in Windber-area mines from 1897 to 1940. Throughout the period, a typical miner worked with a buddy, cutting, shooting, and loading coal in a rather isolated room of a mine. Almost the only time a third person joined two buddies at the working face was when one of the pair had a newly arrived relative from Europe to train.[60]

The one major reorganization of work that did occur in Berwind-White mines in this period seems to have been prompted by the new state laws regarding safety and mandatory worker compensation and by a new industry-wide interest in greater efficiency to meet the huge demand for coal brought on as a result of World War I. That reorganization entailed an increase in the number of foremen and assistant foremen, so that greater supervision could be exercised over the miners and company men. Increasing the quantity of supervisory personnel was one way companies could cut down on accidents, thereby reducing compensation costs, and increase cost-efficient production at the same time, thereby maintaining their competitiveness in an industry noted for a lack of concentration and fierce interstate competition. Moreover, the bituminous code of 1911 in Pennsylvania had added to the duties and responsibilities of the mine foremen, so that compliance often entailed increasing their numbers as a matter of practical necessity. According to the 1893 code, the foreman of every mine, or his assistant, was expected "to visit and examine every working place therein at least once every alternate day while the miners of such place are or should be at work." From 1911 on, foremen or assistants were expected by law to make visits to each working place on a daily basis.[61]

State mining reports indicate that supervision over mining employees was comparatively lax in the formative period, when Windber-area mines were being opened, but that supervision increased afterward. Berwind-White foremen, who supervised 154 mine workers in 1899, directed only 42 miners in 1918.[62]

A miner loading coal by hand.

In 1902 the company had strengthened its supervisory capabilities by adding to the managerial staff a large number of assistant mine foremen who helped the foremen. The need to meet state law and oversee new immigrants, who were widely believed to require more than the customary, almost laissez-faire, supervision, was the ostensible rationale for this change. But nativist prejudices combined with antiunion sentiment played an important part in the personnel change. At the time, it was also widely assumed by coal operators, steel-plant managers, and other industrialists that, in rather childlike fashion, new immigrants naturally deferred to authority and were uncritically loyal to their bosses. Good foremen—those who could inspire confidence and command deference—were thought to be useful in allaying worker discontent and in hindering the development of class-consciousness.[63]

The trend toward increased supervision continued in the following years. From 1902 to 1906, foremen or their assistants supervised an average of 90, not 136, employees. In the next five years, from 1907 to 1910, when the accident rate peaked in Windber, they directed a slightly larger number, 97 workers. From 1911, the year of the new mining code, until 1918, the last year of the war, there was a steady growth of supervision. Foremen oversaw an average of 69 mine workers from 1911 to 1914. The ratio between managerial personnel and employees dropped even further in the period from 1915 to 1918, when each foreman or assistant supervised 47 employees.[64]

Increased supervision probably did help cut the overall rate of accidents, but it did not end worker discontent or allay class-consciousness. The massive strike of 1922–1923, however, apparently prompted Berwind-White to modify its policies. Until then it had no formal training program and selected supervisory personnel from the people who satisfied legal requirements. In 1924, the company departed from custom and opened a night school in Windber for its employees. Practical subjects relating to mining were the focus of the classes. In time, some individuals, including new immigrants or their adult children, could prepare for the examinations required to become state-certified foremen, assistant foremen, or fire bosses. Since union leaders from the 1922 strike were among those whom company officials eventually encouraged to join the classes, the new policy was obviously designed to co-opt existing or potential working-class leaders, as the timing of the innovation itself suggests. If so, the project failed dismally in its goal. Unionization came in the 1930s, with the leaders of the 1922 strike in the vanguard of the effort. Meanwhile, greater numbers of new immigrants began to attend both night school and union meetings.[65]

Berwind-White's recruitment of foreign-born workers and its organization of work in Windber-area mines throughout the nonunion period is incomprehensible without consideration of the company's overall intention to maximize profits while averting unionization. In work, as in the community, fostering ethnic divisions among its diverse work force played a preeminent role in its strategy. Immigrant miners would have to confront the problems posed by inherent ethnic division and conscious company policy if concerted action toward unionization were ever to succeed in Windber. The nonunion policies of low wages, no pay for deadwork, long hours, questionable weights, high accident rates, minimal or nonexistent compensation, arbitrary management, and especially the lack of any representation left the miners with many grievances to resolve whenever they thought the time was right to act collectively.

Implicit in the company's policies was a disdain for the rights of labor—especially immigrant labor. In revealing testimony before the Industrial Relations Commission in 1915, Edward J. Berwind expressed his own underlying condescension toward the very immigrants whose labor enriched him. He complained that his foreign-born work force created special difficulties for him. Because they were undisciplined, drank too much, and overindulged in holidays and funerals, they prevented maximum production from being attained. This statement of his elicited laughter: "For 25 years we have been unable to work our mine at capacity because the men would not work the hours. In other words, we had 26 days work in the month and we rarely could get over 19 days from the men, between various holidays and funerals and pay days and the day after pay day, cut us down in our work."[66]

Berwind went on to explain that, despite his employees' vices, it was in an employer's interest to keep workers satisfied. For this reason, he claimed, his company contributed liberally to his respective towns' churches, schools, playgrounds, and hospitals. But allowing union organizers into his towns or union organization in his mines was another matter. He reluctantly conceded that unions might occasionally be useful in industries where skilled American labor prevailed. However, he had no doubt that they were otherwise undesirable. After all, he maintained, immigrant workers were naive and easily misled. He concluded: "It [unionization] is harmful, especially where the classes of people are not such as to be intelligent and understand your languages altogether and are not educated up to a higher standard." Immigrant workers were better off with the company than with union troublemakers because, in his words, "it is very hard to communicate with them [immigrants] and to make them see what their best interest is."[67]

4

Women's Work

An elderly Carpatho-Russian woman described what she had found on her arrival in Windber, Pennsylvania, in 1910 when she was a young bride of 18: "Men was working real hard. The women was working hard, and men was working hard." She concluded that work then was much more difficult than it is today. "It's different now. It's completely different. The men don't have to work so long, not so hard like they used to. . . . But the greenhorn [who] was coming from Europe was working hard." While men worked twelve or more hours in the mines, women labored just as long and as hard at home, "with the children, having those boarders, and baking bread." Hard work did have some compensations, she explained, because "everybody was like one family." Community made up for a lot of the drudgery: "People in the old age, the people was work[ing] hard. But it was not bad. People be nice to each other, their family, their neighbors." Besides, she added, "When you're young, you don't mind. You work, work, work."[1]

Susie Shuster's reminiscences about hard work were no mere paean to the work ethic, but an accurate statement about the realities of life as she and many other working-class Americans personally experienced them in the first half of the twentieth century. Even in these brief comments, she suggests that her conception of work was a broad one that necessarily encompassed both sexes and included more than wage labor and gainful employment. Mining in Windber—a one-industry company town—was indeed "men's work," but "women's work" was difficult and valuable too.

In this chapter we examine the gender and work relationships that prevailed in this company town during the nonunion period. Berwind-White's labor relations policy not only involved the organization of the work process at the mines but also entailed restricted gender roles that incorporated the work of women and children. Of particular concern are these questions: What role did women's work play in the overall Berwind system, and did the rather rigid sexual division of labor that prevailed strengthen or weaken the coal company's dominance?

Scarcity of Women and Female Employment Opportunities

Windber supports the general picture of gender imbalance found in most contemporary and historical descriptions of frontier mining and steel communities. Characteristic of such places was the abundance of men and the shortage of women. Thirteen years after Windber's founding, males still comprised a far greater portion, 59 percent, of the local population than did females. One female resident who recalled the era explained: "There was a woman shortage in this area. So when a young girl would come, my gosh, there would be a dozen men to visit one girl."[2]

Young single immigrant miners found it difficult to find wives in the town's early years and led many of them to rely on marriage, family, and village networks for selection of their spouses. People interviewed told how men often traveled to previous residences, such as Houtzdale, Pennsylvania, or to large cities, such as Pittsburgh, New York, or Perth Amboy, to select wives from their respective ethnic groups.[3] Women sometimes described simple or elaborate networks that eventually brought them to Windber. Susie Shuster's experience was typical in this regard.

Joseph Shuster and Susanna Gulyásy in Brooklyn, New York, on their wedding day, June 12, 1910.

In 1910, Susie attended a wedding in New York City, where she was then living. At the wedding she met for the first time a Windber miner. Describing the marriage proposal he made at that wedding shortly after they became acquainted, she said: "And we was dancing together, and he asked me, 'Would you like to go to Windber?' I said, 'I don't know.' " Two weeks later, the newly married Shusters arrived to take up residence together in Windber, where they spent the remainder of their lives. Although the two had been strangers until the wedding of mutual acquaintances, their parents had known of the other's family members in Europe and had approved of the marriage. The future Shusters had been born in the same village eight years apart.[4]

In the decade from 1910 to 1920, such networks helped to narrow the ratio between the local male and female populations. Apparently many single men found wives during this period, and many married men, possibly prompted by the Balkan wars, had reunited their families. From 1911 on, the pattern of emigration from Hungary had changed to include greater numbers of wives, very young children, and other dependents. By 1920, males made up 52 percent of the overall Windber population, and females 48 percent. After 1920, males continued to maintain a slight majority, constituting 51 percent of the total number of people in both 1930 and 1940.[5]

In the course of her interview, Susie Shuster offered an interesting explanation for the scarcity of women in Windber: "Well, not very many girls [lived] in here. [Windber] don't have so much work for the girls. Like in the city they have a lot of work for the girls. So the boys take a trip to the city for a wife." She believed that the shortage of job opportunities in Windber was the main reason females were so scarce in the mining community, and she herself had held a variety of jobs in New York City.[6]

The Carpatho-Russian immigrant was perceptive about the lack of female job opportunities in Windber. In the 1920s, a study sponsored by the Women's Bureau of the U.S. Department of Labor found that, overall, employment opportunities for women in bituminous coal-mining communities were very limited. Only 18 percent of working-age females in bituminous families were gainfully employed in the early 1920s, compared to 31 percent of comparable females in anthracite families.[7]

The study attributes the difference in the employment rates between the two coal regions to the location of the mines themselves and the availability of alternative industries. Unlike the more rural areas where bituminous mines existed, anthracite mines were often located in close proximity to urban areas, where many miners lived, and in regions where textile mills and factories were situated.[8] However, the report's focus on location, rather than the economic system, as the source of the problem of scarce female employment opportunities caused it to neglect other possible explanations. "Company towns" were not simply inevitable by-products of rural mining life but deliberate creations, and coal corporations typically restricted alternative industries in the towns they owned.

If the Women's Bureau figures are correct, even fewer women were gainfully employed in the Windber area than were employed in the overall bituminous industry. In 1910, only 15 percent of the town's entire working-age female population, consisting of all classes, held jobs as wage earners. Moreover, Windber was an urban center with high female employment, compared with the surrounding area. The employment rate for females in Mine 38 was 10 percent; Scalp Level 6 percent; and Paint Borough 3 percent. In Mine 37 only 1 of 126 females over the age of 14 was enumerated as employed.[9]

It is also striking that the total percentage of employed females changed little or not at all in the next decades. In 1940 the U.S. Census listed 15 percent of Windber's working-age females as employed in regular jobs. An additional 1 percent, however, were employed in public emergency work programs created for the unemployed during the Great Depression, and another 3 percent were seeking work. Table 8 summarizes the overall employment statistics for Windber in 1910 and 1940. Table 9 lists the specific occupations of Windber women in 1910 and 1940.

Table 8. Employment of Windber women, 1910 and 1940

	1910	1940
Number of Females, All Ages	3,273	4,428
Number over 14 years	1,888	3,326
% over 14 years	58%	75%
Number not in work force	1,599	2,680
% not in work force	85%	81%
Number in work force	289	646
% in work force	15%	19%
% Employed (except on public emergency work)		15%
% Employed on public emergency work		1%
% Seeking employment		3%

SOURCES: Bureau of the Census, Manuscript Schedules, 13th Census 1910: Pennsylvania (1910; Washington, D.C.: National Archives, 1978), microfilm roll 1421; Department of Commerce, Bureau of the Census, *Sixteenth Census: 1940, Population*, vol. 2, part 6, *Characteristics of the Population* (Washington, D.C., 1943), 156.

Berwind-White's monopoly of industry in the Windber area did not allow for many employment opportunities for women and girls outside the home. In the ethnic/class hierarchy that prevailed, professional positions, such as teaching and nursing, and more highly paid jobs, such as bookkeepers and accountants, were exclusively held by native-born or northern European women in 1910. One of Windber's early immigrant

Table 9. Specific occupations of employed Windber women, 1910 and 1940

Occupations	1910	1940
Clerical, sales, and kindred workers	18%	30%
Professional workers	17	28
Domestic service workers, private homes	23	21
Service workers, except domestic private homes	29	12
Operatives and kindred workers	9	3
Proprietors, managers, and officials	3	2
Craftsmen, foremen, and kindred workers	—[a]	1
Semi-professionals	—[a]	1
Occupation, not reported		—[a]
Laborers, except farm		—[a]
	100%	100%

SOURCES: Bureau of the Census, Manuscript Schedules, 13th Census 1910: Pennsylvania (1910; Washington, D.C.: National Archives, 1978), microfilm roll 1421; Department of Commerce, Bureau of the Census, *Sixteenth Census: 1940, Population*, vol. 2, part 6, *Characteristics of the Population* (Washington, D.C., 1943), 156.
[a]Fractional percentage.

residents accurately assessed the situation: "Housework. There was no other work for women, especially for women." With an exception or two, immigrant women who worked outside their own homes were restricted to domestic service, housework, and sales.[10]

In one important respect, Windber's female employment differed markedly from that in other bituminous towns, as reported in the Women's Bureau study. Analysis of the manuscript schedules of the 1910 census indicates that, in Windber, American-born women were more likely to be wage earners than southern and eastern Europeans. Only 3 percent of employed women over 14 years of age were southern or eastern European, but 12 percent of all such native-born or northern European women were.[11]

Nativist prejudices, language difficulties, and a rather rigid pattern of social and ethnic segregation might help explain why American women were apparently preferred over immigrant women for many menial jobs. Domestic servants were an integral part of prosperous middle-class households, and the town's upper echelons generally chose servants from within their own ethnic group. The town's top social structure, composed of native-born merchants, professionals, and company officials, therefore selected Americans or northern Europeans for servants in their homes. In a similar manner, the less prolific and less powerful southern and eastern European merchants chose domestic help from the new immigrant groups.

The same pattern prevailed in business establishments. The majority of hotels were American-owned and employed servants, waitresses, and cooks who were also American-born. Many were daughters of local farmers or lived in the surrounding countryside. Only hotel laundresses were likely to be of new immigrant origin. The Windber census, therefore, indicates a surprising finding about domestic service, which has often been associated with minority ethnic groups and races. In Windber, at least, domestic service was dominated by single or widowed American-born women, and service remained a viable occupational option for them well into the century.[12]

The Women's Bureau study noted that many daughters of bituminous coal-mining families had to leave their families and move to other places in order to find jobs. Such was the case in Windber. Southern and eastern European women interviewed described trips to cities where they became domestics or nannies for families. Apparently there were networks through which new immigrant females found such jobs. Several women described this period of domestic service as a high point in their lives, or at least of their childhoods, because the families they served treated them with kindness and rewarded them with material gifts as well as a paycheck. Others contrasted the quality of this type of life with the bleak, poverty-stricken life they and their families had endured in the mining region.[13]

The case of Katherine Jankosky illustrates the typical work history of a young immigrant daughter in the first quarter of the twentieth century. Katherine's father had been killed in a local mining accident in 1903 when she was six months old. Her widowed mother, left with three dependents, no family in the area, and no resources, had little choice but to remarry, in the daughter's view. The mother did remarry, and Katherine eventually had ten stepbrothers and stepsisters.

At the end of fourth grade, Katherine unhappily left school to go to work. A brief stint in a cigar factory in Johnstown ended because the 30 cents needed for the street car ride was too costly. She then went to work at one of the company stores as an order girl. Traveling from house to house in one of the mining camps, she used her knowledge of various languages to obtain daily orders from the women of diverse nationalities who resided there. Her pay of $30 a month was repeatedly cut at the store, however, and during the strike of 1922 she decided to go to Cleveland with a sister and a friend to find a job.

Through a want ad she obtained a job and was employed for two years as a nanny to care for the three children of a prosperous Cleveland businessman and his wife. Her pay was $30 a month plus room and board and some clothes, and life was pleasant. In 1924 she regretfully left that job and returned to Windber to help her mother, who was ill. She soon returned to the company store, where she met her future husband. Married a year later, she then permanently left the ranks of the official labor force.[14]

The nature of the family economy and the pull of economic obligation toward one's family is evident in Katherine's example, even though she obviously resented her position. She never saw any of the money she earned at the company store, but turned the envelope over to her stepfather. In fact, he never saw $30 in cash either, because the company applied her pay as well as his toward the family's company store bill. A statement merely itemized the pertinent wages and deductions. It is also important to emphasize that Katherine's brothers, who worked as miners, contributed to the family economy in the same manner.

Berwind-White's method of paying out cash wages to employees only after all family debts had been deducted was designed to bind the miner and his family to the company over time. The coal corporation's dominance over jobs, housing, stores, and credit effectively brought the entire family within its tentacles, especially when a mining accident, slack work, or other misfortune caused the family debt to accumulate. One incidental but important result was that the company's policies reinforced and perpetuated the family economy that immigrant workers had come to rely on for reasons of their own.

Katherine Jankosky's contribution to the family economy was not unique but typical. The Women's Bureau study found that 90 percent of gainfully employed females in mining families contributed all or part of their earnings to the family.[15] The primacy of family considerations over individual ones helped to ensure the family's survival and well-being in the midst of the uncertainties of a capitalistic economy, an insecure world, and the absence of a stable welfare state. Hence, as John Bodnar has reminded us, the family economy survived—and characterized—immigrant working-class families into the 1940s or beyond.[16]

However, families were not always the conservative institution they are often purported to be, and a family economy might cut either of two ways.

Depending on circumstances, individual families could opt for security at all costs or become staunch unionists and strikers in an attempt to make a better life for the children. Nor was the family economy synonymous with a harmonious or egalitarian family, as feminists and historians of women remind us.[17] Individual males and females who sought to escape the strict confines of family roles and duties necessarily came into conflict with other family members and with established societal practices. More than one young daughter of an immigrant ran away from Windber to a big city to escape an unwanted marriage to a boarder selected by her father.

The Importance of the Boarding System

The boardinghouse was essential—not tangential—to the development of the mining industry in Windber. Through it the men obtained the food, clothing, shelter, and other necessities of life that enabled them to labor in the mines. This key institution of the community was based on female labor in the way that the mines were based on male labor. Any attempt to assess gender and work relations during the nonunion era in Windber must give "women's work" and the boardinghouse their rightful important place in the overall industrial labor relations system.

As in other bituminous mining towns, the major work from which women derived income came from keeping boarders. In this sense, the census figures about employment are misleading. Only seventeen women were counted in the 1910 census in Windber as boardinghouse keepers, while in reality a majority of the households of new immigrant groups kept boarders.

The many single or recently married young men who came without their wives to work in Windber needed lodging. Consequently, newcomers typically experienced boardinghouse life at least temporarily. Information from the 1910 census indicates that one out of every five people residing in the town limits of Windber was a boarder. But the number of boarders varied significantly, depending on place of residence and nationality group. Americans kept relatively few boarders, and areas such as the town's center and hill district, where many company officials and prosperous merchants resided, contained very few. For this reason, the western section of Windber, which encompassed much of this district, had the lowest overall rate, or 14 percent, of boarders in the area. But surrounding mining settlements with their preponderance of new immigrants and heavily congested households offset such sections. One of every three residents living in the outlying coal settlements was a boarder.[18]

Census information further confirms that boardinghouses were a phenomenon linked to new immigrant groups. Only 13 percent of all American families in the area, and 16 percent of all English families, kept boarders, whereas 57 percent of all Italian, 56 percent of all Polish, and 54 percent of all Magyar families did. The Slovaks were the only ethnic group of the big four southern and eastern European groups to have less than a majority of its families keeping boarders.[19]

Table 10. Families and boarders, Windber area, 1910

Nationality	Number of Families	Number of Boarders	Average per Family	% Families with Boarders
Americans	819	294	0.36	13
English	83	29	0.35	16
Germans	48	34	0.71	23
Swedes	28	16	0.57	18
Magyars	281	647	2.30	54
Slovaks	381	513	1.35	43
Poles	191	395	2.07	56
Italians	171	382	2.23	57

SOURCE: Bureau of the Census, Manuscript Schedules, 13th Census 1910: Pennsylvania (1910; Washington, D.C.: National Archives, 1978), microfilm rolls 1324, 1420, 1421.

Table 10 shows the boarders kept by the top four most populous American and northern European families, on the one hand, and the top four southern and eastern European groups, on the other. Magyars led in the average number of boarders per family, Italians were second, Poles were third, and Slovaks were fourth.

Table 11 summarizes the number of boarders kept by families in each ethnic group. A sizable portion of the Magyar, Polish, and Italian families had from three to ten boarders, but Magyars and Italians led in families that maintained eleven or more boarders. These figures are themselves conservative, because not included in the totals were resident relatives who were identified in the census as brothers, sisters, uncles, in-laws, or other relatives of the head of the household, rather than as boarders

Table 11. Boarders per family, Windber area, 1910

Nationality	0	1–2	3–5	6–10	11 or more
Americans	87%	9%	2%	—%	—%
English	84	13	1	1	0
Germans	77	10	10	2	0
Swedes	82	7	7	4	0
Magyars	46	13	28	11	2
Slovaks	57	23	14	5	—
Poles	44	21	26	8	—
Italians	43	22	22	12	1

SOURCE: Bureau of the Census, Manuscript Schedules, 13th Census 1910: Pennsylvania (1910; Washington, D.C.: National Archives, 1978), microfilm rolls 1324, 1420, 1421.

per se. Moreover, two families sometimes lived together in the same dwelling but no one was classified as a boarder.

Keeping boarders was a way for ordinary people to make money even as it also often entailed aiding relatives, friends, and fellow villagers. The general importance of the income derived from boarding to an immigrant family's economic survival has been well documented in the past. In her classic study of Homestead, Margaret Byington found that many Slavic women supplemented their husbands' wages and that 75 percent of those who kept boarders earned at least the equivalent of the monthly rent. Unskilled laborers simply did not make enough in wages to support a family, even in terms of the lowest standards. In these cases, the boarding income was essential to survival. In a recent study of coal-mining towns in Iowa, Dorothy Schwieder found that a woman could earn the equivalent of a miner's monthly wages if she kept four boarders for the month, and that boarding income helped families survive in hard times.[20]

Whatever the exact amount of money a family derived from the boarding system, those who have studied related incomes and budgets generally agree that the sums involved were not paltry, but rather major contributions to the household economy. The Women's Bureau study of the 1920s concluded that approximately 25 percent of the monthly income of a bituminous coal-mining family derived from the lodging of boarders. In bituminous towns overall, female daughters contributed an additional 15 percent to the family's income, thereby making the total female contribution to family income 40 percent.[21] Given the low number of immigrant women employed outside the home in the Windber area, it is unlikely that daughters contributed that much. However, the prevalence of boarders throughout the area suggests there is good reason to believe that the general finding, the 25 percent boarding-income contribution to the household economy, applies to Windber as well.

The income from boarding can help explain why a small number of mining families were able to purchase a limited number of saleable homes in town. Italians, who lived in town rather than outlying settlements, were especially noted for saving money, and some did succeed in buying their own homes. One Italian woman explained that her parents' home was purchased from the income her mother received from boarding, not from the miner's wages paid her father: "My mother made a living with boarders. She had twenty boarders when she first got married. She thought she was going to be a rich lady in no time. But she saved every penny and then she bought the house. It was $700."[22] In such cases, miners enjoyed more independence from company control than if they had lived in company housing. During the strikes of 1906 and 1922, such home-owning families often put up the families of miners evicted from company houses.

Immigrants kept boarders for financial, personal, and social reasons of their own. Moreover, despite all the problems associated with this congested lifestyle, the boardinghouse could be and sometimes was an educational experience for miners and their families. Migratory Slovak miners from the anthracite region reportedly spread news of the successes of the United Mine Workers through their boardinghouses in the early century, thereby giving great impetus to unionization efforts in Windber in 1906. A leader of the 1922 strike and future officer of one of the miners' locals recalled how his family's boarders had discussed their grievances and brought news of union mines and conditions elsewhere. As a child, he had learned from boarders about the Ludlow massacre and the violent suppression of the Windber strike of 1906. Boarders had been an important source for increasing his class-consciousness, and at an early age he began to appreciate the need for the diverse nationalities to unite if the miners were to have any realistic chance to unionize Windber.[23]

But the boardinghouse system was also encouraged and fostered by Berwind-White because it served other purposes that directly or indirectly benefited the company. Boardinghouses were practical enterprises, of immediate utility, when the town was new and housing for mining employees scarce. Their existence spared Berwind-White money that would otherwise have been necessary to expend on additional housing and facilities for its employees.

Moreover, boardinghouses furnished the company with an additional means of exerting control over employees. It is significant that the male heads of these houses—ordinary miners—were termed "boarding bosses" and that these bosses, as other bosses, were expected to carry out company mandates. Because Berwind-White had monopolistic control over housing, it was able to determine who the boarding bosses were and the terms under which they could keep boarders. When the Eureka Stores, the Berwind company-store affiliate, issued new orders in 1904 that every miner and his family had to buy at the company stores, it was the boarding bosses who were visited and received the decree. Likewise, under the threat of eviction, boarding bosses were expected to see that all boarders, as well as family members, remained at work during the strikes of 1906 and 1922. The right to live in a company house ceased when the renter no longer worked for the company. For Berwind-White, boardinghouses were not what they were to immigrant miners, but an integral part of its larger authoritarian industrial system.[24]

The Significance of Female Labor

As described above, women often made substantial monetary contributions to the family's income, especially through the boarding system. However, whether or not they ever worked outside the home, they were at all times indispensable participants in the overall system of work and industrial relations that prevailed in Windber. Beyond gainful employment, the importance of their role within that system lay within three key areas: (1) the recruitment and retention of the male work force; (2) the maintenance of that work force; and (3) the reproduction of the next generation and the reproduction of work and gender relations.

Recruitment and Retention of the Work Force

The Women's Bureau study of the 1920s concluded that the wives of mine workers were of "peculiar industrial and economic importance" in the coal-mining industry. "The bunk house and the mine boarding and lodging house long ago proved themselves inadequate to attract a requisite number of workers or to maintain a stable labor supply," the study noted. "Only the presence of his family," it continued, "can keep the mine worker in the mining region, and because of the isolation of so many mining operations the mine worker's wife assumes an unusual importance in the industry."[25]

The presence of women and the work they did were indeed important, if often overlooked, inducements for men to come and stay in Windber. The search for wives in other places by Windber's single young miners, and the efforts of married miners to reunite their families, strongly suggest the great importance immigrant workers themselves attached to a domestic female presence. Kinship ties, marriage networks, and the family economy were integral parts of their lives.

Company officials could not have been unaware of the initial scarcity of women, the gender imbalance, or the desire of the miners for this domestic female presence. Moreover, Berwind-White's promotion of a boardinghouse system built on female labor indicated that its officials understood the stabilizing impact that women and women's work brought to the mining locations. Windber's mines were no temporary or fly-by-night operations, and coal reserves were expected to last for many years. For the immigrants, the arrival of large numbers of women, especially wives, was critical in regularizing social, sexual, family, and community relationships. For all their faults, boardinghouses in the context of family settings filled a temporary need for housing and community on the part of many men who were single or whose wives were abroad.

Women and women's work contributed in another important but indirect way in the company's recruitment efforts. Women's labor had proved essential to the development of the local immigrant parishes and other institutions that formed the basis of the town's many ethnic communities. Women established altar societies, cooked for ethnic festivals, and otherwise contributed to a rich cultural life.

As we shall see in the following chapter, the masses of immigrants had hoped to secure a measure of independence and autonomy in their lives by building ethnic institutions and communities that reflected their respective cultures. Their establishment did not presume ethnocentrism or preclude cooperation with other nationalities for common class goals. Berwind-White, however, had another purpose in mind when it promoted the establishment of these communities, and from the outset its conception of them was very different. Keeping the various nationalities isolated, fragmented, and manageable—the old divide-and-conquer strategy—was key to the company's overall antiunion policy throughout the nonunion period. For this reason, it encouraged the development of self-contained ethnic enclaves, easily controllable and loyal to the corporation, rather than the autonomous, potentially organizable, ethnic communities sought by the immigrants. Nevertheless, fostering the development of ethnic institutions of any kind was an implicit acknowledgment of the importance of family and community life to retention of its workers. Into the 1920s, the company continued to use—with some success—the existence of ethnic communities as an enticement for miners to come to Windber. In foreign-language newspapers, it heralded the ethnic institutions and cultures that women had done so much to create.[26]

The presence of women and the work they did directly and indirectly helped Berwind-White attract and maintain a stable labor supply. Given the serious recruitment and turnover problems it had encountered during the mines' most productive years, 1897–1921, it seems reasonable to conclude that women's domestic presence—not steady work or other factors—was of decisive importance in the company's ability to retain male workers in the area.

Maintenance of the Work Force

Women were also an integral part of the overall labor system in that they were ultimately responsible for maintaining the health, welfare, and efficiency of the male work force. They assumed this role, in part, because taking care of one's family's health and well-being was considered a traditional aspect of women's domestic chores, but the role was also thrust on them as a result of corporate greed, public indifference, the lack of unionization, the absence of a welfare state, and the absence of other alternatives. It is significant that the boardinghouse system extended the woman's traditional caretaking responsibility beyond her immediate family and kin to others of the work force who boarded with her. What women had to cope with as part of this caretaking responsibility can be better appreciated by examining their working conditions.

Keeping boarders was hard work, especially in the early decades of the century when company houses had no running water or toilet or bathing facilities. Washing was done by hand, and water had to be fetched from pumps located outdoors. "She washed and cooked for them and kept their beds fixed up," said one elderly man who recalled how his mother took care of the eight to ten boarders she routinely kept throughout his childhood in the early century. "And then your water was outside. Spigots. And if you done your washing or if you needed water in the house, you had to go out there. In the wintertime, it wasn't very pleasant." Spigots froze. He went on: "No water. No conveniences whatsoever. And no washing machines. All by arm power. The [boarding] Mrs.—my mother—she had to do all the washing on a washboard with her arms and haul the water in from outside."[27]

The water problem that women encountered in mining towns was critical and unique. The wives of contemporary farmers often had to haul water from their outside wells and do without conveniences too, but their situation was qualitatively different from that of the miners' wives, who had to contend with coal-laden mining clothes and layers of coal dust, soot, and grease, as well as dirt in their daily environment. The conditions peculiar to mining life required women to expend great arm power or physical strength to perform ordinary domestic chores.[28]

Housing and sanitary conditions in Windber were no worse but certainly no better than those in most bituminous towns. The Dillingham Commission had bitterly complained that open sewage flowed down streets in the outlying settlements of Windber, and sanitation remained a health menace for many years despite sporadic epidemics that prompted cleanup campaigns. The Women's Bureau found that outside privies were still the rule in bituminous homes in the 1920s, that only 20 percent of these homes had any running water, that only 3 percent had bathtubs or showers, and that less than 4 percent had flush toilets. In Windber, few homes outside the urban center had hot water or flush toilets, and miners' families rarely lived in those. Outside privies remained the rule. By the 1920s, women's work was aided by the addition of cold running water inside the house, but work was not easy, and washing machines remained nonexistent or beyond the means of miners' wives. Elizabeth Molnar remembered how she used to help her mother with the washing for the family of twelve and the two or three boarders they kept: "We had water in the house but we had those copper boilers to boil. Imagine! And we had a stick. We used to boil them [the clothes] to make them nice and white."[29]

Washing was an all-week affair, even at this late date. Elizabeth Molnar explained: "So when we would do the wash, we would do, like on Monday, we would do all the white things, like the boys' shirts. Oh, we had a lot of shirts. And then the second day, we'd do the bedding. The third day we would do what—the miners' clothes. Then the carpets, and, oh, God!" She hated to wash the carpets. In order to protect their carpets from the omnipresent coal dust, women often covered them. For this purpose, her mother used to sew together burlap sacks that used to contain feed for her cows. But carpets occasionally had to be washed anyway. She added: "Those things were so big we couldn't do them by hand in a tub. We'd put them on the front porch and scrub them with a brush, and in the wintertime my dress would be all frozen and [I] barefoot, scrubbing those carpets and rinsing them in the tubs."[30]

The water problem also affected another important aspect of domestic and industrial life—bathing. In the absence of company washhouses, showers, and bathtubs, women again routinely hauled and heated water to fill portable tubs for the miners' baths. When the outside pipes froze in the town's early days, however, the men had to go to one of the creeks to break ice to fetch water for their baths. An Italian woman recalled those days and added:

> My mother built a shanty outside in the back. This is where they used to take their baths. My mother put a stove in there. And, well, she carried the water because there was no connection yet. She'd fill these big boilers on the stovetops, get them ready, and everyone, everybody had their own towels when they'd take a bath and their clothes. They'd come in, when you'd see them coming home all black. And, oh, in the wintertime, I used to feel sorry for them.[31]

Wives were expected to assist husbands with their baths. The son of a Hungarian miner recalled:

> They came home, and they'd get the big tub out, see. Then my Dad would wash up, down to his waist, and then they had what they call a *csutak*—Hungarian for a big soap rag. And my Dad would fold it up. It would be like a big pancake, yah, and [he'd] slap it on his back. That was the signal for my mother to scrub his back.[32]

His Slovak wife added that she had to help out with her family's baths, especially the weekly Saturday night full bath: "But that was pretty rough. We girls had to scrub the brother's backs too."[33]

A miner's wife preparing a bath for her husband.

Another major routine maintenance task which women performed was preparing, baking, and cooking food for their families and boarders. At the time, women baked bread, they did not buy it. One company-store clerk who had worked in the store at Mine 35 in the second decade commented:

> Like up at 35 when I was working there, . . . there was so many miners there that, oh, it was pitiful how these women had to work. How they had to wash the clothes on washboards and how they used to work with their arms kneading bread in tubs, not in little pans, but in tubs.

> They'd buy flour by the 100-pound bags. Then make two bakings out of that. Well, they'd make a great big tub. Practically all of them had these outside ovens they built themselves like they had in the old country. They was made out of brick.[34]

Frequent baking in this bulk was an onerous task, although it sometimes provided an opportunity to socialize with female neighbors. Only by appreciating the physical labor involved in this work can we understand why immigrant women were sometimes delighted in later years when they had the chance to purchase less tasty store-made bread.

Women also bore the primary responsibility for tending the gardens and taking care of the animals and poultry that were characteristic of the mining town. Even though they earned little or no cash, they nonetheless contributed to the family's well-being and cut expenses by producing basic necessities of life. Ewa Morawska estimated that such women's services in adjacent Johnstown saved immigrant families an average of five or six dollars a month otherwise spent on food expenditures.[35] An elderly miner recalled those early days:

> The miners always had something to eat. They never went hungry. They had to raise their own food. Coal miners, they had cows and they made their own butter, their own buttermilk, their own cottage cheese. They had chickens, they had ducks, they had geese, and they also raised pigs. You was allowed to raise anything in the borough because it was all right with Berwind-White. So they raised their own food, most of them.[36]

Women's work was critical to survival in those times when the mines only worked a couple of days a week. As one company-store manager put it: "That's how most of them lived that way. I don't believe they could have made it without having that [gardens and animals]. They weren't making enough money, and money was so stinking tight."[37]

As members of mining families, women were already deeply embroiled in the compulsory company-store and credit nexus established by Berwind-White. But their role as shoppers and food preparers brought them into personal contact with a system that bred many injustices and resentments, a subject discussed in a later chapter. Here it is sufficient to note that many women learned from bitter personal experience that they, as consumers, were not free to shop where they wanted to without dire retaliations against their families. Berwind-White had allowed gardens, poultry, pigs, and cows in the borough limits because these were considered necessary to recruit and keep immigrant miners who came from rural backgrounds. This concession to its workers was immediately offset by the corporation's rigid enforcement of compulsory shopping and credit that ensured no loss of profits by doing so. Since the Eureka Supply Company, Ltd., was a comprehensive department store that sold suits, shoes, furniture, and virtually all necessities, dollars not spent on food would be spent on other goods.

For the boarders themselves, food was a major consideration in judging the quality of a boardinghouse. Miners were individuals with their own particular food preferences. They needed nourishment to do their jobs, of course, but they also expected women to honor their preferences because they were contributing to the payment of the household's food bill. A woman had to shop, manage the accounts, and cater to some degree to the boarders' expectations, which were shaped by their peasant and ethnic backgrounds. Otherwise, they might board somewhere else.

Mike Jankosky explained that his family moved from one company house to another because his family's boarders wanted to be closer to the mine where they worked. While his mother served a standard fare, she nevertheless took the boarders' individual preferences into account when she bought and prepared food, which she supplied in generous portions. He said:

> If you had eight or ten boarders, you bought five or six pounds of beef for soup. Soup everyday and a bone with it. Then they bought that soup meat.
>
> You had to cook each man one pound of pork in his dinner pail, besides either [an] apple or banana or whatever kind of fruit, pickle, or whatever they wanted. A pound of pork and bread naturally. She had to bake her own bread. She'd bake as high as eight, ten, big loaves, and cut it, and then put this pork in their dinner pail and an apple or a pickle, and they were ready to go to work.[38]

All buckets might not have been so full, but, as a rule, miners' wives and daughters took special care in preparing the dinner pail that the men took into the mines. In a sense, the pail symbolized basic sustenance, if not life itself, in the mining community. Men valued its contents not merely because their work was physically grueling but, more important, because they had to work long hours at low wages throughout the nonunion era to make a scant living for themselves and their families. Given economic scarcity, some miners subsequently developed a custom that simultaneously showed their appreciation for the food and affection for their families. In the early twentieth century young children often met their fathers when they left the mines after a day's work. On their walk home together, fathers opened up their lunch buckets to give their offspring a treat saved for them.[39]

Women worked long and hard to accommodate the men's needs. Their own laborious work schedules and domestic chores were geared to meet the men's work patterns. Households were economic units, and Berwind-White's pay system and other policies had a direct bearing on the welfare of the entire family. Women's caretaking tasks were made more difficult because of the wages, hours, and working conditions their men endured. The labor and income of both sexes was a necessity of family survival in hard times, but the male was considered the primary breadwinner in that he was normally expected to provide the larger share of the family's income from his wages as a miner.

It was in this context that miners' wives provided for the basic maintenance of the work force. Accommodating the many boarders and male family members who often worked different shifts or irregular hours involved more labor time and arm power than that expended by women in nuclear families. Yet they did more. They cared for accident victims and the sick, including the many victims of "miner's asthma"—a euphemism for the various lung diseases miners contracted from their occupation. A town druggist recalled that in the early decades of the century, people rarely went to hospitals and saw doctors infrequently: "Most of the people in those days went to the drug store, and they had to be their own doctors and nurses. The only reason they had to go to the hospital was surgery. Most of them [operations] [were] for miners, [with] broken legs, broken arms and what not."[40] Once injured miners left the hospital, it was the women in their families and boardinghouses who took care of them.

In an environment in which corporate greed and public indifference predominated, Windber's working women assumed the primary burden of caring for the health and well-being of the miners outside of their workplace. This caretaking task was difficult at all times because of the primitive working conditions in their own workplace—the home—but it was even more necessary to the survival of the family unit because of the low wages, long hours, poor working conditions, accidents, and occupational diseases that Windber's nonunion miners endured.

The Reproduction of Work and Gender Relations

While fathers or other male relatives trained young boys or men to mine coal, mothers and other female relatives trained daughters in all sorts of domestic duties. Gender roles were reproduced. A Carpatho-Russian immigrant explained the rationale: "The boys have to support the family. The girls have to stay home."[41]

At an early age, females learned to bake, cook, sew, wash clothes, and care for the ill. Young girls often assumed major domestic responsibilities for the first time when their mothers had a newborn child, who was usually delivered by a neighbor who was also a midwife. Mothers needed the help of their daughters because families tended to be large, boardinghouses were congested, and "women's work" was never done. Despite some additional schooling and some household improvements, the younger generation of the 1920s had little time to dream of alternatives to domestic, clerical, or sales work, and even fewer opportunities to realize any such dreams in Windber. As in the case of their male counterparts, childhood was still a brief interlude at best. Women interviewed told frequently of their great disappointment when they had to leave school to help out at home. In 1940, some 69 percent of all Windber residents, male and female, over the age of 25 had completed only eight grades of school or less. For both sexes, work began at an early age and remained a fundamental reality.[42]

Women also played an important role in the productive process by reproducing the next generations of mine workers. Children had been employed in the mines throughout Windber's history, but World War I and subsequent immigration restriction in the 1920s made company reliance on miners' families for its future labor force all the greater, even though its demand for labor diminished. The exodus from the local mining region increased during the depression of the 1920s and 1930s, when the mines were fortunate to work two or three days a week, and it intensified during the mine closings of the 1950s. Nonetheless, sons had continued to follow their fathers into the mines throughout these years whenever the sole local industrial employer was hiring.

During the nonunion period of the Windber miners' history and beyond, both the immigrants and Berwind-White accepted a system of work that entailed a rather rigid sexual division of labor and the existence of a family economy, but they did so for different reasons. To the company, the system was an integral part of its larger industrial relations policy. Existing gender and work relations enabled the corporation to exercise and perpetuate its authoritarian leverage over its employees and their families. To the workers, however, rigid sex roles and the family economy were rational measures for coping with the realities of industrial capitalism in Windber, a one-industry town where there were few, if any, job possibilities outside the mines and home. The conditions that shaped industrial policy for one party necessarily shaped the nature of the struggle against that policy for the other party.

Berwind-White had successfully maintained its monopoly of industry in the Windber area throughout the nonunion period by keeping out independent or competitive industries that would have undermined its autocratic control of the workplace and community. As early as 1903, class-conscious miners, disgruntled merchants, and a progressive newspaper editor were crusading for a diversification of industry. They understood that breaking the occupational monopoly was one of the possible roads to unionization, which they considered essential to achieving greater control over their own lives.[43] Their unspoken assumption, however, was that new independent industries would—and should—employ male workers.

Yet the company's stranglehold on job opportunities applied to women as well as to men. We now know that alternatives to domestic and clerical work were almost as scarce for women in Windber in 1940 as they were in 1910. Changing gender relations through large-scale female employment in industries independent of Berwind-White would have necessitated changes in the company's entire system of work relations during the nonunion period because women's work was not subsidiary in importance to men's work, but a bulwark of the overall labor relations system. The Berwind system, dependent on its monopoly position in town, was strengthened by a particular division of labor according to gender, which ensured the recruitment, retention, maintenance, and reproduction of its labor force.

But women's work simultaneously enabled the miners and their families to endure and struggle against the class oppression both sexes encountered in their daily lives. Indeed, their futures were inextricably

Susie Shuster, a union supporter who signed an affidavit against the Berwind Coal and Iron Police, and her children—Katherine, Anne, Joe, and John—during the strike of 1922.

bound together in the family economy. The miners could never have carried out the fierce, protracted unionization battles of 1906 and 1922 without active, broad-scale female support.The sexual division of labor had decreed that women bore the primary caretaking responsibility of home and children. If they had not been behind these strikes, families could never have endured the hunger, evictions, hardships of tent-colony life, and the many deprivations that adversely affected children as well as adults of both sexes.

Moreover, as will be seen in future chapters, women also sometimes took independent actions of their own to support the union cause. During strikes, they often violated injunctions and confronted strikebreakers. By offering these men breadcrumbs and pennies, they effectively shamed them by striking a blow at the very notions of manhood that underlay the miners' family economy.

Women and men had indeed experienced class oppression in different ways, if for no other reason than that their work was different. And their families and households were not egalitarian. Ultimately, however, the autocratic nature of the overall Berwind system gave the men in the mines and the women at home a shared, though not identical, class interest in challenging it.

5

Ethnic Communities and Class

The major dividing line within the community of Windber was simultaneously a class, ethnic, and religious one. In its report, the Dillingham Commission had emphasized the overall dimensions of that division: "The native Americans associate only to a limited extent with the immigrants from southern Europe, while association among the native Americans, Irish, English, Germans, Welsh, and Swedes is very free." Even American Protestant churches were largely indifferent toward new immigrant Protestants and shunned association with them. At the same time, the report commented, "The recent immigrants [from southern and eastern Europe] tend to live apart from the natives and from each other whenever possible."[1]

As already noted, the town's prevailing class structure embodied a particular hierarchical arrangement of ethnic groups in which Americans and northern European nationalities generally occupied the top rungs and were clearly distinguishable from the new immigrant groups in terms of residence, religion, and occupation as well as political and economic power. The result was the emergence of two basically separate, unequal, different societies that cohabited the town. This basic class/ethnic division characterized the nonunion era. The masses of new immigrants had to confront not only class oppression but nativism and systematic discrimination as well.

In this context, new immigrants sought to establish their own ethnic communities early in the century. By creating institutions that reflected their respective cultures, they hoped to gain some degree of power and independence in an otherwise alien, often hostile, environment that accorded them inferior status at best. Narrow ethnocentrism was not the issue. Having been locked out of the dominant society, many local immigrants saw the establishment of these communities as a grassroots means of democratizing their surroundings. Moreover, Windber's diverse nationalities generally recognized that other nationalities had a similar right to use their own languages and create their own institutions. Because they recognized that they were both workers and ethnic people, not one or the other, their desire to build responsive ethnic communities did not preclude cross-cultural cooperation for common class goals, including unionization.

The immigrant miners' desire for greater autonomy in their lives through the establishment of their own ethnic communities inevitably clashed with Berwind-White's alternative conception of ethnic communities as ethnic enclaves that were isolatable, fragmented, controllable, and hence unorganizable. By the early 1900s, the company had actively opted to promote the development of dependent ethnic institutions for reasons of its own. Ethnic communities were an important means of solving its long-term problem of attracting and retaining a foreign-born work force in a competitive labor market. Once established, the communities were a useful mechanism to keep its disparate workers apart and nonunionized. From the outset, the company had rigorously pursued a divide-and-conquer policy in the community as well as in the workplace. It had, for example, successfully separated American miners from foreign-born miners by arbitrarily imposing residential restrictions. Railroad tracks near Mine 35 were used to insulate an enclave of company houses for Americans from a similar enclave for immigrants.[2] In a sense, the fostering of dependent ethnic institutions led by loyal ethnic stalwarts who kept them operating within narrow limits that divided workers of one nationality from those of other nationalities represented the logical culmination of this old antiunion policy. Creating or maintaining divisions in a multiethnic work force was key to Berwind-White's open-shop strategy. "Americanization," even of the corporate sort, never played a role in its antiunion policies.

Both the masses of immigrants and Berwind-White, then, liked ethnic communities, but for different and even conflicting reasons, and they held alternative conceptions about them. Creating independent ethnic communities in the context of an autocratic company-town setting inevitably presented class-conscious ethnic people with very real problems. The result was ongoing class conflict within the established institutions throughout the nonunion era. The purpose of this chapter is to examine the emergence and significance of Windber's various ethnic communities in order to determine how they both hindered and enhanced interethnic class cooperation. In the end, class-conscious miners and their families turned to the ethnically friendly, pluralistic, class-based institution, the United Mine Workers of America, to fulfill their hope of achieving greater democratic control over their lives.

The Importance of the National Parish

In order to assess the role of institutionalized religion as a fragmenting or integrating influence for Windber's immigrants in the first quarter of the century, it is necessary to focus attention on the Roman Catholic population. The vast majority of the new immigrants belonged to this faith, but that should not obscure the fact that a smaller number of immigrants were Byzantine Rite Catholics, Jewish, Slovak Lutherans, Hungarian Reformed, or others. The number of denominations itself highlights the great religious as well as ethnic diversity that existed in the town.

Organized Catholicism came to Somerset County at the turn of the century with the arrival of the new immigrant labor force employed in Berwind-White's mines. The church hierarchy and immigrants quickly established a formal Catholic presence in Windber on the basis of territorial, not national, parishes. By 1900, St. John Cantius Church was serving all Roman Catholics, regardless of nationality, while another church ministered to all Greek Catholics, regardless of nationality. This territorial arrangement worked best in the town's first years when the number of Catholic immigrants was limited. Within a short time,

A First Holy Communion class at St. Mary's Byzantine Catholic Church, Windber, c. late 1920s.

however, the growth of this local religious population began to present practical difficulties, even as the influx of large numbers of Catholics into the neighboring region was simultaneously prompting the formation of a new Altoona Diocese. To meet the needs of the town's expanding population, the church required more space or additional churches. In solving the problem, the head of the new diocese, Bishop Eugene A. Garvey, worked hard to try to preserve the town's original Roman Catholic unity, but ultimately failed. Consequently, St. John Cantius Church of Windber became the mother parish of four other Roman Catholic national parishes in the period from 1905 to 1921.

National or ethnic parishes were the heart and soul of the many immigrant communities that flourished in the United States during the first quarter of the century and beyond. Given the decline of traditional societies, the growth of a capitalistic economy, and the migration experience itself, these institutions facilitated the initial adjustment of strangers in a strange land. Through them, immigrants enjoyed a continuity of

Dramatics Club of St. Mary's Byzantine Catholic Church, Windber, 1932.

religious faith, maintained their native languages and cultures, and participated in a vigorous social life. Immigrant churches thus fulfilled many immigrant needs, and in some cases became more important in the New World than they had been in the Old World.[3]

However, the process by which the different ethnic groups—Slovaks, Italians, Hungarians, Poles, and English-speaking populations—formed separate parishes in Windber tells us much about the subsequent role immigrant churches played in the company town. A review of Roman Catholic institutional history in Windber from 1900 to 1921 makes it possible to draw several important conclusions about potential interethnic, working-class integration.

New Immigrants Founded Parishes

First, it is critical to stress that it was the immigrants themselves who took the initiative to found and build the various national parishes in order to meet their respective spiritual, social, and cultural needs in the new and different environment. Even people who were not particularly religious found it necessary to have some affiliation, if only to partake of the rites of passage in life—birth, marriage, and death. National parishes also helped in the transition to life in a new country. This is important because the early twentieth century was the heyday of immigrant churches in the United States, and the history of Windber's religious institutions—the emergence of a number of national parishes from a territorial, ethnically mixed one—was not unusual.[4] Berwind-White quickly recognized the potential value of these institutions in its industrial labor relations divide-and-conquer policy, but it did not initiate their foundings or mandate their existence. The initiative of the immigrants, their desire for autonomy, and their insistence on the right to use their own languages are illustrated by the history of the first secessions from St. John Cantius Roman Catholic Church.

In Windber, ethnic tensions had threatened to disrupt the original unity of the Roman Catholic church in Windber as early as 1903, when Bishop Eugene Garvey acknowledged parishioners' complaints and publicly recognized a need to harmonize the various nationalities represented among the worshipers at St. John's. His solution was to appoint a priest who he believed could accomplish the task, the Rev. James P. Saas, an American-born priest of German extraction and a talented linguist who could speak four major immigrant languages on his arrival in Windber. Father Saas proved acceptable, and his popularity increased when he rapidly acquired knowledge of four other languages that were important to members of his congregation. Meanwhile, the bishop occasionally sent priests to serve the Italian sector, which wanted to hold separate services in the church.[5]

Father Saas was initially successful in his mission of holding the disparate nationalities together in a working relationship. In 1904 the church was enlarged and the first parochial school in the county's history was established in Windber. At the school's dedication, Bishop Garvey stressed that all Catholics, regardless of nationality, needed to stand together in a common effort for a Catholic education so that the Catholic religion would not fall victim to a growing trend in the world toward religious indifference.[6] Yet ironically, it was this bishop who believed in the commonality of the Catholic faith who later that year precipitated the first open talk of revolt among parishioners of St. John's. By then pastor to more than 1,000 parishioners of diverse nationalities, Father Saas had requested an assistant, and Bishop Garvey, an Irish-

American, responded by sending a young, recently ordained priest of Irish-American extraction.

Windber parishioners accepted Saas because of his linguistic abilities and personality but objected to the new, young priest because he knew none of the major ethnic languages on his arrival. Slovaks and Poles, who claimed to represent 90 percent of the parish's population, were angry. The Slovaks began to talk about forming a separate church, and Poles were publicly quoted as saying they would blow up St. John's rather than give up their language. Their anger was directed at the bishop. Parishioners objected to the English-speaking priest not on theoretical grounds, as a reaction against "Americanization" or the English language, but on practical and democratic grounds. Assistant priests, they argued, should reflect the language of at least one of the parish's major nationalities, not the language of the small minority of Irish and English-speaking parishioners.[7]

This discord quickly abated, but the conflict over the assistant, diminishing space in the church, and desire for their own language and priest made Slovaks determined to establish a parish of their own. In 1905 they organized the parish of SS. Cyril and Methodius under lay initiative and with the bishop's permission. A lay committee of twelve members was chosen to solicit funds from parish families, but solicitations and erection of the church edifice were hampered by the strike of 1906, which dispersed the Slovak population and made resources sparse. Various priests, including Saas, helped to oversee the work, and finally, on November 21, 1909, after four and a half years of effort, the church was dedicated by Bishop Garvey.[8]

Parishes Reflect Ethnic Group Histories

Although new immigrants hoped to increase their democratic participation in religious life by founding national parishes, the first secessions from St. John's indicate that these groups went about the task differently. From the outset, there was no single pattern of church development in the company-town setting, and each ethnic group was influenced by its own particular social, cultural, and religious heritage.

In certain respects, the Slovaks, as a group, were exceptional in Windber. Already, by 1905, they may have been more class-conscious and more suspicious of paternalism than other ethnic groups in Windber. Individual Slovaks are known to have had previous experiences in the anthracite region with coal companies and unions, and they were pioneers in forming working-class ethnic institutions. They had been the first ethnic group in the town to form fraternal societies as well as the first to form a national parish. Slovaks were also prominent activists in the strike of 1906.

Previous experience may explain why the Slovak church is unique among Catholic churches in Windber in that its lay organizers purchased lots in 1905 for the church site at market value from an individual owner and did not ask Berwind-White to donate ground to them.[9] Consequently, the parish was not obligated to the company because of charitable contributions made at the time of its origins. In any event, that SS. Cyril and Methodius was a self-supporting institution created by a broad spectrum of the Slovak population is probably not accidental. Slovak priests, unlike certain other clergy, were consequently more free to exercise at least a modicum of independence during the strikes of 1906 and 1922 when they did not oppose the miners' strike or advise the men to go back to work.

Windber's Italian parish, St. Anthony's, was also started in 1905, but its historical development, which was interrupted and rocky for reasons that went beyond the strike, differed markedly from that of the Slovak church. Rudolph Vecoli has documented widespread anticlericalism among the *contadini*, Italian peasants, along with an indifference to the institutionalized church, which in Italy was considered an exploitative landlord, and a corresponding reluctance of Italian immigrants to initiate and financially support ethnic parishes in the United States.[10] His descriptions of Italian religious culture seem to apply accurately to Windber.

From the outset, it was interested clergy and *prominenti*—Italian labor bosses, men who were associated with Berwind-White's management—not Italian miners, who took the initiative to form a new parish. Father Saas, contractor Frank Lowry, and others thus cooperated in the endeavor. Saas convinced Bishop Garvey to send a priest to organize the parish in the fall of 1905, but this first priest left after only a few months, "on account of the indifference of these people to religion." With his departure, the Italians once again came under St. John's jurisdiction.[11]

The strike of 1906 led to further delays, but late that year Bishop Garvey assigned a second priest, an Italian, the Rev. Nicholas Capaldo, to try to organize a parish. During his brief tenure in Windber, Father

Capaldo encountered so much opposition that he remained only until the church was completed in 1909. In the meantime, the project was advanced by Saas and Lowry. They had persuaded Berwind-White to donate land for a new church, Lowry agreed to build the church foundation free of charge, and a lay committee of prominent Italians was formed to solicit funds. Meanwhile, Italians from various villages and provinces quarreled over which of the many locally favored saints would bear the church's name. Finances continued to be a serious problem, and Capaldo complained bitterly about the parsimony of his own congregation. Even the third priest, who proved satisfactory to the congregation and had a sixteen-year tenure there, had to go door-to-door to collect funds.[12]

The establishment of St. Anthony's did not immediately make Windber's diverse Italians less anticlerical, less indifferent to institutionalized religion, or more oriented to "Italian" than to local loyalties. The formation and support of the Italian parish in Windber was therefore initially more dependent on the patronage of *prominenti* and the Catholic hierarchy than on ordinary Italian miners and their families. Because prominent leaders of this ethnic community were intimately associated with Berwind-White's management, church officials and leaders were not likely to support unionization efforts or strikes.

The Magyars' delay in organizing a Roman Catholic parish was in turn a reflection of the nationality's history of a mixed religious heritage. Unlike the Poles, who enjoyed a more homogeneous religious tradition that unified religion and nationalism, Magyars were hampered by the existence of sizable religious divisions within their ethnic group. Approximately one-fourth of the ethnic group's population were Protestants. The largest portion of these, Hungarian Reformed Church adherents, had formed a parish of their own in Windber in 1905. Hungarian Lutherans had no separate parish, but attended Slovak Lutheran or Swedish Lutheran services. Moreover, language was sometimes as important a consideration as religion. Some Hungarian-speaking Roman Catholics attended Greek-rite services in the town's early years because priests there, unlike those at St. John's, could speak Magyar as well as Ruthenian. On the eve of the world war, when national tensions were intensifying, a few Magyars of Roman Catholic faith continued to prefer the Slovak church to St. John's.[13]

Nevertheless, by 1910, when space had again become a problem at St. John's, where Poles and various other nationalities continued to worship, the Magyars and English-speaking groups began to consider forming separate Catholic parishes. The costs involved in such an undertaking deterred the English-speaking minority, and Father Saas won the bishop's approval to construct a new and enlarged church to meet the mixed congregation's needs. By the end of 1912, however, the Magyars had definitively decided to organize a parish of their own.[14]

Initially, Father Saas, Bishop Garvey, and Berwind-White were cool, if not hostile, to the idea of a Magyar parish. The Magyars' belatedness in organizing counted against them. By 1912, Bishop Garvey was actively promoting adoption of the English language and American customs and was not receptive to the idea of another national parish in the town. Berwind-White officials were also reported to have said that Windber had enough churches and that the company would not donate land to a new one.[15]

Saas, Garvey, and Berwind-White eventually retracted their initial opposition to a Magyar parish. The clergy were greatly displeased when Hungarian-speaking priests from other dioceses began to visit Windber to challenge their jurisdiction, and when local Hungarians threatened to take their case for a new parish to Rome if necessary. In the end, Garvey agreed to support the effort, but only on condition that Father Saas supervise the work. Berwind-White did donate land, and after seven years of strife and the retarding impact of a world war, St. Mary's Roman Catholic Church was dedicated on September 3, 1920.[16]

The departure of the Magyars from St. John's precipitated the founding of a fifth Roman Catholic parish, Holy Child Jesus, in July 1921. This new English-speaking parish left St. John's exclusively to the Polish nationality. A new church, Holy Child Jesus, was formally dedicated in October 1926 and became popularly known as the "Irish" church—which was erroneous because the Irish were few and membership included English-speaking Catholics of all nationalities. The national "Americanization" campaigns of the war years, and growing evidence of nativist sentiments and legislation, could also have contributed to the timing of this last division. In any event, a number of prominent ethnic businessmen from new immigrant nationalities, those individuals who had become "Americanized" enough by the 1920s to prefer English to a foreign language, transferred parish allegiances.[17]

Windber's immigrant groups founded national parishes over a span of years and under different circumstances, but in each case the group's previous experiences and aspirations influenced church development.

The Role of Berwind-White

While the national parishes met many of the needs of their members and reflected their groups' histories, there is no question that Berwind-White encouraged their development by making crucial donations of land and charitable contributions, because it saw them as convenient mechanisms for its divide-and-conquer policy. To guarantee its monopolistic control, the company both fostered ethnic segregation and sought to establish long-lasting control over ethnic institutions. Through a certain degree of paternalism, it sought to inculcate sentiments of loyalty among the miners, but it never considered sentiments sufficient to guarantee perpetuation of the open shop. In the end, legal restrictions and the involvement of leading company officials in parish affairs were more reliable control mechanisms. The company saw institutionalized religion—bolstered by ethnic divisions, legal constraints, loyal clergy, and "safe" ethnic lay leaders—as a conservative force that inhibited class-consciousness.

It was standard procedure for the company, whenever land was sold, leased, or donated to anyone, to include provisions in the deed that gave Berwind-White or any successor company perpetual rights to mine, remove, and carry away all coal and minerals beneath the lots. After farmers had brought numerous lawsuits against the company for injuring or damaging the surface, premises, or buildings as a result of mining, a new provision that released Berwind-White from any liability for such things was incorporated in deeds. Deeds conveying land donated to churches or charitable organizations, however, contained additional restrictive clauses, which gave the company perpetual control over how the properties would be used.

The formation of the Italian church furnishes us with a good example of how Berwind-White sought to maintain control over churches and other organizations through restricted land donations. Bishop Garvey was so displeased by something in the original deed granting the Altoona diocese land for an Italian church that he returned it, rejected, to the Wilmore Coal Company, Berwind's real estate subsidiary. Ultimately, however, he accepted a second deed that contained a number of restrictions. The 1908 deed for St. Anthony's land contained the following clause: "It is expressly understood that the said lot #1055 is to be used for church purposes only." Therefore, if the church rented out rooms for union purposes or used its premises to promote social causes or political movements that the company opposed, Berwind-White would have legal recourse. Provisions like this guaranteed that churches would never become meeting-places for miners who were seeking to organize.[18]

Despite subsequent changes in American society and the establishment of the union in Windber in the 1930s, the company still sought to perpetuate its influence by restricting the use of donated properties to specific purposes, such as "church purposes only" or "cemetery purposes only." In 1942, when the Italian church obtained an additional lot, a clause not only stipulated that the property be used "for church purposes only," but also added: "In the event that the lot is abandoned, or if the premises are used for purposes other than church, the land could revert to The Wilmore Coal Company, or its successors, at the option of the said Wilmore Coal Company." St. Anthony's got a release from such restrictions only in 1974; St. John's obtained a release from similar restrictions on leased properties in 1984.[19]

This focus on Berwind-White's restricted-deed land donations to Roman Catholic churches should not obscure the fact that, from the town's inception, the company treated other churches and denominations in the same manner. Deeds conveyed to trustees of the First Presbyterian, Methodist Episcopal, Brethren, Mount Zion Evangelical Lutheran, Hungarian Reformed, and Gustavus Adolphus Swedish Lutheran churches contained restrictions similar to the ones described in the deeds conveyed to St. Anthony's.[20] Legal provisions circumscribing the independence of churches and all other potentially independent organizations were part of the company's arsenal for maintaining autocratic control in the community.

The company also sought to continue to exercise control of both Catholic and Protestant religious institutions through other mechanisms, especially financial contributions and the presence of loyal company officials on church boards. Consequently, celebrations of church anniversaries were characterized by banquets where Berwind-White officials were honored and the company was routinely praised for its philanthropy. But priests and ministers were also expected to support the company's policies in the pulpit and elsewhere. In 1912, for example, priests were called on by the company to preach calm in order to forestall a mass exodus from town that it feared would occur when the Democratic presidential candidate it had bitterly opposed and aroused fears about unexpectedly won the election. And people who dared to oppose company policies were harassed by the company. This was the case, for example, with the Rev. G. W. Pender of the Methodist Episcopal church, who publicly declared his support for the rights of working people during the strike of 1922 and ultimately left town in the face of company harassment.[21]

The company even expected certain priests and ministers to serve as active agents for it. Berwind-White had found that immigrant churches, the basis of the town's ethnic communities, were invaluable instruments in its recruitment drives. In want ads placed in Hungarian-language newspapers in 1920, the company utilized the Hungarian Reformed church and the new St. Mary's Hungarian Roman Catholic Church as reasons for miners to locate in Windber. Prospective Hungarian miners were encouraged to seek out the town's Hungarian minister and priest to verify the company's advertising claims.[22]

Berwind-White's attempts to control immigrant churches meant that the immigrants' attempts to use churches as meeting-places for purposes opposed by the company would meet legal obstacles. Like the miners or other residents, clergy who exercised the right of free speech on behalf of the labor movement, or otherwise supported it, would normally not survive long in nonunion Windber. In 1906, when immigrants were only beginning to form ethnic parishes, their ethnic priests supported the union and the strike. In 1922, when the churches were well established, most ethnic priests supported the company and were ousted by the miners. In the end, leadership within the immigrant communities was a crucial factor. Although the constituents of national parishes were predominantly working-class, their leaders and clergy did not necessarily represent their class interests.

The Limits of Fragmentation

On the whole, the Catholic church in Windber was not an integrating force in Windber or an important assimilating or "Americanizing" influence in the first decades of the twentieth century. With reluctance, the Catholic hierarchy bowed to the new immigrants' autonomous movement to increase their democratic participation in the church through the establishment of national parishes and use of their own languages. Thus, the church recognized and sanctioned cultural divisions that neither it nor the company created. However, recognizing and tolerating different languages and traditions—pluralism—is not synonymous with unifying diversity in a larger purpose.[23]

Nonetheless, the fragmenting tendencies implicit in the formation of national parishes, which Berwind-White supported for its own reasons, were muted by other tendencies, which had implications for potential interethnic, working-class integration. Among these were the relatively friendly manner in which church separation was effected, ongoing working-class relationships and grievances, and the community's major class/ethnic/religious division.

The manner in which Windber's national parishes came into being determined that subsequent relations between the dissolving ethnic constituencies would be relatively amicable. It is not surprising that divisions surfaced within St. John's ethnically mixed congregation, but it was not inevitable that separation occurred with so little conflict and bitterness. The absence of serious hostilities upon departure muted the fragmenting or segregating tendencies inherent in the establishment of national parishes. Windber's Catholic immigrants interviewed often expressed a pragmatic and democratic acceptance of these institutions which dated back to their origins. The same impulse that found it just that St. John's have an assistant priest who spoke the language of at least one of the parish's major nationalities decreed that disparate nationalities had the democratic right to have their own churches and languages. National parishes in Windber were born of tolerance and pluralism, not intolerance and hatred.

Today, few of Windber's Catholic population who can trace their family back to the early century have only one church in their background. Parents or grandparents were typically members of the ethnically mixed parishes of St. John's Roman Catholic or St. Mary's Greek Catholic before their dispersion. The following comments from an elderly member of St. Anthony's illustrate this mixed religious heritage. Pearl Leonardis was born in Windber in 1905. Her parents had been married at St. John's when it was the only Roman Catholic church in town. In 1910, the family began to attend St. Anthony's but continued to visit St. John's upon occasion.

> When they built that church [the new enlarged St. John's in 1913], everybody went wild because it was such a beautiful church. We used to visit, and . . . my mother used to go to it because she was [once] a member of that church. Well, she was married there, and she had children—my sister, a brother after me, another sister, and I—[who] were baptized there. Only one of the boys, when our

> church [St. Anthony's] was built in 1910, was baptized up here, at St. Anthony's.[24]

Late in life, Mrs. Leonardis obtained a passport by documenting her birth through a baptismal certificate from St. John's, and when Pope John Paul II became pope in 1978, she chided Windber Poles: "I always tell these Polish when they say something about the Pope [to indicate ethnic pride], I say, 'Shut up. I'm Polish, too. Don't say nothing about it. I was baptized there [at St. John's], you know.' "[25]

The relatively peaceful and friendly secessions from St. John's stand in sharp contrast to the town's one really bitter religious schism, which occurred in the 1930s *within* a single nationality: the Carpatho-Russians. In 1936, St. Mary's Greek Catholic parish divided over the enforcement of papal decrees on Latinization of the rite and over priestly celibacy. The conflict led to a new local church, lawsuits, the formation of an independent denomination, and the creation of the American Carpatho-Russian Orthodox Greek Catholic Diocese of U.S.A.[26]

Then too, ethnic divisions were muted by the class experience in Windber. The grievances that concerned miners at the workplace—wages, honest weights, pay for deadwork, the length of the work day, and most of all, democratic representation and grievance procedures—cut across ethnic lines, and these grievances remained remarkably constant throughout the nonunion period. Accidents that injured or killed large numbers of new immigrants did not respect ethnic boundaries either.

Other class experiences muted ethnic divisions in the community. Regardless of nationality, miners and their families were required to shop at the company store or risk losing their jobs and houses. Foremen or other company bosses told them how to vote at the polls. No church or organization was exempt from Berwind-White's control.

Nor was housing segregation as rigid as such terms as "Little Italy" or "the Polish sector" suggest. Certainly there were sections or streets in which one nationality predominated, especially near the center of town. However, census records show that none of these areas was entirely homogeneous. In addition, a great variety of nationalities resided as neighbors in the outlying populous mining camps. Miners walked to work and wanted to live near their workplaces. They necessarily came into contact with one another. Women of various nationalities met at the company store or outside water pumps.

Moreover, for many new immigrants, the experience of ethnic diversity that they encountered in Windber was not entirely new. Various nationalities and religious groups had coexisted in the Carpathian Mountains and adjacent regions, from which many Windber immigrants had originally emigrated. While many had had little or no formal education, they could nevertheless speak a variety of languages they had learned in Europe or picked up in multilingual Windber or similar places in the United States. The sons and daughters of miners, some of whom were store clerks and order girls for the company stores, had to know numerous languages to do their jobs. Miners who held positions that required them to move about the mines in the course of their jobs often came into contact with other nationalities and learned some of their languages. It is probably not accidental that mobile motormen and spraggers from various nationalities were catalysts of the local 1922 strike.[27]

Because national parishes were, by and large, the creations of new immigrant groups in Windber, it is easy to stress the obvious linguistic and cultural differences between them, to make working-class integration appear to be an impossible dream. But to do so is to overlook the critical fact that the major religious division in the town did not occur within these nationalities at all and that the primary demarcation was simultaneously a class, ethnic, and religious one.

The gap between the masses of ethnic Catholic workers and the company's dominant Anglo-American Protestant establishment was indeed a wide one. The Dillingham Commission had first noted this division in the early century, but it remained long afterward. Protestant churches were the bedrock religious organization of the town's upper-class structure, which included local company managers and borough officials. Certain Protestant denominations were simultaneously bastions of nativism, and it was American-born Protestants who belonged to nativist fraternal associations, such as the Patriotic Order Sons of America, and who were leading advocates of temperance and sabbatarianism. In the nativist 1920s several prominent local Protestant clergy publicly endorsed the Ku Klux Klan, which was waging its antilabor, anti-immigrant, anti-Catholic campaign in the town and in the region.[28]

The result was that even the fragmented institutional Catholic church occasionally served as a unifying force for the various nationalities. This integrating tendency was most clearly evident in the mid-1920s, when Catholics of all nationalities were attacked by nativists. In the absence of the influence of the United Mine Workers of America, which had temporarily unified the disparate ethnic and religious groups in the course of a defeated strike in 1922, the Catholic churches were forced by circumstance to unite and mobilize for the purpose of defense.

The Real Problem

Immigrant churches both enhanced and hindered working-class integration in Windber. To the extent that immigrants mobilized to create these religious bodies in order to meet a need of members and give them some voice in an otherwise autocratic and often nativist society, they were progressive institutions. National parishes inevitably reflected existing linguistic and cultural differences that did not automatically preclude broader interethnic class cooperation. Ethnic diversity always posed practical problems for aspiring unionists, but such problems were surmountable in certain circumstances when ethnic miners saw themselves as both "ethnics" and "miners," not one or the other.

Class-conscious miners in Windber were not deterred from organizing by the existence of national parishes per se, a common phenomenon, or by ethnic diversity, which in any case was a given. For them, the real problem was not ethnic institutions but company domination, which existed within the local churches as it did elsewhere, and the disloyal ethnic and religious leaders who allied themselves with Berwind-White to keep miners separated and unorganized. Ethnic unionists would have to provide an alternative working-class leadership, mobilize to seize control of their own churches, and transcend narrow ethnic or religious boundaries. They visibly did so at key moments, such as during the strike of 1922, when they took control, ousted the clergy who had preached to them to go back to work, and joined with miners of other nationalities in an ethnically diverse, class-based organization, the United Mine Workers of America.[29]

The emergence and role of immigrant churches in Windber raises other interesting questions, because despite their prevalence, national parishes were not universal. Even all coal-mining towns with large diverse ethnic populations did not have them. Why not? Did the prior existence of a class-based industrial labor organization, such as the UMWA, inhibit ethnic fragmentation in Nanty-Glo, Pennsylvania, and other places?[30] Were national parishes less prevalent in towns where industry was diverse or monopolistic company control was nonexistent? Only future comparative community studies can answer such intriguing questions.

The Importance of Fraternal Societies

Fraternal societies were the material counterparts of national parishes in immigrant communities. Their proliferation in the late nineteenth and early twentieth centuries, not only in the United States but also in Europe, symbolized the attempts of working people to protect themselves from the insecurities and hardships imposed by the disintegration of traditional economies and by the rise of unregulated industrial capitalism. The most important purpose of the fraternal society was to provide minimal death and sickness insurance benefits to victims or their families. Immigrant workers found these organizations indispensable. The American middle classes also formed such societies, but they had other options for security that working people lacked. Miners and millhands were denied coverage by an emerging profit-seeking insurance industry because of the high death and accident rates associated with certain occupations. Moreover, at the turn of the century only exceptional corporations willingly assumed any responsibility for employee welfare, and few mining or steel companies compensated accident victims or their families, even when these companies were clearly at fault. The first worker compensation law, long sought by the United Mine Workers, was not passed in Pennsylvania until 1915.[31]

Miners in Windber remember that in these early years the custom for aiding accident victims was for a fellow miner to stand outside the entrance to a Berwind-White mine as miners went to work. With hat in hand, this person collected donations from fellow workers for the wife and family of a man killed or seriously injured in that same mine a day or two before. In the absence of family aid and fraternal insurance, such collections were the only assistance given to bereaved dependents. The state and company contributed nothing.[32]

The key role that ethnic fraternal societies played in the lives of working-class immigrants in industrial America is also reflected in the fact that the establishment of such societies often preceded that of national parishes or even of some ethnically mixed churches. Three different major Slovak fraternal societies, and one that was Carpatho-Russian Greek Catholic, are known to have existed in 1899, before either of the two ethnically mixed Catholic churches had been built. In addition, the first fraternal society of Windber Hungarians, established in 1901, preceded the formation of the Hungarian Reformed parish by four years and the Hungarian Catholic parish by about twelve years. The

Italian Workmen's Beneficial Association received its charter in February 1904, a year or so before talk of separation of the Italian Catholics from St. John's Catholic Church is known to have occurred. Of the major nationalities in Windber, only the Poles formed an initial ethnic fraternal society after their church—ethnically mixed St. John's—had been erected.[33]

By spring 1904 all of Windber's four major nationalities had set up one or more of these ethnic fraternal aid organizations. The charters of all these associations included a clause similar to the following one contained in the local Polish society charter:

> The said corporation is formed for the purpose of . . . rendering aid and comfort to its members in case of sickness, accidents, or distress, and of assisting and giving pecuniary aid to the widows and orphans of deceased members the funds for these purposes to be raised by assessments upon the members by dues, by voluntary subscriptions and the receipts of entertainments, etc.[34]

Proliferation of fraternal societies among Windber's four major nationalities was such that at least fifty were organized from the first decade of the twentieth century onward. The Slovaks alone had eleven or twelve societies that habitually marched in Fourth of July parades, and these did not include women's organizations. Nor were these organizations ephemeral. The jubilee history of the Slovak parish in Windber still listed twelve fraternals connected with the church in 1956, and all had been founded in the first quarter of the century.[35]

Mutual assistance was the one feature common to all fraternal societies, but these organizations often had additional stated purposes that varied from one association to another. These purposes often also reflected the diverse composition of the respective ethnic groups, whose members formed separate associations based on intraethnic differences. Local fraternals were thus founded on the basis of one or more of the following: religion, occupation, gender, nationalism, citizenship, and local or regional origin.

Beyond nationality, religion was the basis of division that most characterized the development of local fraternals. Greek Catholic societies, composed of Carpatho-Russians and Slovaks whose national identities were sometimes confused, had helped establish Windber's Greek Catholic mixed-nationality parish. Carpatho-Russians formed branches of the

Greek Catholic Union, while Slovak Greek Catholics joined it or one of two chapters of the Pennsylvania Slovak Roman and Greek Catholic Union. The names and composition of these Slovak and Carpatho-Russian societies in Windber's early years indicates that the lines between religion and nationality were not as rigidly drawn as might be presumed, and reminds us that these associations were originally linked to ethnically mixed parishes.[36]

On the other hand, Slovak secularists and Slovak Lutherans were most likely to join the Windber branch of the National Slovak Association, whose national organization was originally designed to be secular and nationalistic rather than religious in orientation. While Slovak Roman Catholics might join an association with Slovak Greek Catholics, more often than not they became members of one of the two Windber-area branches of the First Catholic Slovak Union, or Jednota. One of the largest organizations was St. John the Baptist, Jednota branch 292, which sought "to uphold, preserve intact, and perpetuate the doctrine of the Holy Roman Catholic faith." On the national level, Slovak fraternals had divided into religious groupings in the 1890s, and Windber's organizations simply reflected this course.[37]

Whenever immigrants established national parishes in Windber from 1905 to 1921, they usually founded fraternal societies directly associated with the parish if no suitable ethnic ones already existed. Thus, in due course, the Hungarian Reformed Beneficial Society of Windber, the St. Anthony Catholic Beneficial Society of Windber, and the Windber and Surrounding District St. Mary's Roman and Greek Catholic Hungarian Church and Sick Benefit Society were formed. In 1902, Poles had limited membership in their first fraternal society to Poles who were also members of St. John Cantius Church. They preempted the church name to establish the St. John Cantius Beneficial Society, which was never the ethnically mixed organization that the parish was. Membership in the Hungarian church societies, however, was never restricted to church members, because language and Magyar nationality were considered by clerics, laypeople, and the Hungarian government to be more important than particular religions.[38]

Immigrant fraternal organizations were working-class institutions, although in Windber the relatively small number of ethnic businessmen, contractors, company foremen, and professionals joined the masses of miners in the same associations and often became leaders of them. Given the large mining population, these societies differed very little in terms of the occupations of their members. Yet there is one exception to the general pattern. In 1907 the Italians organized a band, which became a prominent addition to the Italian community. Band members soon formed a new fraternal aid society: the Italian Literary, Musical and Beneficial Association of Windber Borough. According to its charter, in addition to insurance and mutual aid, the association was formed "for the purpose of having a hall for its members in which they can meet to read, practice music and accumulate a fund." While the Slovaks had had a local band since 1902, the miners who were band members had joined one of the existing fraternals instead of creating a new one, as did the Magyars, who formed a Hungarian band in 1913.[39]

As a rule, women's fraternal societies were founded later than men's on the national and local levels. However, as the scarcity of women diminished, these societies increased in number and importance in the Windber area. The usual pattern was to form separate women's branches that paralleled male organizations. Thus, Slovak women formed two branches of the First Catholic Slovak Ladies' Union in Windber early in the century; Carpatho-Russian women soon set up St. Mary's Greek Catholic Women's Society #354 and Women's Sobranye #172. In 1914 the constitution of the National Slovak Society provided for direct membership of both sexes for the first time, but on somewhat different terms.[40]

By their very existence, ethnic fraternal societies perpetuated the language and culture of the homeland and provided occasions for members of a nationality to socialize. Dances and drama events were common but some societies sought to promote national unity and national causes, such as Polish independence or Slovak autonomy in Europe. The Polish Falcons and the National Slovak Society were important in this regard in Windber, but almost all associations fostered some interest in European affairs. National foreign and American holidays were usually celebrated by all the societies of a particular ethnic group. Often other nationalities joined the celebrations or sent their bands to perform. In 1909, Slovak societies in Windber assembled for the National Slovak Society's dedication of a new Slovak flag; Polish organizations paraded and celebrated the 118th anniversary of the Polish constitution; and Italian societies planned and carried out a massive parade, program, and dance for the first local commemoration of Columbus Day. Such events were even bigger in subsequent years.[41]

The war inevitably heightened ethnic consciousness, because immigrants had originally come from countries that were now at war with one another. In 1914 many still had relatives and friends in Europe, or expected to return there. After Woodrow Wilson proclaimed the right of self-determination, and after the United States entered the war, foreign-born nationalists vied for influence in determining the postwar situation. In Windber, Slovaks seem to have been the most active interested parties in this regard.[42]

A national federation of Slovak organizations had been formed in 1907 to promote Slovak autonomy and freedom from Magyar oppression, and Windber had formed branch #106 of this organization, the Slovak League. During the war, the local group became very active and brought prominent Slovak leaders, including Albert Mamatey, president of the Slovak League, to speak in Windber even as he was trying to influence postwar decisions about Slovakia. Slovak immigrants contributed financially to national Slovak causes and just as generously to Liberty Bond drives.[43]

The native-born and the local American press monitored which foreign societies and nationalities contributed what to the war effort, and foreign societies joined in the drives. It is interesting to note that the *Windber Era* openly denounced only one ethnic group once the United States was involved in the war. Somewhat ironically, it found German farmers, who had had a presence in Somerset County since at least the 1830s, less American and less generous in their support of bond drives than Hungarian, Slovak, Polish, and Italian miners.[44]

American citizenship was another cause that a number of fraternal associations promoted. The National Slovak Society paved the way in this regard, because it required members to become citizens within six years after their initiation into the organization. Immigrants typically saw no contradiction between advocating nationalistic movements in Europe and citizenship in their new land. Perhaps Slovaks were far ahead of other nationalities in Windber in helping people of their ethnic group to become citizens because they had come to America earlier. Unique among Windber's ethnic groups, the Slovaks had formed an educational club and a couple of Republican party clubs by 1907. The Dillingham Commission also cited their pioneering role in citizenship and political work.[45]

In the second decade of the twentieth century, a few other such associations came into existence. Polish leaders founded the American Polish Educational Association of Windber to teach adults about the principles of the U.S. government and the English language in order to prepare Poles for citizenship, and in 1912 the Italian-American Citizens' Beneficial Association was formed for similar purposes. On the national level, Magyars are known to have lagged notably behind other nationalities in this regard, and Windber confirms the general picture. The legal incorporation of fraternal societies required that most, if not all, of the prospective incorporators be American citizens, and the charters of Hungarian organizations usually listed a greater number of noncitizens as incorporators than did the charters of other nationality societies.[46]

The effectiveness of these local organizations in promoting citizenship, however, was limited. The vast majority of foreign-born in Windber became citizens only after the United Mine Workers had successfully organized area mines and established a citizenship school in the 1930s. The 1920 census indicated that a large portion, 73 percent of Windber's foreign-born male population of voting age, were not citizens by that date. Still, the 27 percent who were naturalized more than doubled the 1910 census figure of 11 percent. The point here is that any increase in American citizenship that occurred on the local level during the nonunion period was largely the result of the work of fraternal societies—not of Americanizers, Berwind-White, or the public schools.[47]

Localism continued to be a prominent feature of Windber's Italian community in the second decade of the century, and the names of fraternal organizations founded during this era sometimes indicated the provincial origins of their members. Thus, the Abruzzi Workman's Beneficial Society of Windber was organized in 1913, and the Sicilian Italian Mutual Beneficial Society in 1918. By contrast, a year later, in 1919, a local lodge of the broader-based Sons of Italy also came into being.[48]

This description and categorization of the diverse purposes and memberships of local ethnic fraternal societies is by no means exhaustive. Slovaks were the first of several ethnic groups to establish Sokols, which stressed physical well-being, athletics, and sporting events. Selected members of the Abruzzi society formed a cooperative for Abruzzi members only. The range of purposes beyond mutual insurance conveys an accurate idea of the multifaceted usefulness of these societies to people in the immigrant communities. These institutions provided immigrants with sociability, continuity, and preservation of cultural traditions, but they also assisted workers confronted by the harsh realities of industrial capitalism in a strange land.[49]

Windber's fraternal associations also had an importance of their own in national immigrant circles. Delegates from the area attended national conventions of the various societies, and individuals participated in national affairs. Early in the century, the Rev. D. D. Polivka, of St. Mary's Greek Catholic Church in Windber, served as spiritual adviser to the national Greek Catholic Union. Steve Molnar and the Rev. Steve Borsos, two of the eight Hungarians who obtained a charter for the Hungarian Reformed Federation of the United States of America from the U.S. Congress in 1907, were residents of the borough at the time.[50]

Although Windber was much smaller than the cities that usually hosted national conventions, the United Greek Catholic Societies of the United States chose it in 1915 for the site of its ninth biennial convention, which was held with great pomp and ceremony. Occasionally, national figures, such as Albert Mamatey, came directly to Windber to speak. More frequently, they visited the city of Johnstown, where delegations from Windber and other mining towns might also attend lectures and events.[51]

The Limits of New Immigrant Ethnic Conflicts

There is no denying that new immigrant groups in Windber had occasional conflicts with one another or that fraternal societies were sometimes involved. The most spectacular incident of this kind took place on July 4, 1900, when a group of Magyars tried to take over a picnic held by St. Stephen's Archangel Society of Greek Catholics. When the Slavs of the society refused to change their music to Hungarian national songs, some Magyars fired guns. One Slav died, and another was seriously wounded. Eventually eleven Magyars were arrested. The fight was widely viewed as a classic case of Magyar-Slovak antipathy. Old World antagonisms were occasionally carried into the New World.[52]

Yet such nationality-based conflicts and such overt violence were rare. Court proceedings and police reports published in area newspapers suggest that most crimes, serious or minor, occurred within given nationality groups. Weddings and christenings, which lasted for days and involved large-scale drinking, did produce fights. Congested boardinghouses filled by homogeneous ethnic groups also bred hostilities. Italians feared the various gangs of other Italians who harassed or extorted money from them but left other nationalities alone.[53]

Moreover, while national and religious divisions within and between new immigrant groups were real, the lines between them were not always rigid. Diverse fraternal societies of the same nationality group cooperated with one another for social, national, and religious events, and bands composed of the various segments within ethnic groups performed for other nationalities. Foreign societies of disparate nationalities joined together to celebrate American holidays in an orderly manner. Disparate ethnic groups occasionally cooperated with one another in such common ventures as the founding of a joint meeting hall or mutual support of miners' strikes.

Then too, ethnic conflict between these nationalities was limited by the existing broader occupational and social ethnic hierarchy, which was itself reinforced by discriminations against the foreign-born and the attitudes of some Americans, who considered all new immigrants, regardless of nationality, to be "Hunkies." Nativist organizations, such as the Patriotic Order Sons of America and the Ku Klux Klan of the 1920s, remind us once again that the main ethnic division in Windber was still the one between the American-born and the foreign-born.[54]

As previously mentioned, class experiences at the workplace and in the community in Windber also muted ethnic hostilities. That new immigrants saw themselves as workers and not just ethnic people, and their institutions as working class and not just ethnic ones, is clear from the names they gave to a number of fraternal societies. Certain Slovak, Hungarian, and Italian organizations, including the town's very first Italian and Hungarian fraternals, used "workers," "workingman," or "miners" in their names. Thus, Italians had the Italian Workman's Beneficial Association, the Hungarians had St. Michael's Hungarian Miners' Sick Relief Society, and the Slovaks had the Slovak Workingman's Benevolent Association. Class-consciousness existed, but only a broad-based organization, such as the United Mine Workers, could integrate all the various working-class segments into a common body for action.[55]

Berwind-White's Role

Berwind-White tolerated and often encouraged the development of ethnic fraternals initiated by the immigrants for the same reasons that it promoted national parishes and segregated ethnic enclaves. Moreover, fraternal aid rendered by workers to fellow workers generally allowed corporations (and the state) to escape financial responsibility for the welfare of employees.

The company's approach to fraternal societies is best understood as part of broader labor management policies that were designed to keep foreign-born miners satisfied, unorganized, and separated from American miners. In 1903, Will Hendrickson, the *Windber Journal* editor, accused the company of "nursing" the foreign element in order to divide miners and arbitrarily deny workers' basic rights. A U.S. Department of Labor investigator reached the same conclusion in 1919. To placate new immigrants, the company routinely allowed immigrant miners to take time off to attend fraternal events and occasionally even closed the mines for foreign national and religious holidays. Greek Catholics adhered to the old Julian calendar and always celebrated their holidays accordingly. At such times, the mines worked less than a full week or with less than a full work force. In public testimony before the Industrial Relations Commission in 1915, E. J. Berwind chided his foreign-born miners for a propensity to take days off, but hinted strongly that company toleration of such absences was considered necessary to keep his miners content, which, in his worldview, meant nonunionized and less likely to migrate elsewhere.[56]

This policy met with the general favor of the immigrants, who did not have to fight Berwind-White for their holidays. Yet even this practice led to unforeseen consequences because of the general arbitrary behavior of company officials and foremen. One leader of the 1922 strike recalled that his commitment to the union crystallized because of an incident that had occurred in 1918 when he was a young man of 16. Joseph Zahurak had been working as a trapper when he was ordered by a foreman to take over a spragger's job. The spragger had the day off to celebrate one of his ethnic group's holidays. Zahurak was not trained for the specialized and dangerous job, for which he was also underage according to state law.[57]

Nevertheless, the foreman threatened that if he did not do the work his father would not get any cars. (Miners could not load coal without cars to put it in, and lack of cars therefore meant no pay.) Zahurak was injured on his first day as a spragger and subsequently had to undergo several operations. Unable to work for a year, he found worker compensation laws inadequate and had to seek aid from an attorney and a priest to obtain another job in Windber. On April 1, 1922, he was one of a number of men who led Berwind-White miners in Windber out on strike.[58]

Fraternal societies were also expected to be loyal and subservient to the company. On the twenty-fifth anniversary celebration of the founding of the first Polish fraternal society, which was celebrated four years after the strike of 1922 had been defeated, prominent speakers included the town's burgess and two top company officials. The burgess presented the foreign society with a new American flag and lectured the foreign-born on the need to love, honor, and defend it. General Manager E. J. Newbaker then briefly expressed his pleasure to be there, but R. S. Baylor, who directly supervised St. John's members in the mines, revealed the standard by which the company measured such societies. Baylor reported with delight "that he never had troubles with *Polish* [emphasis mine] workers." So long as nationalities and their fraternal societies did not create "trouble" or act independently, Berwind-White considered them useful organizations.[59]

But fraternal societies did sometimes act independently, and whenever they did they came into conflict with the company. They faced many of the same difficulties and legal restrictions that churches did. Berwind-White controlled access to ethnic halls through subsidies and property restrictions and relied on a handful of ethnic leaders and spies to locate potential troublemakers and keep fraternal societies operating within narrow limits. The best example of the struggle by these ethnic institutions for greater autonomy and the countermeasures taken by the company to squelch independence and interethnic cooperation occurred in 1914.

On April 5, 1914, Slovaks and Carpatho-Russians from various fraternals gathered together to consider and approve the possibility of organizing a new association that would erect a building for the fraternals' common use as a meeting-place and social center. The first practical difficulty organizers faced in proceeding with the project was to locate and obtain a suitable site for the building, and hall association members immediately formed a committee to find out about appropriate lots and land prices. They also tacitly recognized the power of Berwind-White in the town when they simultaneously selected another committee to inform the company about their plan. Some members hoped that the corporation would donate land. But the company's immediate response to the new project was hostile, and subsequent negotiations proved nasty and highly unsatisfactory to the association. Thomas Fisher of Philadelphia, Berwind-White's general manager, refused to donate or even sell a piece of land outright. He would offer use of company land only under stringent conditions, and he rejected a proposal to sell a lot in return for the association's guarantee that nothing would be done against company interests in the building.[60]

Fisher instead proposed terms that were roughly comparable to those ultimately accepted by another new organization, the Slovak Workingman's Benevolent Association. By contractual agreement with this fraternal in 1916, Berwind-White allowed use of some land for a nominal sum, but the company retained the right to reclaim the land at any time and for any reason for ten years. It also retained the right to reclaim the land and any buildings erected on it for the next twenty-one years, provided that it paid the fraternal society some money, not to exceed $7,000, for any buildings erected in the meantime. The National Slovak Hall Association members rightly concluded that similar company proposals provided them and future generations with no independence, no security from eviction, and no guarantees whatsoever. Consequently, they rejected such terms and purchased a lot from Will Barefoot, a private citizen, instead.[61]

The company countered the association's independent land acquisition with other steps designed to ensure its control. On June 25, 1914, the Rev. Francis Jevnik, priest of SS. Cyril and Methodius Church, came to the association's meeting with the latest telegram received from Fisher about the hall project. Father Jevnik read the communication and explained that Berwind-White was threatening the officers of the Roman Catholic societies with loss of their jobs in the mines if they went ahead with the building project. The company was also demanding the right to name new officers for the association and claimed that it would not negotiate with the present ones. Berwind-White treated fraternal society committees as arbitrarily as it had treated miners' committees elected to present grievances to it in past labor conflicts.[62]

John Hritz, secretary of the association, reported that hall association members then debated whether to go ahead with the project and that traitors had come to the June 28 meeting during one such discussion. He himself proposed continuation of the project, but Michael Washko presented an alternative motion against building. Hritz won the debate, and the project went ahead.[63]

In the ensuing months, hall association officers made repeated attempts to pacify the company. In September they sent another committee to go to it to explain that the project was not directed against it. Officers made similar public pronouncements. After many obstacles, six

The National Slovak Band, Windber, in front of the contested project, the Slovak National Hall Association Building, c. 1918.

shareholding organizations, including the First Catholic Slovak Jednota No. 292, Greek Catholic Union No. 151, National Slovak Society No. 304, Pennsylvania Slovak Jednota No. 68, National (Slovak) Sokol No. 60, and Sobranie Russkich Bratsvo No. 26, received a charter for the new Slovak National Hall Association in September 1914. In June 1915, the building, more modest than originally intended, was completed by outside contractors. But Berwind-White seems to have won the immediate battle over independence. It successfully imposed its established custom of paying subsidies to the owners of halls. In return for a small sum of money paid to the hall association, it secured the right to make decisions about who would meet there. Shortly before the strike of 1922, members voted to discontinue such subsidies and to exercise the right to decide for themselves who could or could not rent the hall.[64]

The Slovak hall project indicates that, even in an era of competing nationalisms and the onset of world war, there was considerable movement toward cooperation by fraternal societies and ethnic groups on the local level in Windber. Ethnic rivalries and religious differences were less an obstacle to fulfillment of the original hall proposal than the artificial constraints imposed by Berwind-White.

Leadership again emerged as a key issue, as it had in the churches. Officers might ally themselves with Berwind-White or with their working-class memberships. Both types of allegiances existed within immigrant communities. Although the top officers of the town's various fraternal societies often included company foremen and ethnic contractors, there were ordinary miners in these ranks as well. For individuals sympathetic to the plight and welfare of their mass membership, fraternal societies were a school in class education. No institution, no leader, no member was immune from Berwind-White's interference in their lives and activities. It is not coincidental that Slovak organizations generally supported the miners in the strike of 1922 or that the Slovak Hall was the first place in Windber to allow its rooms to be used for union meetings in town in the 1930s.[65]

The Problem of Ethnic Leadership

It is generally recognized that ethnic leaders served as intermediaries between their respective immigrant populations and the dominant American nationality. Non-English-speaking people, recent arrivals from Europe, may have relied primarily on family and kinship networks, but they still needed a minimum of assistance from individuals with some power and influence in American society to obtain information, transportation, jobs, housing, legal documents, citizenship, necessities, or amenities of life. By virtue of their ability to provide such necessary intermediary services, those individuals inevitably occupied a middle-class position. Recent research has tended to confirm that while a few ethnic leaders may have become self-made entrepreneurs in the United States, most had enjoyed a middle or higher class position in their homeland.[66]

Possession of a modicum of wealth, occupational rank, political power, and connections distinguished middle-class ethnic leaders from immigrant workers. But to be effective functionaries, intermediaries also had to have a fair knowledge of English and practical, if not formal, knowledge of American life and institutions. That immigrants appreciated the value of seniority in length of residence in the United States is generally acknowledged but not always understood. Veteran immigrants of any class had certain advantages over "greenhorns." Fraternal societies and churches were chartered by individuals who had emigrated earlier than others because American citizenship was required for most legal proceedings. Consequently, a certain degree of power and influence gravitated toward the ethnic leaders who played an important role in establishing the institutions around which immigrant working-class communities were constituted.[67]

Nonetheless, ethnic leaders who could wield economic power acquired an inordinate degree of influence in most immigrant communities. The ability of such individuals to procure jobs and housing for recent arrivals, whose main purpose in emigrating had been to find work and higher wages than could be earned in Europe, was important everywhere, in the big cities and small towns of America, but it was critical in company towns like Windber. Only the ethnic leaders who had a special relationship with Berwind-White were in a position to serve an immigrant clientele in an industrial mining center with one major employer. From the outset, Windber's immigrant communities were not homogeneous societies.

A number of select southern and eastern Europeans, who differed from the masses of their fellow compatriots in terms of their class position and loyalty, had come to Windber in the town's earliest years. Such individuals occupied the top positions in the stratified ethnic communities. Given their power and influence, it would not be inaccurate to term these people the "aristocracy" of the ethnic communities, even though they were middle-class in terms of the dominant American society and industrial capitalist economy. The prominent ethnic aristocrats and leaders of Windber included immigrant bankers and steamship agents; ethnic contractors; mine foremen and their assistants; and certain parish priests. Given the particular context of this company town, the first three of these four categories of people either worked directly for Berwind-White or had contracts or other cooperative arrangements with the company. They therefore personally derived substantial material benefits from company patronage, while they themselves acted as patrons to respective immigrant clients who worked for the company. Obligated to the coal corporation, they upheld the existing system of company control. Many parish priests performed this same function.

The ethnic owners of various small-scale businesses, such as modest groceries, are not included in this group. Such businessmen occupied a class position that was more ambivalent. Often they were only one step above the miners, had recently emerged from the mining ranks themselves, or had family members who were miners.

An examination of the nature of ethnic leadership in Windber, therefore, necessarily focuses on the small two or three percent of the ethnic population who constituted an ethnic aristocracy. Kurt Lewin has termed such people leaders on the "periphery" of the ethnic group and contrasted them with ethnic leaders who emerged from the ethnic group's "center."[68] A description of a few leading ethnic individuals on this periphery and of the mediating role that they played in the community is instructive.

The first group, immigrant bankers and steamship agents, initially played an important role in the local immigrant communities. These men not only provided traditional banking services, arranged travel, and sent remittances to the homeland, but also served as labor recruiters and notary publics in the United States. Windber's immigrant bankers typically had official ties to steamship lines, the Pennsylvania Railroad, and Berwind-White, corporations that had interlocking interests of their own. Periodically they traveled to New York to banquets and conventions, to meet with officials from the various companies they represented as agents.[69]

While various immigrants were employed in the steamship and banking business over the years, two men were the most successful. Andrew Zemany came to Windber in 1900 with impressive credentials for an ethnic community. Zemany was the son of a nationally prominent Slovak leader, Michael Zemany, who had founded the Slovak Evangelical Union in Freeland, Pennsylvania, in 1893, and a national Slovak newspaper as well. The younger Zemany immediately established what became a flourishing steamship and banking business in Windber, and the range of activities he was involved in beyond his immediate business suggests his central importance to the community.[70]

Zemany not only participated in fraternal affairs but also notarized documents for the formation of a number of the fraternal societies. He became an important contributor to the Greek Catholic church, participated in a Slovak Republican club, and took immigrants of various nationalities to Somerset and witnessed for them in citizenship hearings. Of the 119 Windber-area people who became citizens from

Andrew Zemany

Bankový, šífkartový a notársky obchod.

Peniaze posielam do kraja. — Predávam šífkarty do kraja a z kraja sem. Vystavujem plnomocenstvá a vybavujem starokrajové záležitosti. — V každej potrebe obráťte sa na mňa

ANDREW ZEMANY,
AGENT,
P. O. Box 375 Windber, Pa.

An advertisement by Andrew Zemany of his Immigrant Banking and Steamship Agency in Windber, 1909. From *Národný Kalendár*, 1909.

August 1900 to December 1903, he personally was a witness for twenty-four.[71]

Ultimately Zemany's banking business met the fate of most such immigrant enterprises and went out of business. A combination of bad business acumen and outright theft contributed to the bank's demise. When Zemany fled Windber in 1912 with his bank's funds, many foreign-born small depositors of the Zemany bank lost a total of $85,000.[72]

Albert Torquato, Zemany's Italian counterpart, had founded his bank and steamship agency in 1905, but his business really prospered after Zemany and a number of other competitors went out of business around

ANDREW ZEMANY
WINDBER, PA.
Notary Public
STEAMSHIP TICKETS
REMITTANCES MADE TO ALL PARTS OF EUROPE.

(5525-A, April, 1908.)

No. Windber, Pa., Oct. 15. 1910.

Received from Feodor Gerula at Windber, Pa

Dollars

For (Foreign Currency) 615 Krone

To be remitted to

Residence

ANDREW ZEMANY

per

This remittance will be forwarded to the payee named, subject to the rules and regulations of the various Post Offices used in making the Remittance.
No claim covering non-delivery or erroneous delivery of this remittance, or for failure to produce receipt, or for postage deposited for return of payee's receipt, will be considered after three months from date hereof.

Receipt for cash remittance Feodor Gerula sent abroad through the Zemany Agency, October 15, 1910.

1912. Although it too eventually failed, it survived until the 1940s.[73] Torquato was probably second in importance only to Frank Lowry in the Italian community. He was the founder and president of the Italian-American Citizens' Beneficial Association, which helped Italians become citizens. He had close ties to the Italian consulate and personally resolved occasional conflicts involving Italians in Windber. Respected by some as a *prominenti* and denounced by others as a gang leader, he was nevertheless well known to all Italians. It was he who had originated and organized the first local celebrations of Columbus Day.[74]

Ewa Morawska found that ethnic leaders in Johnstown did not confine themselves to participation in a single society or type of activity, but took part as leaders in a wide range of ethnic affairs. Zemany and Torquato illustrate that ethnic leaders in Windber also engaged in multiple leadership roles. Moreover, it is significant that Zemany was the only new immigrant to run for political office in Windber before the 1920s, and that Torquato ran for tax collector in 1925. Both men ran as Republicans and lost their races to other Republicans or Prohibitionists. Their defeats show how difficult it was, even for prominent new immigrants, to gain acceptance into the American political mainstream. Success in an ethnic community and success as mediators between immigrants and the dominant society did not translate into acceptance or integration into American society.[75]

The second group of ethnic aristocrats—ethnic contractors—were fiercely loyal, and beholden, to Berwind-White. Again, two individuals stand out because of their importance. Frank Lowry, the Italian contractor who came to Windber with Berwind-White and built many of the area's railroads, streets, sewers, and building foundations, has already been mentioned. After retirement from construction in 1915, he operated a hotel and later owned a local brick factory. He had been instrumental in founding St. Anthony's and had built that church's foundation as well as St. John's. He had also been a charter member of at least one Italian fraternal society as well as the Eagles home. Indeed, Lowry had his labor gang and devoted followers. The *Johnstown Democrat* reported that American leaders in Windber considered him "king" of the Italians. Yet Lowry was not universally liked, was frequently involved in legal disputes, and on two separate, well-publicized occasions, years apart, was shot by fellow Italians who had had excellent personal reputations before the incidents.[76]

Stiney Roden was Lowry's counterpart in the Polish community. A Lithuanian rather than a Pole, Roden was another Berwind-White contractor who supervised gangs of men from about 1900 until the 1930s. Roden was active in parish affairs, Polish fraternal associations, and a Republican club. That the company considered him important is suggested by newspaper accounts. He and Lowry were the only two new immigrants invited to attend top local Berwind-White social functions in the early century. Miners interviewed frequently said that Berwind-White had one or two key recruiting agents per nationality in Windber, and one man commented: "Stanley Roden—he was the Polish getter." A union leader described Roden's importance to the ethnic community and the company in the following manner:

> The Polish people had a great leader in Stiney Roden. He was like a godfather to the Polish people. Now, he received slush money from Berwind-White all the time, and he'd treat all the members in the Polish club. They followed his rule, what he wanted, what Berwind wanted, and they followed him, and they believed in him. So they [the company] kept those people separated that way.[77]

Berwind-White relied on key ethnic contractors and other privileged employees to keep miners "satisfied" and nonunionized, according to a U.S. Department of Labor conciliator who had come to Windber in January 1919 to investigate charges against Berwind-White. Ethnic leaders apprised the company of the miners' grievances, diverted discontent, and reported potential troublemakers. Yet, beyond repression, the system worked in part because contractors hired their own men, whom they considered clients. The son of a Carpatho-Russian contractor explained, "They [the company] could depend on my Dad because, well, he had leadership with his clients." When his father hired men, the contractor told them, "You work with me and I'll take care of you." As for the United Mine Workers and union activity, the father told the son, and presumably others, "Keep your nose out of it."[78]

The third group of ethnic aristocrats, a handful of ethnic mine foremen and their assistants, also occupied positions whereby they directed work and maintained loyalty to the company. They had a stature and power that extended far beyond their small numbers. These men not only hired and fired, but also enforced company policies at the workplace, in the polling place, and in the community. During the strike of 1922, such men initially dared not oppose the strike for fear of losing all influence with the men. Certain parish priests were the final link in the Berwind system.

Such ethnic leaders frequently tried to gain acceptance—for themselves—from the dominant American society, which generally disdained new immigrants. Frank Lowry's real name was Lorenzo Quagliarillo, but he had it legally changed in 1904. When he died in 1935, his funeral took place at the English-speaking church of Holy Child Jesus, not at St. Anthony's or St. John's. Albert Torquato may have appeared as the Genoese navigator, Christopher Columbus, in a float in the first Columbus Day parade in 1909, but by July 4, 1928, Stiney Roden appeared on horseback in the holiday parade as George Washington. Flanked by a costumed General Thaddeus Kosciusko on his right and a costumed General Casimir Pulaski on his left, Roden led the Polish societies, which marched behind him.[79]

The masses of immigrant miners and their families inhabited stratified ethnic communities, but they also were workers. To them, ethnic worlds and working-class worlds were not separate entities but intersecting dimensions of human experience. Ambivalent sentiments about Berwind's ethnic leaders were commonplace, but in times of overt labor unrest a choice would have to be made. A female Italian union supporter described her feelings toward an elderly Italian leader and company loyalist who had treated her like a daughter in her youth: "I used to tell him. I said, I love you very much, but I hate what you do. I said, that isn't fair for those other men. I says, who in the hell do you think you are? . . . I says, rather help our people." This same ambivalence surfaced when people were asked about the early system of obtaining citizenship, which many disdained. Citizenship was linked to suspect ethnic leaders who were viewed as mere company functionaries because they immediately registered new citizens to vote Republican and support Berwind-White candidates.[80]

Throughout the nonunion era, Berwind-White and class-conscious miners fought a never-ending battle for the hearts and minds of the miners within the ethnic communities of Windber. Class-conscious miners had to contend not only with the practical problems of ethnic division and hostile external enemies—the company and all its powerful resources, an unsympathetic state, nativist prejudices and legislation—but also with hostile internal obstacles, spies and ethnic leaders who did not represent the class interests of their respective constituencies. The argument is that a small minority of strategically placed, prominent ethnic leaders were the linchpin that held the Berwind-White open-shop system in place during the nonunion era and that, to succeed in unionizing on the local level, class-conscious ethnic miners had to provide a new alternative leadership to challenge the system. As we shall see in Part Two, such leaders became evident in those critical historical moments, such as the strikes of 1906 and 1922, when particular circumstances and events convinced the masses of miners that it was time to take a stand and try to unionize.

On June 26, 1915, the leading coal operators' journal in the country editorialized that the prime necessity of the day was to destroy class-consciousness. In its view, that task could only be accomplished by creating class interdependence, and class interdependence in the mines could be secured only by retaining permanent reliable foremen who had gained the trust and loyalty of the employees. "The agricultural workmen from Europe who come to our mines hardly know what a 'company' is," *Coal Age* commented. "The most successful companies are those retaining men in office whose personality is so marked that the men work for the official and not for the company."[81] Beyond repression, strategically placed ethnic leaders, foremen, and contractors were Berwind-

White's way of trying to gain and maintain the loyalty of immigrants while keeping them separated and unorganized in an autocratic setting.

Berwind-White liked immigrant communities because its owners and officials saw them as a means of keeping workers divided. That view is implicit in advertisements for miners which the company placed in diverse ethnic newspapers in July and August 1920. These ads are interesting for what they reveal about the corporation's assumptions about immigrants and about the Berwinds' labor policies.

Because immigrants were widely believed to prefer steady work above all else, the ads carried such headlines as "Miners Wanted for Steady Work" and "Where the Mines Work Every Day." Good wages and coal seams of reasonable proportion were then described. Because prospective miners came from rural backgrounds, the ads noted that miners could keep gardens, cows, chickens, and ducks. Ethnic churches and fraternal societies were cited as inducements, and occasionally specific priests or ministers were named as authorities to verify company claims. But the most interesting thing about the ads was how they differed.

Immediately after the headline, the body of these ads referred to Windber, but in different terms. In *Dongó, Szabadság, Amerikai Magyar Népszava,* and *Magyar Bányászlap,* Windber was a Hungarian mining city. In *Ameryka Echo* it was a Polish mining city; in *Keleivis,* a Lithuanian mining city; in *Slovák v Amerike, Jednota,* and *Amerikánsko-Slovenské Noviny,* a Slovak mining city; and in *Italo-Americano* and *Il Progresso,* an Italian mining city.[82] It is significant that no terms denoting "Americanization" of any sort ever appeared in the ads, and that Windber was never described as an American town or as one composed of diverse nationalities. Instead, Berwind-White had portrayed Windber as a single ethnic community to appeal to prospective foreign-born miners. Unlike the immigrants who had initiated them, the company had perceived and promoted ethnic communities as isolated and self-contained enclaves. As long as loyal ethnic leaders continued to help the company keep the communities segregated and their members' activities enclosed within narrow limits, the large coal corporation hoped that class-consciousness could be muted, company control maintained, workers divided, and unionization averted.

But as we shall see in Part Two, Berwind-White had seriously underestimated immigrants and immigrant communities, which were continuously producing ethnic leaders with working-class loyalties even as the company, work experiences, and life in a company town were producing class-consciousness among the general mining population. These alternative leaders fully appreciated the need to integrate miners of disparate nationalities into a class-based industrial union, such as the United Mine Workers. To do so successfully, they understood, required displacing the company's ethnic aristocrats and mobilizing the masses of miners to take charge of their own ethnic institutions—national parishes and fraternal societies. In 1922, Windber's immigrant communities exploded as foreign-born miners and their families fought a long, hard, and bitter strike and temporarily accomplished these objectives. A more permanent displacement of the company's old ethnic stalwarts had to await the establishment in the Windber area of several permanent locals of the United Mine Workers in August 1933, after passage of the National Industrial Recovery Act. Nevertheless, the long and uneven path to unionization, ethnic diversity, and the central importance of external obstacles in impeding organization should not obscure the fact that the masses of immigrants had sought to form ethnic communities for their own autonomous, largely democratic reasons and that immigrant communities had had internal class divisions throughout the nonunion era and had been arenas of class conflict.

PART TWO

Struggles and Strikes

6

First Stirrings

The miners and others who came to Windber in its early years set foot in a company town that Berwind-White had consciously created to ensure its ongoing dominance in the workplace and the community. From the outset, the autocratic structure of the town and the work process were givens. The skilled, semi-skilled, and unskilled workers who streamed into the area from other parts of the United States and Europe had emigrated for the plentiful jobs that existed there, not because they desired or were willing to submit to unlimited, undemocratic restrictions over their lives. With them they brought givens of their own—their respective cultures, experiences, and hopes. By 1899, less than two years after the town's founding, the diverse working class that Berwind-White had recruited and brought together in the Alleghenies made its first direct challenge to authoritarian company rule. Windber miners established a brief-lived but important local of the United Mine Workers of America.

The UMWA was the one viable, class-based alternative to company rule that existed for the miners in Windber, and they were drawn to it in large numbers at critical moments throughout time. That Berwind-White successfully smashed the short-lived local in 1899, and all subsequent unionization efforts until the New Deal of the 1930s, should not diminish the importance of these organizing efforts. Indeed, it is impossible to understand the history of Windber and its people without this appreciation. The nonunion era was far from quiescent, uneventful, or devoid of ephemeral unions or labor-capital strife. As McAlister Coleman pointed out in his *Men and Coal* more than fifty years ago, it may be possible to study the history of some industrial workers and not bring in unionization until fairly recent times. However, such is not the case with coal, where "whether or not they belonged to the U.M.W.A., the influence of that organization upon the life and fortune of every mine worker has been so preponderant as to make its story the bulk of any adequate chronicle of American coal mining."[1]

It is significant that the first stirrings of union activity in Windber dated from the town's earliest days—not 1906, when a major strike and dramatic events made local support for the UMWA manifest to all. That Windber-area miners were already protesting and trying to organize by 1899 revealed the falseness of the company's publicized claims that it had created a new "model" town and workplace. This chapter examines the miners' initial organizing drive.

To understand what happened on the local level, it is crucial to look at the broader context. The founding of the new town and the opening of new mines had coincided with the revival and growing strength of the district and national UMWA organizations after the disastrous depression of the 1890s. To the first local unionization campaign in 1899, both the outside contending parties—the UMWA and Berwind-White—brought particular policies and strategies that had been shaped by previous experience.

The United Mine Workers of America

The United Mine Workers of America came to the Windber campaign as a relatively new but important labor union that had deep roots in the past. It had been founded in January 1890 when representatives from two competing national labor organizations—the Knights of Labor and the National Progressive Union—recognized the futility of the internecine warfare they had been waging against each other in the coal fields. They met together in Columbus, Ohio, to bury the hatchet and create a single national organization in order to educate miners about the need for "demanding and securing, by lawful means, the just fruits of our toil."[2] The young organization consisted of approximately 17,000 members at the time of its founding. The unionized miners of District 2, which encompassed Central Pennsylvania, including what would become the Windber area, lent critical financial, moral, and numerical support to the fledging national organization.[3]

From the outset, the UMWA was a singular organization that adopted progressive policies that would subsequently be important in organizing Windber. Unlike the craft-based American Federation of Labor of which it was a member, it was an industrial union that welcomed all miners who worked inside or outside the mines, regardless of skill or method of payment. Moreover, from the Knights of Labor, which had held a strong position in Central Pennsylvania before 1890, the new union also inherited experienced leaders and policies that were, in theory if not always in practice, ethnically friendly and pluralistic. Like the Knights, the UMWA employed foreign-speaking organizers and made important documents available in a number of languages. The new UMWA constitution incorporated the principle of brotherhood by prohibiting discrimination based on race, nationality, or creed.[4]

As Victor Greene has shown, the UMWA's inclusive principles regarding new immigrants were not always followed. English-speaking miners did sometimes take nativist actions against new immigrant miners.[5] There were limits to racial toleration as well. Yet, in general, major UMWA leaders considered the problem of the influx of new immigrants into the coal fields in the United States in the 1890s to be a practical one: organizing and incorporating them into the organization. Because immigrants from southern and eastern Europe (and blacks) were often unwittingly introduced into an area first as strikebreakers, they had to struggle to offset operators' attempts to arouse the fears of English-speaking people or otherwise exploit national or racial divisions.

In October 1891, District 2 President James White led such efforts in Central Pennsylvania when he took a nativist, anti-immigrant, anti-Catholic editorial from the *Houtzdale Observer* as the basis for subsequent attacks on local coal operators for their policy of encouraging and exploiting national and religious prejudices for their own benefit.[6] In a similar vein, in his annual address to the fourth national UMWA convention in April 1893, President John McBride felt compelled to comment:

> It is with deep concern, and the most profound regret, that I have noticed during the past year the danger which threatens our organization from the growing tendency to divide mine workers on the question of nationality and religion, and I feel that it is time for those who are interested in the miners' welfare to speak in denunciation of the same in unmistakable terms.[7]

During a later successful drive to organize anthracite miners in 1902, National UMWA President John Mitchell again warned miners: "The coal you dig isn't Slavish or Polish or Irish coal, it's coal."[8]

Prejudice existed, but sooner or later many English-speaking miners in the various regions discovered that Hungarians, Italians, Poles, Russians, and later blacks, were not averse to unionization when properly approached, preferably in their own languages. Andrew Roy, historian, miners' advocate, and onetime Ohio state mine inspector, claimed that the strike of Connellsville coke miners in 1891 exposed the myth of the new immigrants' docility to bituminous miners; Victor Greene has dated the end of the myth in the anthracite region to the Lattimer massacre of 1897. Like others who found immigrants receptive to UMWA efforts, Thomas W. Davis, District 2 leader, reminded English-speaking miners that the UMWA knew no creed or nationality and suggested that they set the example for foreign-speaking workers who would join the UMWA if the organization was not spasmodic but stable.[9]

The 1890s proved to be hard times for the new UMWA, however. The greatest depression to date, the depression of 1893, took a heavy toll in coal fields everywhere in terms of mine closings, wage cuts, massive unemployment, and weakened organization, even as what Frank J. Warne termed the "Slav invasion" extended into bituminous regions. In an attempt to forestall further wage cuts and regain some semblance of control, national leaders called a national strike, which received broad rank-and-file support, in April 1894. However, the leadership's decision in June to accept a controversial compromise, a partial settlement, undercut the possibility of forging a national or interstate agreement, thereby leaving individual district organizations on their own to negotiate wage scales for their respective competing fields. The national strike quickly proved to be a notable failure that in many regions set the stage for even greater destitution, exploitation, and employer offensives against all forms of labor organizations.[10]

The failure of the 1894 strike in the midst of depression tested the durability of the national union but almost dealt a fatal blow to district organization in Central Pennsylvania. After the June settlement, District 2 was unable to negotiate a compromise settlement with regional operators, and district leaders were forced to call off their ongoing strike in August.[11] Companies knew they had gained the upper hand and were not content with mere wage reductions. Repression, blacklisting, and mass discharges were the order of the day, and many union members were forced to tramp the highways and leave the region. William Lockyer, a regular correspondent to the *United Mine Workers Journal*, traveled through the district and concluded that operators were determined "to stamp out every vestige of organization that looked so prosperous a few months ago."[12] Subsequent events showed that Central Pennsylvania operators and mining superintendents generally considered 1894 a decisive victory in their struggle against unionization. Meanwhile, conditions in this region were so bad that a committee of the Pennsylvania state legislature began to investigate and document cases of starvation in the coal fields.[13]

The once-active and relatively prosperous District 2 organization held a convention in January 1895 only because it was financed by the national, which was in debt itself. The numerous checkweigh associations, along with the locals and assemblies that had supported them, were, with an exception or two, casualties of the strike. In their place, local mine committees arose spontaneously in a vain attempt to maintain some sort of unified action, but their leaders, once discovered, were discharged by their employers.[14]

District 2 was also conspicuous by its absence from the national UMWA conventions in 1896 and 1897; it had no funds to send delegates. At the seventh national convention, in April 1896, while alluding to criticisms of national officers, Secretary-Treasurer Pat McBryde noted the district's absence with regret and commented that, since the strike, the suffering of miners in Central Pennsylvania had been greater than that of miners in other parts of the nation.[15] On the eve of the UMWA's national strike in 1897, District 2's organization was, at best, a mere shadow of its former self. It was from this deep abyss that the regional organization had to climb when the national union initiated its first successful nationwide bituminous strike in 1897.

The founding of Windber in 1897 coincided with the youthful UMWA's first successful national strike, a turning point that reversed the union's devastating decline in the midst of the depression-ridden 1890s. Only one of thirteen Windber-area mines was operating in September 1897, when the strike settlement was reached. Although the Windber miners (and many others in Central Pennsylvania) took no part in the landmark bituminous strike, the strike itself signified the revival of an assertive national organization that neither the company nor local miners would long ignore. The inclusive progressive policies of the UMWA offered both new immigrants and the American-born a viable vehicle for intra-class organizing across ethnic and other lines.

Berwind-White and Collective Bargaining

What type of policies did Berwind-White pursue during these years? The company's virulent antiunion sentiments and practices, especially evident in the history of Windber and Somerset County, have tended to obscure an earlier history of occasional, if begrudging, cooperation in arriving at collective bargaining agreements with the organized labor movement. Even though E. J. Berwind earned a disputable but widely proclaimed reputation as a "rugged individualist," during the 1880s and 1890s the Berwind brothers occasionally entered into joint wage-scale conferences with other operators and with mining representatives of National Trades Assembly 135, Knights of Labor. In her study of Maryland miners, Katherine Harvey, for instance, points out that in March 1891 Berwind-White was the only company to accept a UMWA invitation to attend a joint conference in Cumberland but that its representative left when no other operators appeared.[16]

Paul Pritchard, a biographer of William B. Wilson, a Knights of Labor organizer who later became a District 2 UMWA president and the nation's first secretary of labor, characterized an agreement Wilson negotiated with the Berwind-White company in 1889 in Horatio, Pennsylvania, as the "greatest progress that had ever been made in the Punxsutawney field" and *the* standard that other miners soon sought to emulate in

Central Pennsylvania, if not the nation. According to Pritchard, this lauded agreement provided not only for checkweighmen on all tipples but also for company deduction of wages from paychecks for the purpose of supporting union checkweighmen. It was this novel use of an old feature—the "checkoff"—that, he claimed, made this particular agreement "the most important step in the development of unionism among miners since the beginning of the joint wage-scale conference" and a guarantee of the UMWA's future financial stability.[17]

In the late 1880s and early 1890s, then, some of Berwind-White's largest mines in Horatio, Houtzdale, and DuBois, Pennsylvania, had some sort of organization, and in its pre-Windber history the company negotiated settlements with organized labor in some of its twenty-nine mines. It did so at times when workers and coal were in demand, when company contracts were waiting to be met, and when organized labor had the strength and prestige to enforce its demands. It did not do so because of any love for unionization or belief in the inherent rights of workers to organize or bargain collectively.

The subsequent history of places like Horatio and Houtzdale showed that when conditions promoting joint agreements, conditions such as those cited above, were absent, the company had no compunction about refusing to enter into agreements or about using its immense wealth and power for repressive purposes and union annihilation. Nor were signed contracts deemed sacred. Soon after the famous Horatio agreement was reached, for example, union members and organizers complained about the Berwinds' attempts to violate the terms and prevent fulfillment of the checkoff provision.[18]

Horatio itself was a financial and moral bulwark of district organization, until the debacle of 1894. The town, located about fifty-five miles northwest of Windber, in Jefferson County, had been built in the 1880s as a company town owned and controlled by Berwind-White. Knights and UMWA organizers habitually cited it first as a negative example of how terrible nonunion company towns and mines could be and then as an example of how organization could radically improve both. Although other mines were larger, Horatio was in the forefront of every major UMWA effort in the district in the early 1890s, and even seems to have successfully enforced a closed shop in Berwind-White mines there. Leaders such as District 2 Secretary-Treasurer Dan Lennon held Horatio up as a model for others to follow: "Every man in the Horatio Checkweigh Fund is a regular contributor to the United

Mine Workers of America and do it without much murmuring; in fact the Horatio miners lead all others in Pennsylvania as far as organization and generally favorable conditions in their work is concerned." After the destruction of the local in 1894, District 2 sorely missed Horatio's militancy and its annual contribution of $600 to the district treasury.[19]

The depression and the failed strike of 1894 had given Berwind-White and other companies a golden opportunity to destroy all existing organization, thereby dictating wages and working conditions for years to come. During the strike itself, the Berwinds indulged in the same antilabor tactics commonly used by other coal operators. They imported southern and eastern European immigrants obtained from an employment agency in New York, hired their own police and Pinkertons, sought court action against strikers, and evicted miners' families from company housing.

In the aftermath of the strike, their repressive policies, which differed only in degree from those of other operators in the region, earned the company a reputation as one of the most ruthless in the region. Writers to the *United Mine Workers Journal* told of Berwind-White's many discharged and blacklisted men, of its new order requiring home-owning miners to move into company housing as a condition of employment, and of its revitalization of the dreaded "pluck-me," or company-store, system.[20] Gone were agreements negotiated with organized labor. Gone also were checkweighmen and the checkoff provision, pay for deadwork, and other hard-earned gains. The irony was that Berwind-White miners in Horatio, Houtzdale, Windber, and other places had to fight and refight battles well into the twentieth century to achieve wages and working conditions comparable to those in Horatio in the early 1890s.

The repression that occurred in Horatio and in other Berwind-White mining towns after the 1894 strike served as an important lesson for contemporary and future labor organizers who had no illusions about the company's philanthropy or antiunion policies. William B. Wilson had participated in the original organizing effort in Horatio in the 1880s; Thomas Haggerty, another important Knights of Labor leader who had helped found the UMWA in 1890, had lived in town and mined coal there. Both of these union men were later active in the first attempts to organize Windber. Also, after the 1894 strike debacle, an important future UMWA leader, John Brophy, lived and worked in Horatio as a young boy. In his memoirs, he pointed to this experience as an additional motivating force and useful guide when he led the effort to organize the Berwind mines in Windber during the strike of 1922.[21]

The Rebuilding of District 2

The victorious strike of bituminous miners that began on July 4, 1897, lasted twelve weeks and reversed the declining fortunes of the national UMWA while inaugurating a new era in labor's militancy just as the depressed economy was beginning to show some signs of improvement. By early 1898, the nation's bituminous miners had won wage increases, paved the way for the revival of the interstate joint conference system, and secured the eight-hour day in a historic agreement in Chicago.[22]

These successes led directly to a revival of the Pennsylvania state and District 2 organizations, and indirectly to the first organizing campaign in Windber. On November 23, 1897, the national union's President, Michael Ratchford, and national organizer Chris Evans attended a meeting in Altoona at which the state organization was founded and a permanent district organization was established. Delegates from Berwind-White mines other than Windber were among those who came to this meeting. By December a few locals were actively reorganizing in such places as Dunlo and South Fork, where no organization had existed since 1894, and district delegates asked the national to send organizers into Central Pennsylvania to help get the district back on its feet. In mid-February 1898 permanent district officers were elected at the Pennsylvania state convention held in Altoona.[23]

Rebuilding the Central Pennsylvania organization and getting regional operators to take part in a district-wide joint conference system of agreements were difficult tasks to accomplish. Pragmatic district leaders took careful steps toward these goals while trying to avoid disastrous mistakes and strikes that might threaten the precarious existence of the rebuilding organization. But by 1899, under the leadership of President William B. Wilson, the district organization had succeeded in reestablishing the joint conference system in Central Pennsylvania, and growth was such that, as of January 1, 1900, District 2 had 38 locals and 3,854 paid members. Throughout the year, sporadic strikes occurred and were successful in winning on such issues as wage increases that were in accord with the district scale, enforcing state laws that required operators to pay wages semi-monthly, and allowing miners to have checkweighmen on tipples.[24]

Meanwhile, the national made a commitment to help the district, and both cooperated in a joint effort to reach the unorganized and to establish new locals. District 2 was indeed getting on its feet in 1899, and in a private letter to an editor of a Joliet, Illinois, newspaper, John Mitchell, who had recently succeeded Michael Ratchford as president of the national, took credit: "Their victory is entirely due to efforts made by me. I have kept men there, paid their salaries from the National Treasury, and have practically paid the salaries of the District officers."[25]

Formation and Demise of an Ephemeral Local

The organizing campaign that took place in Windber in 1899 grew out of these larger efforts and resulted in the establishment of an ephemeral union local. Because its records have not survived, we unfortunately know very little about the local itself, its membership, or its leaders. Our knowledge and perspective is therefore limited, but we can glean some understanding of the organization's brief history through national union records and other outside sources that have documented its existence.

National organizer Chris Evans had come to Somerset County in February 1899 to attempt to resolve a lengthy strike over wages paid for machine-mining and loading in Elk Lick, Pennsylvania. While settling the strike in March, he visited Windber, Scalp Level, and South Fork. Miners employed in the new Berwind-White mines at Windber and Scalp Level had expressed interest in the union, and the reorganized South Fork local subsequently became an important magnet of union activity.[26]

Meanwhile, the district's rebuilding efforts were succeeding. Officers issued a call for a district convention of organized and unorganized miners to meet on March 23, 1899, in Tyrone, Pennsylvania. They stressed that the national would keep a force in the field as long as miners continued to respond and join the UMWA. Delegates to the important convention at Tyrone elected William B. Wilson president and passed sixteen resolutions, which were disseminated throughout the district. The most central resolutions affirmed the priority of establishing uniform wage rates and ordered a suspension on April 1, 1899, at mines that had not yet met the 45-cent wage scale. The national and district organizations expressed willingness to commit people and resources to achieve this immediate and limited goal.[27]

April 1 was not only the date set for possible suspension but also the date set aside by the national UMWA convention to celebrate as a general miners' holiday to commemorate the one-year anniversary of the eight-hour day achieved in bituminous fields by terms of the Chicago agreement. It was in the context of these events and a new assertiveness that Chris Evans announced the welcome formation of a new local in Windber on April 1, 1899.[28]

Within three months of this announcement, Evans's initial optimism about the Windber local had turned to a concerned lament. In June he reported that he and fellow organizer Ed McKay were busy in Windber "trying our best to save as many souls as we can from everlasting punishment."[29] Area miners had not been the only ones to respond to UMWA initiatives. Berwind-White, along with other open-shop companies, quickly took active steps to forestall organization. Among these steps were measures of both reform and repression.

Timely wage raises and a massive public relations campaign were among the tools used by the company to undermine the union's support. Between April 1, 1899, and January 1, 1900, Berwind-White announced wage increases affecting Windber miners on three separate occasions. In each case, local newspapers generously praised the company and treated the raises as if they were isolated, voluntary events. Informed people, union officials, and organized miners quickly understood that each pay raise had been prompted by the union's rebuilding efforts and the district's recent successes.

It was no accident that the first of these increases was announced by the company on April 1, 1899, on the same day and shortly after Evans had made known the existence of the new local. In order to counter the union's momentum, Berwind-White quickly posted notices of a general 8 percent wage increase effective April 1. Word of the raise reached Evans in Elk Lick, where he was attending a miners' celebration, and caught him completely by surprise. Initially he took the advance as a sign of the company's goodwill. Newspapers praised the "voluntary" raise. Few had noted that one month earlier, during the strike at Elk Lick, the superintendent of mines there had cited specific figures showing that Berwind-White paid much less in wages than his own company did. That the company announced the wage increase in terms of percentages rather than in terms of dollars and cents was also not accidental. The percentage method tended to obscure the amount really being paid and allowed the corporation to make broad claims, which were harder to refute. The wage increase, a direct response to the local union's emergence in Windber, apparently did not apply to the company's operations in DuBois, Pennsylvania, where Berwind-White miners walked out on strike on April 1 in an attempt to obtain the district scale.[30]

Berwind-White normally exerted authoritarian controls of all sorts in its company town, but it did so in extraordinary ways shortly after the appearance of the Windber UMWA local. The company was able to take advantage of an outbreak of smallpox to completely shut off the town from outside union organizers for a critical four- to six-week period in April and May. The fledgling organization was thereby deprived of important outside assistance in its formative days, and local union supporters were isolated.

A few cases of smallpox had first appeared in the town in March, but in April, after the local union had publicly emerged, wild rumors of undetermined origin about an epidemic of major proportions in Windber began to spread throughout the region. To deal with the outbreak, Berwind-White sought and quickly received permission from understaffed state and county authorities to deal with the local smallpox crisis as it saw fit. The company had absolute authority. The result was that it prohibited entry into and out of Windber during the height of the crisis and rigidly restricted it for about four weeks. Quarantine guards were removed from the Windber and Scalp Level roads on May 5, and by mid-May the crisis was considered at an end.[31]

Smallpox was a deadly disease, and quarantines were routinely imposed in the early century to keep such diseases from spreading. Yet the gravity and extent of the epidemic in Windber is highly debatable. Despite vague reports of numerous, perhaps hundreds, of minor cases, only two victims ultimately died; both had been given special nursing treatment in a pesthouse established for that purpose. On one occasion, the *Johnstown Democrat,* mystified by the simultaneous occurrence of boom and epidemic in Windber, noted that foreigners there were not suffering from the disease at all, because they had been vaccinated during the immigration process. After the crisis was over, newspapers commented that fears had been greatly exaggerated but that safety had been the primary concern.[32]

Whatever the exact dimensions of the epidemic, the important consideration for union organizers was that Berwind-White, not the state or any neutral authority, had control of access to that town, so that the company was able to unilaterally enforce the quarantine as it saw fit. The result was an uneven, partial enforcement. Company officials and a privileged few could come and go, while others could not. On April 7, 1899, for instance, Clark Duncan, prominent businessman and owner of the Windber Hotel, visited the Mansion House in Johnstown during the earliest and most hysterical days of the panic. He had come to buy a pair of horses for his business. While in Johnstown, he raved about the growth of Windber and noted that Berwind-White had recently placed an ad in the *Johnstown Democrat* for 500 miners in hopes of attracting area miners who went on strike on April 1. Duncan claimed that the ad had brought 1,000 new people to Windber in the last few days and ridiculed any suggestion that smallpox was a serious concern there.[33]

Company officials, such as Superintendent James S. Cunningham, who had charge of the epidemic, and leading businessmen, like Clark Duncan, may have been able to bypass quarantine regulations, but union organizers certainly could not. On April 16, Chris Evans noted that the quarantine was retarding work in Windber. A week later, he reported that Windber's organization needed "rounding-up," that "the Berwind-White people have full charge of the smallpox quarantine," and that "the quarantine against invaders is complete."[34] In 1899, for the first but not the last time, critics of the company charged that it had unfairly and arbitrarily enforced a quarantine to fend off a progressive movement. Businessmen would make the same charge in 1919 during the much more substantiated flu epidemic there.[35]

Meanwhile, the end of the smallpox scare coincided with the company's announcement of the second wage increase in two months. By June, when organizers returned to Windber, they faced a situation that was quite different from the situation in March. Union organizers had never been welcome in Windber, but the epidemic had enabled the company to control information as well as access to the town. The newly founded local, not yet a solid organization in Evans's view, was deprived of important outside assistance in a critical period of its development. At the same time, the company's ability to advertise plentiful jobs and to recruit large numbers of workers even during the supposed epidemic created an unstable, inchoate, and amorphous work force.

Americans and immigrants poured into Windber for work. Under the circumstances, disintegration was probably inevitable.

Berwind-White's second raise, effective June 1, 1899, was not only a response to events in Windber but also a clear reaction to district success in achieving an important goal: the revival of the joint conference system in Central Pennsylvania. At the end of April, an agreement between miners and operators of the Beech Creek region of the district was effected in a joint conference held at Clearfield, Pennsylvania. Moreover, the signed scale agreement raised the basic wage from 45 cents to 50 cents a ton. In their convention, miners then passed a resolution calling for a suspension on June 1 for mines that were not paying the new scale as of that date. District policy that all mines of a given company had to be included in any settlement was reiterated. It was in this context and under threat of a strike at some or all of its mines that Berwind-White quickly announced it would pay the 50-cent district scale in all its mines, not just in Windber.[36]

While it is possible that the smallpox outbreak presented Berwind-White with a merely fortuitous opportunity to cut off Windber-area miners from access to union organizers for almost two months, the repression instituted by the company when the organizing drive resumed after the ending of the crisis was neither accidental nor random. By June, returning union officials and visiting organizers were reporting serious instances of harassment and discrimination directed against themselves, even as they began to expose institutionalized abuses and repressive company practices directed against the miners. Windber's new importance to the company and to coal production in the district, along with the company's key role in Central Pennsylvania, had led the UMWA to send its district president, district secretary, three board members, and various organizers to help in the renewal of the organizing effort there in June and July. In recognition of the need to communicate with the growing multiethnic work force in their own languages, District 2's English-speaking officers also employed a number of foreign-speaking aides.

Because our information about this initial drive in Windber comes exclusively from English-speaking organizers who dealt primarily with English-speaking miners, the sources are heavily biased in this direction and unlikely to tell us much about the foreign-born. Nevertheless, these sources say enough to refute the notion that new immigrants were not interested in the labor movement or inherently unorganizable, as some labor historians have suggested. Indeed, immigrants in some numbers had responded positively to the union's message, attended meetings, and suffered repression in the same manner as the American-born. On June 29, 1899, for instance, Chris Evans indicated that a number of foreign-speaking miners had just been discharged—they had attended the last union meeting. Company officials had reacted to the drive by attending union meetings, observing which miners came, and discharging those who did the following day.

The result was that, by July, Evans reported that repression was so severe that miners dared not make any sign of recognition to organizers passing on the street for fear of what he called the phenomenon of "tramping ties." If seen talking to an agitator, a miner was sent tramping down railroad tracks to new pastures. If anyone dared to complain about working conditions, he was called into the company's office, discharged, and told to seek relief from Charley McKay or some other union officer.[37]

Company repression was supplemented by the hostile reception that leading businessmen and other citizens gave the union organizers. William Warner was one person who found that Windber's hospitality left something to be desired. On June 29, 1899, he wrote T. W. Davis, a union friend in Gallitzin, Pennsylvania: "The circumstances at Windber cannot be described. Every business man and influence of every kind is arrayed against us." Spies, detectives, discharges abounded. "I believe if we were to return now, as they know us," he continued, "we would be refused a meal or lodging in the town."[38]

Elizabeth Catherine Morris, private secretary to the national UMWA president, John Mitchell, conveyed essential messages to him when he was away from headquarters on trips. In one such instance, on July 1, 1899, she summarized the contents of several pessimistic letters received from organizers in Windber. In one, Ed McKay described how organizers' steps were dogged by company spies who had successfully prevented them from holding any meeting. In another, President William Wilson informed Mitchell, in Morris's words, that "they are meeting with much opposition from the Berwind-White Company, that their footsteps are dogged, and that it is almost impossible for them to make a move that the company has not already been apprised of." A second letter from Wilson reported that eighty-seven men had recently promised to join the union at Windber but that twenty of these eighty-seven had already been discharged for unionizing efforts.[39]

The portrait organizers painted of conditions in Windber differed markedly from Berwind-White's description of Windber as a model town. In particular, Chris Evans ridiculed the notion of the company's philanthropy. Miners told him and other organizers of their many grievances. Dishonest weights, the absence of checkweighmen on the tipples, compulsory car-pushing or having to haul heavy cars long distances, and no pay for deadwork were frequent complaints that often negated any wage advances. The recent discharges in June and July for union activity were, in Evans's view, "philanthropy with a vengeance."[40]

Chris Evans was the first of a subsequently long line of organizers and outside observers who over the years compared the status of Windber-area miners to that of pre–Civil War slaves. In a letter written on July 21, 1899, he described Windber as "growing in population of that class of men (color excepted) that gave Abraham Lincoln so much trouble and cost the country so many lives in former years." Furthermore, the company's manipulation of power was such that other workers were tantamount to European serfs: "But," he said, "coal miners are not the only willing victims here. Carpenters, bricklayers, stone masons, plasterers and a host of other serfs are also numbered with the throng, and so far as the scale of wages are concerned of their various trades are even worse."[41]

Repression resulting from this first systematic organizing campaign in Windber was such that, in July, Evans advised the district and national organizations to decrease their mobilization there. He believed that a reduction of outside forces would enable Windber-area miners to take greater charge of the movement. As it was, the presence of UMWA outsiders greatly increased the likelihood that they would be discharged. On July 1, therefore, he suggested new tactics that relied on local initiative to continue the battle, which, he claimed, Windber miners would soon win.[42]

The UMWA adopted the strategic retreat advocated by Evans. During the next few years, district and national organizers occasionally passed through the Windber area and tried to maintain a continuity of interest in organization there, but a massive commitment of people and resources to a new organizing campaign had to await events and developments that revealed serious internal discontent, local initiative, and some promise of success.

On January 1, 1900, Berwind-White announced its third wage increase in a year. Again, neither the advance itself nor the timing of its announcement was accidental. A District 2 convention held in Clearfield on December 12–14, 1899, had pressed hard for an additional wage increase by January 1 in the mines like Berwind-White's whose operators had never signed the scale agreement, and by April 1, 1900, in the mines whose operators had signed. District officers persuaded the workers to defer a strike on January 1 pending the outcome of the forthcoming national UMWA convention and joint conference. They were still hoping to avoid costly strikes by convincing noncomplying district operators to attend the regional conference and take part in collective bargaining.[43]

Berwind-White's granting of three wage increases in the space of a year reflected the growing strength of union organization in Central Pennsylvania. Without the revival of the national and district UMWAs, it is highly unlikely that Windber-area miners would have made such advances, despite returning prosperity. Moreover, the wage increases or reforms served other purposes for the company. Pay raises conceded "voluntarily," without any need for Windber's miners to strike or even organize permanently, undercut union support there. Also, because extensive new mines were being opened throughout Central Pennsylvania as well as elsewhere in the nation as the depression ended, the company was in reality forced to pay wages in relative, if not actual, conformity with the district scale. In order to keep up with competitors, the company had to attract and maintain the labor force needed for its own expansion, and wages were an important inducement. Finally, the reputation it fostered and gained as a liberal employer and generous benefactor among the broader public made it even more difficult for union activists to unmask nonunion wage scales and other conditions. The benefits to Berwind-White of responding to the new initiatives of the union by advancing wages were therefore considerable. Minimal wage concessions, combined with a great deal of repression, were a small price to pay to maintain a prosperous business and an open-shop policy.

Meanwhile, the importance of Berwind-White and Windber to the district was impossible for officers to ignore. Unorganized miners were a drag on the entire district and an exasperation to organized miners in the region. The *Lilly Signal*, a pro-union newspaper in a nearby town, expressed such sentiments. On some occasions it tried to be informative and demonstrate that Windber's actual wages were 10 to 20 percent below the district scale. On other occasions, as when Berwind-White "voluntarily" raised wages in response to district gains, the paper blasted Windber-area miners as "scabs" who derived the benefits of organization fought and paid for by others.[44]

The Failure of Organizing "From Above"

Because the first attempt in 1899 to organize Windber "from below" had met fierce opposition and failed, district and national UMWA leaders turned to a new strategy early in the new century. What they could not gain "from below," despite considerable effort, they now sought to win more easily "from above." The new conservative method was in accord with the class-collaborationist and corporate liberal philosophy espoused at the time by Mark Hanna and the founders of the National Civic Federation (NCF). National UMWA President John Mitchell hoped to use his position as a charter member of the NCF to win E. J. Berwind's acceptance of the United Mine Workers and of collective bargaining.

In so doing, Mitchell was reacting to initiatives "from below," specifically to complaints about Berwind-White mines and to repeated requests from the new District 2 president, Patrick Gilday, in November 1901 for help from the national in organizing Windber. His response to Gilday's requests was not to commit men and resources there but to write Ralph Easley, head of the National Civic Federation. On November 23, 1901, he urged Easley to get E. J. Berwind into the NCF and to get him to take part in the joint conference system:

> The Berwind-White Coal Company is the largest interest in Central Pennsylvania which does not treat with our organization and at present the indications are that trouble will ensue unless we can induce Berwind to grant the same conditions as prevail in competitive mines. I think that if you could bring pressure to bear upon him to take part in the meeting on the 16th and 17th in New York, good results would follow.[45]

E. J. Berwind did not attend the NCF meeting in New York, and UMWA organizers in the field, like Ed McKay, were sorely disappointed that Mitchell had not been able to confer with him. On December 27, 1901, McKay wrote Mitchell: "I had hoped that you would have succeeded in getting more thru peace than we will be able to force from him." McKay continued to urge Mitchell to meet with Berwind, if at all possible, especially after a conference he had held with Superintendent James Cunningham in Johnstown. The organizer had urged Berwind-White's top official in Windber to attend the next joint conference at

Altoona. Cunningham replied that he could not do so without the senior Berwind's explicit approval.[46]

From time to time, district officers contacted company officials and planned trips to Philadelphia or New York to issue operators like the Berwinds special invitations to attend joint conferences. The strategy of organizing Windber "from above," however, proved a complete fiasco, and Berwind never did join the NCF. In a letter to McKay on December 23, 1902, Mitchell recognized this failure:

> You know I expected that he [Berwind] would be induced to go on the committee which was formed by the National Civic Federation, but notwithstanding the fact that an urgent invitation was extended to him in a communication from Mr. Easley, he failed even to make a reply. . . . I am of the opinion—and so is Billy [Wilson]—that the only way to force him into line is to inaugurate a strike at his mines.[47]

Organizing Windber would not be easy. As one miner put it, "The Berwinds said we'll sign the union contract when grass grows on the railroad tracks."[48] E. J. Berwind's conception of labor relations was closer to the open-shop philosophy of David Parry and the National Association of Manufacturers than it was to Mark Hanna's or John Mitchell's. The initial contests between the company and the union seem to have made the national leadership wary of taking on Berwind-White or of committing sparse resources to the costly task. Given corporate resources and a hostile state, local miners could not hope to succeed without a lot of outside help. From the outset, District 2, which had an immediate stake in what happened in Windber, proved to be more reliable and supportive of organizing efforts in Windber than the national.

The Anomaly of Windber's Position in the District

By the end of 1900, Windber was rapidly becoming an anomaly in a district that was aggressively signing up union members and winning contracts from local operators. In the next few years, given the district's successes, the gap between union and nonunion mining conditions widened and became even more obvious. Despite Berwind-White's attempts to isolate its workers within the strict confines it imposed in its company town, it could not totally isolate the miners from the environment that surrounded them. The more successful the district organization was, the more likely it was that Windber miners would make another serious attempt to organize, and the more likely it was that they might succeed.

As part of its overall strategy, District 2 increased its special efforts to attract the many new immigrants who continued to pour into Central Pennsylvania. It placed a foreign-speaking representative on its executive board, employed more foreign-speaking organizers, and made literature available in a variety of languages. Unlike Samuel Gompers and other national labor leaders who advocated immigration restriction or the adoption of a literacy test, district miners took the lead in opposing most forms of nativism. At district conventions they passed resolutions that referred to immigration as a positive force and as an argument for reform. One such resolution demanded the eight-hour day because immigration had led to such a plentiful supply of labor that operators' objections to the shorter workday were no longer valid. But there were clear limits to the toleration and inclusion practiced by the UMWA. District leaders and locals were willing to extend the UMWA's policy of knowing no race, creed, or nationality to southern and eastern Europeans and possibly blacks, but not to the Chinese. They enthusiastically endorsed a petition drive inaugurated by the national organization in 1901 to continue excluding the Chinese.[49]

The historic anthracite strike of 1902, which led to unprecedented government intervention and the creation of an arbitration commission that granted some but not all of the anthracite miners' demands, presented District 2 with problems and opportunities. Bituminous miners were not involved in the strike, which had an ambiguous impact in Windber and Central Pennsylvania, where miners worked throughout the strike.

Among the beneficiaries of the strike were the various coal companies operating in the bituminous fields of District 2 and the Pennsylvania Railroad, which hauled the product. District coal was like anthracite in that both had supplied eastern markets before the strike. With anthracite competition gone during the strike, this bituminous output captured new eastern markets, often permanently. Indeed, Central Pennsylvania coal mines boomed during the strike, and despite frequent complaints of labor and car shortages, production reached an all-time high. Moreover, numerous anthracite miners left their region and came to these bituminous fields to work. Simultaneously they spread news of strike events and educated miners about unionization.[50]

Under normal circumstances, the new prosperity that Central Pennsylvania enjoyed would have been unreservedly welcome, but district and national UMWA leaders quickly recognized that the region's increased production was potentially, if not actually, harmful to strike efforts in the anthracite region. They debated what to do. Their first response was to initiate a policy aimed at restricting output. At the end of June, district leaders ordered miners not to work on Fridays and Saturdays, thereby enforcing a four-day week, which they considered sufficient to meet the demands of existing bituminous contracts but insufficient to invade anthracite markets and damage the strike.

Although the order affected 40,000 district miners, most of whom had recently participated in an anti-injunction general holiday on May 17, 1902, Windber remained unaffected. In hopes of gaining the cooperation of its miners in the restriction effort, Gilday, other officers, and organizers entered Windber en masse on 4 July, ostensibly to attend holiday festivities there but really to agitate. Their efforts succeeded in setting up a special Sunday meeting attended by 600 miners. The *Lilly Signal* optimistically concluded: "Had it been possible to secure a hall for the meeting there is no doubt the men of Windber would have organized." Getting a place to meet had not been the only difficulty, according to the paper. "All the Windber officials, from the managers down, were present for the purpose of intimidating the men and preventing action being taken to enforce scale rates and justice at the tipples." Still, the responsiveness of Windber-area miners was taken as a hopeful sign for the future.[51]

Ultimately, the district's ordering of a four-day work week proved unpopular and unenforceable not only in Windber but throughout the district. There was some regional sentiment that bituminous miners make the ongoing strike a general one by joining the anthracite suspension, but this option, vigorously opposed by Mitchell because it would violate the sacredness of signed contracts and endanger the bituminous joint conference system, was defeated in a special national convention in July. Because restriction efforts had clearly failed, however, delegates supported a plan that levied special assessments on all working miners to aid in strike relief. The UMWA still had some difficulty collecting the assessments, but organized District 2 miners could now salve their consciences and work a six-day week.[52]

District success reached new heights in 1903. A financial report in May indicated that District 2 had grown to 113 locals and a paid-up membership of 19,201. The organization's strength had paved the way for an important agreement reached in Altoona on March 20, 1903, during a joint conference of miners and operators. This agreement, a landmark for the district, established the eight-hour day for the first time. It also granted miners an average wage increase of 11 percent and made provision for deadwork and overtime pay, as well as for mandatory checkoffs, holidays, and funeral attendance. William B. Wilson, then the national UMWA's secretary-treasurer, claimed that miners had given up a demand for even higher wage increases in order to get the much-desired eight-hour day, already achieved in other organized fields.[53]

Getting the signed agreement was not the final word, however. Throughout the year, district miners in various places were forced to go on strike to achieve or maintain the new district scale, and on October 22, 1903, district officers found it advisable to issue a special circular in five different languages. The circular traced the history and explained the importance of the eight-hour gain and argued that unless miners limited the hours they worked, district officers could not hope to get remuneration for deadwork.[54]

The new district agreement had placed Windber-area miners in a position clearly inferior to that enjoyed by area miners who worked in organized mines. In typical fashion, in late February 1903, before the joint conference met, Berwind-White had announced a "voluntary" but unspecified wage increase effective April 1. Speculation and media praise abounded, but notices of a general wage raise of 10 percent for manual labor and 12 percent for machine labor were not posted until one week after the district settlement had been made.[55] Then the company tried to prevent local newspapers from publishing the real union scale. Windber miners continued to work for ten or more hours a day, continued to receive no pay for deadwork, and gained none of the other benefits contained in the district contract. The anomaly of their position could not go unnoticed for long.

The first stirrings of Windber-area miners to organize in 1899 and the early 1900s established some of the basic contours of a labor movement that persisted throughout the entire nonunion period. Three important conclusions can be drawn from this early history.

First, the struggle of Windber's miners to democratize the workplace and community began in the earliest days of the company town's existence, not later, and involved new immigrants as well as English-speaking people. In 1899, miners were already citing local grievances that they would continue to cite until permanent unionization was achieved in the 1930s. Such issues as low pay, no pay for deadwork, dishonest weights, and the general lack of free speech and civil rights were constant complaints of the nonunion era. Given the abundance of new jobs, the inchoate development of the local working class, the corporation's resources, and the authoritarian context, it is not surprising that Berwind-White was successful in destroying the first UMWA local in Windber. What is surprising is that an ephemeral local ever emerged at all. The local's short-lived existence is a testimony to the aspirations of diverse working people—the American-born and new immigrants—for social justice.

Second, the fate of the labor movement in Windber was intimately linked to the successes and failures of the district and the national union. The national UMWA's resurgence in 1897 directly spawned District 2's revival in 1898, which in turn indirectly spawned the organization of a local in Windber in 1899. Despite Berwind-White's efforts to quarantine its miners from outside influences, Windber miners could not be isolated from their larger environment. Their first organizing campaign in Windber revealed how costly and bitter the struggle for unionization would be. As pragmatists, they understood the immense difficulties and need for allies in the uphill organizing battles they waged. They realistically and rationally weighed their options. Injustices and grievances were perennial, and it made sense to take the great personal risks involved in organizing on the local level in periods when the UMWA was on the move in the district and the nation. Then there was some chance of success. The converse was also true. In periods of the larger labor movement's weakness or quiescence, Windber miners would reject taking risks for what seemed certain to be a losing battle.

Although a nonunion bastion, Berwind-White was not immune from the UMWA's successes in the surrounding region either. As indicated by the timing of company pay raises, Windber's nonunion miners were often indirect beneficiaries of the gains made by unionized miners in District 2. In 1899 and throughout the nonunion period, the company, whenever it was challenged, made blatantly untrue claims that it habitually honored the union scale and working conditions. The fact that it felt compelled to make such claims, to suppress contrary evidence, and to conduct massive public relations campaigns reflected the growing influence of organized labor in the region. Bituminous mining was a competitive industry, and miners were migratory people.

Third, this first organizing drive indicated to the national UMWA that Berwind-White would be a formidable adversary. In the company-town setting it had created, its powers of repression were great, and as a nationally prominent corporation with ties to other open-shop interests, it could mobilize enormous economic and political resources against organized labor. John Mitchell would be the first, but not the last, UMWA president to hesitate and then refuse requests "from below" for an all-out commitment by the national to support a sustained drive in Windber for fear of the costs that such an effort would entail for the larger organization.

At the same time, the unionized miners in District 2 rightly perceived that the continued existence of the Windber-area mines as nonunion entities threatened their own organizations. The new Berwind-White mines were big coal producers. Other major enterprises, such as the steel mills in Johnstown, were also nonunion. Unionized miners would not be able to maintain their standards if nonunion employers dominated the landscape. From the outset, informed district miners understood that they had an immediate stake in ending the anomalous position Windber occupied in District 2.

7

Friends and Enemies

Working people of many nationalities and races participated in economic and political struggles in the United States in the early century. Miners themselves had pioneered an industrial form of union organization that had successfully incorporated unskilled new immigrants. Whether these workers ultimately supported radical organizations such as the Industrial Workers of the World, or more moderate industrial unions such as the United Mine Workers, they typically confronted the opposition of open-shop employers and a repressive state apparatus.

But labor struggles were fought on many fronts, especially in company towns. In communities with large foreign-born populations, the struggle to establish a union base necessarily intertwined with a struggle against nativist discriminations. Moreover, the success or failure of organized labor to take hold in such places affected the livelihoods, status, and activities of other classes. Businessmen, borough councils, and reformers who functioned in the environment of closed company towns could not ignore the wishes of the dominant employer and did not enjoy the relative independence their counterparts enjoyed elsewhere. As a result, it became clear to many in the early century that the opening up of towns such as Windber to basic constitutional liberties, competitive businesses, and progressive reforms was integrally linked to the success of organized labor. For that reason alone, the relationship in such towns between workers and other classes was often complex. Herbert Gutman discovered that labor sometimes found unexpected allies among other classes in small towns during the Gilded Age.[1] In Windber, immigrant miners had a similar experience at a later date.

This chapter examines the support that Windber's union-oriented miners found on the local level among other classes in the early century. Although organizations and individuals changed over the years, the complex interclass relationships that developed in the formative years of the town set a pattern that lasted throughout the nonunion era. That these allies included certain Americans is significant because immigrant miners searched for friends in the midst of many enemies, but always within the context of a deep-rooted nativism that pervaded the predominantly American-born, Protestant middle classes.

The Nativist Context

Southern and eastern European immigrants were compelled to wage labor struggles in the midst of a new or resurgent nativist environment. While the national triumphs of nativist legislation—passage of the literacy test and immigration restriction—did not occur until World War I and the 1920s, they had been preceded by earlier efforts and by important discriminatory measures on state levels.

Pennsylvania was one of the leaders in this regard. It was also the birthplace of a number of prominent nineteenth-century nativist organizations, including the Order of United American Mechanics, the Junior Order United American Mechanics, and the Patriotic Order Sons of America (POSA).[2] These secret fraternal organizations had been active in Central Pennsylvania before the founding of Windber, but the arrival of large numbers of new immigrants into Somerset County, which had been inhabited predominantly by native-born Protestants, prompted a revival of such organizations late in the century. Once Windber itself was established, its segregated ethnic and class configurations provided a milieu in which nativism could flourish at the local level.

In the early 1900s the most active nativist organization in Windber, as well as in many other Pennsylvania towns, was the local chapter of the antiradical and anti-Catholic Patriotic Order Sons of America. If newspaper reports are correct, Windber contained one of the largest chapters in the state. The society's national platform, published in Philadelphia in 1894, sought to exclude "foreign speculators or adventurers, who do not wish to become citizens" and to crush and prohibit from entry "that foreign element which comes here to advocate communism, and nihilism, and which does not identify itself with our country, and does not respect our flag." Since, according to its preamble, sons of America loved the country as no others could, its membership was limited to "native-born citizens who believe in their country and its institutions, and who desire to perpetuate free government, and who wish to encourage a brotherly feeling among Americans. . . ."[3]

Little is known about the activities of Camp 640 of the POSA except that it was quite active in the Windber area during those early years and that Protestant churches and public schools regularly turned out at the flag dedications and patriotic addresses it sponsored. By August 1904, Camp 640 was important enough to have sponsored a women's auxiliary whose charter membership was thirty.

Meanwhile, the Pennsylvania organization, whose membership counted 95,000 in 1909, remained the bulwark of the national organization, which firmly advocated immigration restriction. In 1910 it supplied the Dillingham Commission with a resolution endorsing adoption of a literacy test, an increased head tax, and strict enforcement of laws designed to exclude undesirables. Its views stood in sharp contrast to those of the National Liberal Immigration League, which sent a comparable statement opposing restrictive measures to the commission. At the national level, coal and steel interests, including Berwind-White, supported the efforts of the latter organization, not those of the POSA.[4]

Some of the most important manifestations of nativism at the local level occurred in the arena of the legal system. Pennsylvania itself subsequently gained a notorious reputation for the law enforcement system that existed in its remote mining towns. During the first decades of the century, private police, public police, and the state constabulary were routinely used to repress strikes and labor-organizing activities. But day-to-day law enforcement at the local level also led to discriminatory treatment fueled by the nativist sentiments of an exclusively American-born police establishment.

On June 4, 1903, after an incident in which two foreigners were severely beaten by a policeman, the *Windber Journal* carried an editorial about the problems of local nativism and police brutality. According to the editorial, the police seemed to believe that they were above the law and could beat prisoners, especially foreign-born prisoners, at will. The newspaper stated: "We are opposed to resistance to arrest on the part of offenders, but we are also strongly opposed to the customary treatment accorded foreigners when they get in the toils—that is the beating that some officers delight in administering to the unfortunates." It went on to describe nativist sentiments: "They [certain members of the police force] seem to forget the fact that foreigners are human and possessed of feelings, and if they can boast of having split open one or two men's head with a mace each week, the more elated they seem to be."[5]

Pennsylvania's antiquated legal system also made a discriminatory nativist enforcement of laws more than a matter of psychological or sentimental recompense. Indeed, interested parties could and did financially benefit. For example, justices of the peace, who were not required by the state constitution to have any legal qualifications, received remuneration out of the fines they collected from fellow citizens. Emily Balch and others have cited foreigners' complaints about Pennsylvania's archaic system and specific abuses of power, such as the readiness of some law enforcement officials to instigate fights in order to collect revenue from the fines.[6] Unsuspecting innocent or guilty foreigners, intimidated by unfamiliar legal procedures or their inability to speak English, were at best severely disadvantaged in criminal disputes. At worst, they were easy prey for corrupt officials. The fines that were routinely imposed on Windber-area residents for drunkenness, fighting, and other minor crimes accumulated to sizable sums.

But justices of the peace were not the only interested participants in the fining system at the local level. Fines imposed by the burgess were also an important source of borough revenue. On March 1, 1904, the burgess reported to the Windber council that he had collected monthly fines of $203.96, nearly half the borough's balance of $420.33. Such collections were not unique. The *Windber Journal* reported that 500 defendants had appeared before the burgess in a ten-and-a-half-month period and had paid fines and costs totaling $1,500. These fees not only paid the salaries of the burgess, the chief of police, and other police for the period entailed, but also contributed $777.50 to the council for general borough use.[7]

Certain state laws and fines applied exclusively to foreigners. A nativist law passed by the state legislature in 1903 required that all unnaturalized foreign-born Pennsylvania residents who wanted to hunt pay a special fee, not applicable to Americans, for a hunting license. Until then, Windber-area immigrants had hunted in nearby woods without the need for any license. The new law, unknown to many foreign-born hunters, quickly netted a handsome income for the local government and law enforcement officials, who benefited personally from the fines. Fines were typically $25 for each case of illegal hunting. In one week in October 1903 the police chief single-handedly collected $500 worth of fines for the offense. Violations were so frequent and costly that the editor of the *Windber Journal* performed a public service by publishing the law in its entirety in order to inform unsuspecting foreigners of the new regulation.[8]

Hunting restrictions had an additional effect, which benefited Berwind-White indirectly. Immigrants who had once furnished meat and game to their boardinghouse keepers became much more dependent on the company store. Subsequent state laws prohibited foreigners from hunting and from possessing firearms of any kind.

If newspaper reports are accurate, most of the fines that enriched the borough coffers and individual pockets came from the area's large foreign-born population. This observation is of more than passing interest because the sizable penalties involved meant that the taxes on the local American population and coal corporation were correspondingly reduced. At the time, no income or sales taxes were in existence. Instead, license fees, fines, and real-estate taxes were the basis of revenue for local governments. Because Pennsylvania state law excluded propertyless unnaturalized foreigners from direct taxation, the burden of local taxes fell on area property owners who were primarily American-born. Corporate taxes were low or nonexistent. In a sense, then, and in this context, fines amounted to an informal and indirect method of taxation that fell mainly on the foreign and transient population.

For that reason, the fining system enjoyed popular support in English-speaking quarters, and for that reason councils and newspapers, conscious of its relationship to taxation, publicly congratulated burgesses for banner months when impositions were large. Burgess Thomas Delehunt was accordingly commended when he retired in 1909 after three years in office. He had turned over nearly $5,000 to the borough's treasury during his tenure as burgess.[9]

Taxation was also an important controversial political issue on the state level. One economically discriminatory law that had passed the Pennsylvania legislature in the late 1890s as the result of the lobbying efforts of American-born miners in the anthracite region ultimately backfired and hurt both American-born and foreign-born miners.[10] The law required that coal companies deduct from foreigners' wages a special tax based on the number of unnaturalized coal miners they employed. Although the law was quickly declared unconstitutional, coal companies used it afterward as a precedent to make unauthorized and illegal deductions from wages in the name of taxes.

Berwind-White apparently indulged in this unlawful practice. On November 19, 1903, the *Windber Journal* reported that, for the purpose of local taxes, the company had deducted $2.65 from the current semi-monthly wage statements of every Windber-area miner. In an attack on the company and this practice, the newspaper noted: "It has been customary for the company to make wholesale collections of taxes from the unnaturalized foreigners. This is unlawful, but that fact does not concern the czars on Somerset avenue." While the newspaper believed that taxes should be uniform and that foreigners should be taxed in some manner, no laws to that effect existed at the time.[11]

Besides, foreigners were not the only victims of the illegal imposition. Americans who had already paid their taxes were taxed a second time, and minors and nonresidents, who were not obligated to pay borough taxes, were taxed as well. Shortly after the company's action, disgruntled miners had assembled at the company office to get corrections on their statements. They did not receive any satisfaction, and the corporation called in police to disperse the crowd. In another instance, in 1906, Daniel Ott, auditor of Paint Township, formally complained to state authorities that Berwind-White had collected taxes from its employees but that the revenues had not been turned over to public officials.[12]

Corporations and politicians sometimes tried to exploit resentment at the foreigners' exempt tax status. In the early century and long afterward, Pennsylvania tax collectors were elected and paid out of the amount of taxes they collected. In the company-town context in which the directorship of town government was interlocked with that of Berwind-White, those with important positions at the coal company often served as tax collectors and members of the town council as well. In

1909 the head accountant at Berwind-White and secretary of the Windber council ran for the office of tax collector. In an ad that was designed to appeal to the native-born and was reminiscent of the company's illegal deductions, he promised that he could enrich the treasury by making exempt foreigners pay taxes: "My occupation brings me in close touch with these people [foreigners] and places me in a position to collect taxes from hundreds who would otherwise pay nothing."[13]

Although fines and illegal or discriminatory levies were profitable sources of town revenue, Pennsylvania's legal system contained within it inherent difficulties that plagued borough government in Windber. There was no guarantee that justices of the peace or burgesses would report all cases arbitrated or all revenues collected. Throughout the years, the Windber council had periodic trouble in making key officials financially accountable to it. In an extreme instance, at a meeting on August 13, 1902, the council asked Burgess Herrick Thomas to resign. He had ignored previous requests, not properly accounted for fines, and was at least $400 in arrears to the council. In another instance, Justice M. E. Sell, who had been given previous warnings, was told by the council in 1906 to make his report on fines within thirty days or "suit was to be instituted." From time to time, tax collectors proved to be as difficult as burgesses and justices to oversee and hold accountable.[14]

The Limits of Nativism

Nativism of a limited sort on the local level served the interests of Berwind-White and the dominant American-born middle classes who were in large numbers allied with the company. Because Americans and northern Europeans occupied the highest positions in the company, town enterprises, and local government, the town's major ethnic/class line and the company's autocratic dominance were mutually reinforced by any discriminations that maintained the existing hierarchies and kept the masses of "Hunkies," "Polaks," and "spaghetti eaters" in their place. Although the company took great pains to divide its disparate immigrant nationalities from each other, its primary concern was always to maintain this larger ethnic and class division. In the broadest sense, national and racial prejudice could seriously inhibit the development of class-consciousness among workers, and as long as the new immigrant miners could be kept apart from American and English-speaking ones, the chances of successful union organization were slim. Nativist sentiments were thus a valuable tool in fending off unionization. But to take no chances, the company also arbitrarily isolated these two groups from each other by mandating company housing in two segregated residential enclaves that were rigidly separated by railroad tracks.[15]

From the outset, Berwind-White's policies toward its immigrant work force were caught in a maze of contradictions. At the same time that it tolerated or fostered local nativism to reinforce its dominance, it also sought to attract and retain new immigrant miners to work in its mines. In various ways, it thus tried to placate foreigners and eventually turned to a policy of promoting separate segregated ethnic enclaves and institutions. Its success in maintaining this contradictory policy resulted from ensuing segregation, whether voluntary or imposed, and ethnic stratification. While the town contained a plethora of fraternal societies and clubs, those with an exclusively native-born membership coexisted alongside those with an exclusively ethnic one. Rarely, if ever, did the twain meet. Except for public school events, where a captive audience included immigrant children, the Patriotic Order Sons of America spoke exclusively to its already converted, native-born constituency. Meanwhile, the numerous foreign societies conducted meetings and social affairs of their own in their respective native languages.

However sympathetic certain segments of the American middle classes were to nativist sentiments, the company's and the town's dependence on foreign-born labor set practical and de facto, if not formal, limits on the exercise of nativist practices in the Windber area. Thus, the American-born and foreign-born segregated societies occasionally participated together in a number of community enterprises. For instance, national holidays such as July Fourth were celebrated by massive parades through town. But in these instances the parade formations, determined by town leaders, also indicated the nature of Windber's stratified and segregated social structure. Parades typically led off with the Windber Fire Company, the POSA, or another bastion of the American-born population, while contingents of the many ethnic societies, often decked out in colorful uniforms, marched in formation behind.[16]

That Windber's large foreign population set some practical limits to nativism even as it was being discriminated against is clear from borough employment policies concerning public work projects. From the outset, Berwind-White's numerous companies and various allied contractors had performed many of these services. Initial contractors had included prominent foreigners, especially the Italian leader Frank Lowry, whose ethnic work gang had built many of the town's roads and sewers. The scarcity of American workers in the area had contributed to dependence on immigrant labor for much of the unskilled nonmining labor as well as mining labor.

On the state level, American workers, fearing immigrant competition for their jobs, had successfully lobbied to pass a law in the late 1890s to exclude aliens from public works. Several years later a similar movement arose in Windber. In 1905 fifty interested signatories petitioned the Windber council and asked that citizens, not foreigners, be employed in the borough's work. The council's response was to adopt a policy that seemed to satisfy borough needs and American discriminatory sentiments. The council did not formally exclude foreigners from borough work, but gave preference to citizens for street commission and committee work.[17]

Windber's large foreign population also imposed limitations on the temperance and sabbatarian movements, which were popular crusades in some American sectors during the period. Indeed, Somerset County had many organizations that espoused these causes. But the movements' identification with middle-class Americans and Protestant churches undoubtedly set reformers on a collision course with new immigrant industrial workers, whose religious and drinking customs were significantly different. To be sure, Windber had its own prohibitionists and sabbatarians, but they immediately confronted the realities presented by their coexistence with large numbers of immigrant miners and ethnic businessmen. Personal preferences aside, local reformers typically sought to regulate and limit the drink trade and Sunday business activity, rather than abolish them entirely.

Nativism quickly manifested itself in the battles to procure the limited number of lucrative liquor licenses in Windber. Licensing in Somerset and Cambria counties could not be taken for granted because elected judges were conscious of the growing middle-class sentiments for temperance and prohibition. Even American-born hotel keepers had to refute remonstrances drawn up by competitors and active prohibitionists. However, immigrant businessmen had to contend not only with these sorts of difficulties but also with nativism. On several occasions, American businessmen tried to restrict, if not entirely eliminate, the foreign-born from the competition for the licenses, and ethnic clubs and societies, which had traditionally served alcohol on special occasions without a license, were reported to the courts, which called ethnic leaders to account. In a noteworthy instance, in February 1903, the county court received a general remonstrance from Windber asking that the court refuse all foreign retail and wholesale applications from the town. The alleged reason for the retail discrimination was the propensity of the foreign population for lawlessness, which, the remonstrance claimed, resulted from drink and made life unpleasant for the "better class of people." The one local license denied that year was to a pair of new immigrants whose hotel went bankrupt shortly after they lost their liquor license.[18]

In later years, especially after nativists blamed the strike of 1906 on drink and lawless foreigners, foreigners continued to be the main reason cited in requests to deny liquor licenses. Thus, the *Somerset County Democrat* supported remonstrances against the granting of a wholesale liquor license in 1908 in Windber because "the petitions recite grievances which no industrial community dependent on large foreign labor should be allowed to suffer." On this one occasion, Berwind-White, through its superintendent Walter R. Calverley, who was unique among the company's top officials in that he personally favored prohibition, supported such remonstrances because the wholesaler's "miserable tactics" encouraged "this weak foreign element to drink."[19]

The issue of Sunday closing also arose periodically in Windber during the first decades of the century. Because miners typically worked six days a week, many businessmen, especially small businessmen who were barely surviving, preferred to keep their doors open on Sundays whenever possible. Moreover, Jewish merchants observed a different Sabbath and therefore closed their stores on a different schedule. Nevertheless, sabbatarian laws did exist, and occasionally a burgess attempted to enforce the closings. Such efforts usually produced strong resistance from interested merchants of all nationalities, and the typical result was a compromise. For example, in June 1905 fifty merchants met with Burgess S. H. Mills to dissuade him from rigidly enforcing the existing laws. Priests and doctors attended as well. According to a newspaper report, most people at the meeting opposed enforcement of the outdated so-called "blue laws" because they believed that such laws did not suit Windber. Some even cited Scripture to strengthen their case. Ultimately the burgess satisfied the merchants by extending the list of "necessities" that could legally be sold on Sundays.[20]

In future years this list was sometimes restricted and sometimes expanded.[21] Still, throughout the first decades of the century, desecration of the Sabbath was a fineable offense subject to vast differences in interpretation, and American law enforcement officials tended to use it to discipline foreigners, to enrich borough coffers, and to supplement their own income.

The YMCA Experiment

In theory, if not in practice, there were in Windber certain mediating influences that tried to bridge the gap between the town's American-born and foreign-born populations during the first decade of the century. Chief among such influences was the Young Men's Christian Association (YMCA), which was implanted in Windber in 1903 by the state association's newly founded bituminous department. This department originated as a by-product of the anthracite strike of 1902. Religious leaders and committees of leading coal operators had sponsored the expansion of YMCA activity into bituminous mining towns as a means of harmonizing social relations and allaying class conflict.[22]

A serious and prolonged effort was made to ensure the success of such organizations, but the fate of Windber's chapter paralleled that in many other mining towns, and despite an auspicious beginning it seems to have always had a precarious existence. It almost ceased to exist in 1906, when the strike affected every local business and organization, but it did enjoy a brief resurgence afterward, before its final demise and disappearance in 1909. Throughout the years, reports of dramatic periodic membership campaigns and finance drives, along with frequent reopenings, new starts, and revivals of the organization, suggest that its difficulties were inherent.[23]

At Windber's YMCA housewarming in December 1903, the Rev. C. E. Smith described the objectives of the YMCA and said, "This is an opportunity for practical Christianity to reach the lives of the people." He issued a cordial invitation for all nationalities, creeds, and classes to join the organization. Newspapers welcomed the association's existence, and the *Windber Era* commented that it considered the YMCA to be second only to the church in its efforts to better the religious, moral, and physical life of a community.[24]

Engineers approached General Manager Thomas Fisher, who persuaded Berwind-White to donate to the YMCA quarters the corporation vacated when it moved to a new and larger building. Before long, the local chapter was conducting regular Bible classes and devotional services, sponsoring lecture and entertainment series, and organizing a basketball team and league. It had also opened a reading room, set up a gym, and formed a women's auxiliary. But by 1905, references to unspecified YMCA problems were cited in area newspapers, and the local association requested a paid secretary to help in the efforts to solve them.[25]

On the eve of the 1906 strike, there were 225 nominal members and a much smaller number of active members in the Windber YMCA. Its rebuilding in 1907, after the hiatus in 1906, was delayed while the structure it occupied was remodeled. In 1908, in an effort to solve the association's indebtedness problems, Berwind-White donated $1,000; its subsidiary, the Alpha Construction Company, gave another $300 for building repairs. Meanwhile, for a short time, the local chapter ran an English class for foreigners. By 1909 the company's superintendent and other officials were actively serving on the financial and educational committees of the YMCA, but the association apparently did not survive the year. Similar YMCAs in other mining towns disappeared about the same time.[26]

A number of reasons help explain the eventual failure of Windber's YMCA. From the outset, officers and directors were exclusively prominent employees of the Berwind-White Coal Mining Company, American-born or English-speaking, and Protestant. Its appeal was to engineers, merchants, professionals, and the middle classes in general. Foreign-born miners, who were predominantly Roman or Greek Catholic, found other entertainments and sporting activities. Their own ethnic leaders and fraternal societies, not Americans, assisted them when they wanted to become citizens. Americans too had alternative organizations, including the fire department, the POSA, veterans' groups, political parties, church groups, fraternal organizations, and social clubs. In short, the prevailing pattern of ethnic and class segregation proved impossible to overcome, and in the end neither a cadre of dedicated volunteers nor a paid secretary could provide the stability needed to make the local prosper.

Although the attempt to implant a YMCA in Windber would not have occurred had Berwind-White been hostile to the idea, the company's relationship to the organization was somewhat ambiguous. Despite its expressed sympathy and the donation of its old quarters, the corporation did not provide significant aid in the initial effort to successfully establish the YMCA, and on the whole seemed surprisingly indifferent to its fate. Indeed, as the Dillingham report indicated, the company ignored welfare work and lagged behind many other corporations who were currently experimenting with work of a "humanizing" nature.[27]

The 1906 strike and a change in Windber's local management in 1907 did seem to spark a new interest by the company in the YMCA and its efforts to uplift, harmonize, and mediate ethnic and class relations in the community. But in the context of the aftermath of the bitter clash of 1906, Berwind-White's visible new level of support for the organization may even have doomed it. Immigrants had never patronized it, but it is also possible that independent progressive elements in the YMCA and in the community were hindered, stifled, and demoralized by the constraints within which the organization had to operate in an open-shop company town. During the strike of 1922, Windber citizens told a New York reporter that Berwind-White had destroyed the YMCA along with other progressive institutions in town.[28] The real question was whether any independent or semi-independent force or organization could thrive in such a situation, especially in a period that seemed to preclude unionization.

An Unexpected Alliance

Nativism and the autocratic setting of the company town, however, did not automatically silence all independent criticism of the company, even from the unexpected quarter of the American middle classes. In fact, during Windber's formative years a courageous crusading newspaper editor not only publicly exposed company policies, thereby shedding much light on how Berwind-White exercised authoritarian control, but also helped initiate a major UMWA organizing drive in the winter of 1903–1904. The unlikely alliance between progressive middle-class Americans and new immigrant labor thus transcended narrow class and ethnic boundaries.

At first glance, Will F. Hendrickson was an unlikely friend of immigrant miners and the United Mine Workers. The young American-born progressive newspaper editor who arrived in the spring of 1903 to join the staff of the *Windber Journal*, one of the town's two newspapers, was no radical. Nor was the paper he joined known for taking critical stances. In fact, the first seven months of the journalist's tenure in town were uneventful. However, this seeming quiescence ended abruptly in November 1903, when he and his newspaper broke publicly with Berwind-White in a pathbreaking article, "Time to Call a Halt." Angered by long-term and recent events, Hendrickson came out actively in support of the miners' right to organize.[29] He argued that "organization is the only alternative" for the miners to free themselves from conditions resembling slavery and for the town to free itself from company dominance. For the next five months the *Windber Journal* was a vigorous local pro-union organ that courageously exposed company abuses and lent what support it could to active UMWA organizing activities there.

What awakened the editor's conscience and emboldened him to act was Berwind-White's arbitrary dismissal of ten miners. On Saturday, October 24, 1903, Ben Stevens, a popular 22-year-old American spragger, had been killed on the job in Mine 35. Shortly after the fatality, ten motormen and spraggers ignored the mine bosses' objection and left work a half day early in order to call on the young man's parents as a delegation to inform them of their son's death. On the following Monday, when they returned to work, they found that they had been suspended indefinitely.[30]

Hendrickson was outraged. While he acknowledged that the company had a right to set reasonable rules for work, he believed it had no right to use oppressive tactics. Moreover, he claimed that there were many precedents for miners to leave work early. Showing rare insight into the company's contradictory divide-and-conquer ethnic policies, he claimed that large numbers of foreigners were frequently allowed to leave work early to attend fraternal society events or celebrate their national holidays. In his view, the company habitually sought to "nurse" the foreigners, even as it denied Americans basic rights. In the Stevens case, the company had decided to make examples of independent-minded young men and discipline them with suspension or dismissal. When other miners subsequently objected to the treatment of the ten, they too were suspended, in what the editor termed a "punishment for exercising the privilege of free speech."[31]

The immediate issues were rather quickly resolved. As *Journal* pressure mounted, the company gradually took back the suspended men a few at a time. In fact, it temporarily closed the mines for part of the day on the day of Stevens's funeral so miners could attend it en masse. Hendrickson claimed that the company would have faced outright revolt had it not done otherwise. But by then the die had been cast, and the resolution of the immediate grievance did not prevent other grievances from rising rapidly to the fore.

It was in this context that company representatives visited the *Journal* to offer what Hendrickson considered lame explanations for oppressive policies. He recalled that, during the previous spring, the company had tried to prevent the newspaper's publication of the Altoona wage scale but that the *Journal* had not been intimidated. Its publication of the union mining scale at that time had made it clear that Berwind-White's working conditions and wages were not up to district standards. Nonetheless, Berwind-White continued to insist that its mines worked on the union scale. If so, retorted Hendrickson, the company should have no hesitancy about recognizing the UMWA. All parties understood that the real battle was now being waged over the issue of labor's right to organize. The *Journal* took the position that miners could not get their rights within the existing system, that justice could be attained only through organization, and that right should replace might.[32]

In the next few months, grievances discussed in the columns of the newspaper included the company's failure to live up to the Altoona agreement, the lack of checkweighmen in the local mines, illegal deductions from wages for taxation, the general lack of free speech, specific oppressive policies such as compelling miners to vote according to company wishes for fear of their jobs, the collusion of the Windber council and the company to keep out independent businesses, and the evils of the company-store system. This impressive range of issues indicates that the miners' fight for unionization was not—and could never be—narrowly confined to the workplace. Nor were the issues applicable only to miners or to new immigrants. Indeed, the miners' cause was inevitably linked to other fundamental rights and to other interests in the community.

The ensuing UMWA-*Journal* campaign was fought in the context of extraordinary threats that went beyond the usual practice of discharging union-minded miners. When company officials publicly stated that they would close Windber's mines if the miners organized, Hendrickson called the threat "silly twaddle" and pointed out that profits mattered more to capitalists than mere sentiment. Berwind-White's purchase of eleven special riot guns, another intimidation, prompted the editor to urge the miners to seek an injunction in court to eliminate them. No miners' organization could, he claimed, have taken a similar action without the company's seeking an injunction. Moreover, he argued that the company's resorting to armed intimidation revealed its relative weakness and lack of popular support within the larger community. Meanwhile, however, spotters and spies continued to harass miners and union organizers who could not secure a hall for union meetings. The company's men sat on the boards of foreign societies, and board approval was necessary for hall use. Berwind-White owned or rented most other suitable places, and according to Hendrickson, paid off the owners of any other possible alternatives.[33]

Hendrickson's courageous stand came as a pleasant surprise that prompted the UMWA to renew its organizing efforts in Windber. The *United Mine Workers Journal* (*UMWJ*) welcomed the *Windber Journal* as an unexpected ally in the struggle, but from the start the miners' organ placed primary responsibility for the drive on the miners themselves, not on Hendrickson or others. It urged the miners to assert their rights like men, to correct wrongs, and to change the slavery that existed in the so-called model town. At the same time, it asked miners to support the *Windber Journal*, whose very existence was at stake as a result of its contention with a company that was considered a mere subterfuge for the powerful Pennsylvania Railroad.[34]

Soon, the *Journal* and the *UMWJ* were receiving letters from former Windber residents who supported many of the charges Hendrickson had made. For several months, *Windber Journal* columns on the local situation were reproduced verbatim in the *UMWJ*. Meanwhile, the Windber paper and editor won the admiration of many other regional newspapers, who regardless of their views on organized labor understood the contest as one between David and Goliath. By late December, according to Hendrickson, eleven organizers were active in Windber, and the *Coal Trade Journal*, an operators' paper, reported that a serious effort was being made by district officers to organize Berwind-White mines there.[35]

Local support existed for Hendrickson's efforts even before outside organizers and interpreters arrived on the Windber scene. Immigrant miners rapidly became aware of the struggle of an English-language newspaper on their behalf, and one unusual piece of evidence indirectly illustrates popular support for the *Journal* in foreign quarters at the height of its attacks on Berwind-White. The Slovak societies held a masquerade ball on Thanksgiving Day Eve 1903. The Slovak band played for the event, and during the night, prizes were awarded for the best costumes. First place was captured by a woman who dressed as a facsimile of the *Windber Journal*. Given the battle currently being waged, the costume itself was a statement, as was the awarding of the prize. Nor was Hendrickson immediately ostracized by all Windber businessmen and professionals. During his tenure in town he had been active in various organizations, and during the course of the unionization campaign, in December 1903, he was elected vice-regent of the Royal Arcanum fraternal society.[36]

In sharp contrast to the *Journal*, the *Windber Era* always adamantly supported the company and for decades of its subsequent history never reported anything negative about it. Thus, it printed nothing about the UMWA-Hendrickson unionization drive except for one comment in January 1904, that could be considered a hostile allusion to it: "Conditions in Windber are all that could be desired—mines working full time, employees earning good wages—labor and capital being at peace."[37]

The joint UMWA-*Journal* campaign in late 1903 and early 1904 did not succeed in the immediate goal of unionizing Windber-area mines, but it did serve important educational purposes that paved the way for action in 1906. Contrary to the satisfaction and smugness reported by the *Era*, miners had been made more aware that Windber's position was an anomaly in the district, where unionized mines benefited from eight-hour days, checkweighmen, and a different wage scale. The campaign also brought out into the open many aspects of the company's control of a company town, thereby revealing the severe limitations it could impose on personal independence and basic rights. The *Journal*'s coverage of events outside Windber broadened the horizons of area miners and educated its readers. Moreover, unlike the *Era* at this time, it regularly carried news of naturalization proceedings and printed the text of important laws to inform Windber's immigrant populations about matters that directly affected them.[38]

From the moment of its initial attack on the company, however, the *Journal* was engaged in a life-and-death struggle for its own survival. Even after the dramatic organizing campaign ended in Spring 1904, it continued to serve as a critic of the company and as a vehicle for progressive causes. Sometime during the strike of 1906 it succumbed and went out of business permanently. Hendrickson himself had left the paper a year earlier in March 1905, immediately after his departure from town to attend the inauguration of President Theodore Roosevelt.[39]

After the demise of the *Journal*, no independent-minded newspaper was ever published in Windber again.

Windber miners had found an unexpected ally, although the alliance forged between Hendrickson and the UMWA was basically a conservative one. The editor consciously identified himself with Mark Hanna's variety of progressivism. He believed not only that workers had an inherent right to organize, but also that the organization of labor was a necessity of the times, given the existing organization of capital. An advocate of arbitration, he did not endorse strikes—except as a last resort, when grievances could not otherwise be adjusted—and he deplored violence of any kind.[40] For him, the UMWA was an example of a responsible union under intelligent leadership. While he quoted the pro-union statements of various individuals and cited Pope Leo XIII's encyclical on labor during the course of the campaign, his sympathies were closest to Hanna's. In fact, he reproduced in the *Journal* the senator's "Socialism and the Labor Unions," in which the author described the nonradical motivations behind his support for organized labor. Hanna had written: "My plan is to have organized union labor Americanized in the best sense and thoroughly educated to an understanding of its responsibilities, and in this way to make it the ally of the capitalist, rather than a foe with which to grapple."[41] In short, a conservative-oriented labor movement would end class conflict and defeat the menace of socialism.

Many UMWA leaders espoused philosophies similar to Hanna's and Hendrickson's. While the national union did contain socialists and radicals within it, its president participated in the National Civic Federation and adhered to a pragmatic, business union philosophy of trade-unionism. Out of respect for Senator Hanna's support of organized labor, the national UMWA even asked organized miners to pay him a final tribute and refrain from work on the day of his funeral, February 19, 1904.[42] However, Berwind-White was no more willing to accept business unionism and progressive reforms than it was to accept their radical counterparts.

The Weakness of Independent Business

Business people in a coal-mining company town occupied a peculiar intermediary position that in times of conflict between labor and capital sometimes led them to support miners' organizations, as happened in many Pennsylvania towns during the anthracite strike of 1902. In other instances, merchants threw their lot in with the coal companies or supported affiliated antilabor groups, such as Citizens' Alliances. Evidence suggests that business people in Windber were especially divided over the issue of unionization.

In fact, Windber merchants were not a homogeneous lot. In the mid-1900s and afterward, more than a hundred town businesses were listed in the county's mercantile appraisement lists, and the merchants represented many different nationalities. Thus, the dependence on one basic local industry, along with its company stores, did not preclude the existence of a multitude of smaller stores, ethnic specialty shops, or larger businesses, whose activities were linked to company patronage. But whatever the size and nature of the enterprise, Berwind-White could not be ignored. Miners and others interviewed have suggested that some businesses survived because the company ignored tiny, noncompetitive businesses whose hours often extended beyond company-store hours, and that if the company had chosen to do so it could easily have put such places out of business. Other businesses thrived on company patronage or catered to the nonmining middle classes, who unlike miners were free to shop where they wished.[43]

That some businesses were unhappy with the company's monopoly is reflected in early efforts of merchants to form an independent business association in town. After the failure of the short-lived Windber Protective Association, progressive business interests waged a new campaign to form a Board of Trade in March 1903. The *Journal* recommended that one of the organization's main objects should be to secure new industry because "that community which has the greatest diversity of industries is the most independent and also the most prosperous," and it argued that labor disturbances were more harmful to businesses in general when one industry predominated.[44] By June 1903, Hendrickson was citing the fact that coal would someday be depleted and that providing alternative manufacturing industries would be advantageous in the future as well as a healthy community addition in the present.[45]

But genuine industrial diversity and competition had been precluded at the time of the town's founding when Berwind-White's monopolistic controls were first established. The original pattern set by interlocking directorates, borough council collusion with the company in matters of business restrictions, and the awarding of lucrative borough contracts to company enterprises or favorites continued in subsequent years. Miners interviewed often cited the company's success in keeping other industries out of the town. Even when a rare industry that did not compete with Berwind-White's enterprises succeeded in establishing its presence in town, it often faced numerous impediments and acts of harassment. The minutes of Windber council meetings indicate that such was the case for the Windber Brewery Company. Berwind-White had written the council a letter explicitly asking it not to grant the brewery approval of siding rights until the company could make its own arrangement with the concern. Brewery development was systematically stalled for months, and once siding was finally approved the new enterprise faced other obstacles and discriminations, such as a sewer-tapping fee that, it claimed, was unusual and exorbitant.[46]

The Windber Council continued to award lucrative borough contracts to Berwind-White or its affiliates. Outside competitive bidding did not exist. In June 1904, for instance, the council granted the Alpha Construction Company, a Berwind subsidiary, the borough's profitable heating contract, even though the company had refused repeated requests to make even minor concessions to the town in return for the franchise. In another noteworthy incident in 1912, the council revealed its abject subservience to the company. The council had advertised for bids for a borough contract to provide electricity for the town. Competition was nonexistent, and the sole bidder, the Windber Electric Company, a Berwind subsidiary, had included in its bid conditions and terms markedly different from those the council had advertised and presumably preferred. Because the terms differed, the council rejected the initial bid. However, it then acted immediately to amend the terms of the proffered contract so that they would match those presented by the Berwind company. At a subsequent meeting, the Windber Electric Company, sole bidder on the new contract terms, was formally granted the valued franchise.[47]

Important outside capital was also systematically discouraged from coming into the community through other means. For example, company officials, who directed the new Windber Building and Loan Association, kept control of the concern by prohibiting outside investment and limiting the ownership of shares to local, wealthy prominent persons who could be trusted to follow company wishes.[48]

Business people who wanted a diversification of industry and genuine competitive opportunities were therefore in a weak position in a company-town setting. Business was not monolithic, and Berwind-White had its supporters. In fact, the only point of agreement that seems to have united the many small mercantile enterprises was the request, cited in council minutes, that the burgess refuse to license itinerant peddlers and that he take steps to keep such people out of town.[49] This action merely contributed to the closed nature of business competition and reinforced the company's dominance. Clearly, if independents had any possibility of achieving greater diversity and independence, they needed allies. Under the circumstances, their fates were integrally linked to that of the miners and the success of organized labor.

The Tyranny of the Company Store

The issue that subsequently brought together miners who were trying to organize and progressive businessmen who wanted greater independence from the company's control over business and the town was the tyranny of the company-store system. During the Hendrickson-UMWA organizing drive in the winter of 1903–1904, a newly founded independent business association made a futile attempt to overcome the company store's stranglehold over the miners and thereby increase competition. Its efforts reveal the weak position independent business enjoyed in the Windber setting, and its failure paved the way for the demise of the association. In the future, if not before, progressive businessmen would understand that competition, business diversity, and town independence could not be achieved without the unionization of Windber miners.

To miners and their families, perhaps nothing symbolized the oppressive nature of a company town better than the company-store system that prevailed in Windber. "You had to shop at the company store" was a theme frequently recited in interviews because, company propaganda to the contrary, miners were not free to shop where they wanted to. The issue of the company store was a perennial grievance persisting into the 1930s or beyond, when automobiles, technology, and unionization changed the context in which the company store existed. Miners who initially came to Windber early in the century, in part because of the credit advanced by Berwind-White and the Eureka Supply Company, often later rued the existence of the "pluck-me" stores. Many business people did also.

Windber-area miners' complaints about the company-store system were similar to complaints wherever this type of monopoly existed. Miners and their families especially resented the compulsion inherent in the system. Although they were wage earners and consumers in a capitalist society, they were not free to shop where they pleased. Compelled to patronize the company store out of fear for their jobs, and deprived of the full value of their cash earnings through a compulsory company-mandated credit system, they bought goods for higher prices than could be obtained in comparable, competitive stores. In trying to free themselves from the grip of the dreaded system, Windber miners joined a long historical procession of workers who struggled to destroy such compulsory exploitative institutions.

Wherever company stores were established, protests had followed in their wake, especially in Pennsylvania, where many existed. A number of state laws had been passed in an attempt to restrict or eliminate them, but such legislation was declared unconstitutional, successfully evaded, or just not enforced. Pennsylvania employers avoided an 1891 law that prohibited mine owners from operating company stores by incorporating their stores as separate enterprises and linking them to the parent company via interlocking directorates. Hence, Windber had both the Berwind-White Coal Mining Company and the Eureka Supply Company, Ltd. Thus, in testimony before the Industrial Relations Commission, E. J. Berwind could technically deny that he owned any company stores, even though his brothers and his coal company's managers constituted the affiliated store's board of directors.[50]

Company stores served many useful purposes for employers. Company stores that could supply credit gave their owners an advantage over other employers in recruiting prospective miners to places such as Windber. Employers and coal operator journals also recognized that the same stores and credit mechanisms helped to maintain a dependent, stable labor supply, especially when there were lulls in work. The Dillingham Commission had reported that immigrants and miners in Windber and other places were migratory creatures who were not loyal to any particular town or company. From the outset, Berwind-White had encouraged and allowed gardens, poultry, pigs, and cows in the borough limits because these were considered necessary to recruit and to keep immigrant miners who came from rural backgrounds. The corporation lost nothing in the exchange. Since the Eureka Supply Company was a comprehensive department store that sold suits, shoes, furniture, and virtually all necessities, dollars not spent on food were usually spent on other goods. Moreover, Ole Johnson has pointed out in his study of industrial stores that company stores of this early era took much less risk in granting credit than did independent stores, which had no control over jobs and wages. Operators themselves have suggested that company stores were a certain source of revenue and sometimes more profitable for companies than the actual mining of coal, which faced a competitive market.[51]

The Dillingham report substantiated the miners' case against the Eureka Supply Company. It noted the interlocking managerial interests of Berwind-White and the store, and the fact that employees hired by Berwind-White were required to sign an agreement that allowed the coal company to deduct Eureka store purchases from biweekly wage statements. In effect, miners had no choice but to trade there on credit advanced for the current two-week pay period. While the commission stated that dealing with the company store was not "absolutely compulsory," it nevertheless concluded that "a good customer of the store is less likely to be discharged should occasion arise than is one who deals at other places." Moreover, in outlying settlements, the only stores that existed were company stores.[52]

At the same time, the commission found that people who were not employed by Berwind-White were regularly charged a lesser amount for the same items purchased by miners and their families. When asked about the matter, the company attempted to justify the discrimination by claiming it had to make up for credit expended in the miners' behalf.[53]

Of course, it was the company, not the miners, who required the use of credit by linking jobs, passbooks, and deductions. Nor were the profits from the stores inconsequential—in fact, they were the most important source of revenue for the company besides marketed coal. The commission had information on the number of dollars deducted by the company for all categories of services in the month of August 1908. In all, more than $27,000 had been taken out of employees' wages for rent, blacksmithing, the hospital, store, and miscellaneous items. Of these, deductions for the store headed the list. Over $19,000 was given back to the company indirectly via the store deduction.[54]

Company stores, or successful store managers, ran profitable businesses, not altruistic or charitable enterprises. Company officials always denied that compulsion of any sort was involved in the miners' patronage of the Eureka stores, and claimed instead that miners were grateful for the institution. Evidence to the contrary is overwhelming. Of course, some miners and their families occasionally overused credit, and during hard times many workers were glad to be able to get credit for food from any source for their families. But generosity was minimal, when it existed at all.

One manager interviewed told of how he had always maintained an eye on profit margins and control over the apparatus of credit, while miners reported that when work was slack, credit was extended to them only on a very limited basis. Managers contacted foremen to see how many hours or days individual miners were working, and an amount of credit in accordance with that work schedule—and no more—was granted. In later years, Berwind-White and other corporations were prohibited by state law from taking all a person's wages in deductions. Consequently, coal companies had to leave an individual with at least $2.00 in cash after deductions. However, credit required or extended could also be withheld. Consequently, store clerks collected passbooks during the strikes of 1906 and 1922 because the company did not intend to extend credit to strikers during a suspension. Managers also knew that they had to show profits or risk being replaced, and one manager explained that he had been very concerned, presumably for his career, during the strike of 1922 because his store had lost thousands and thousands of dollars.[55]

Miners bitterly resented the compulsion inherent in the company-store system, and they and their families protested against the system in a variety of ways. Some tried to cheat the store by forging figures on their statements and increasing the amount of credit due them. A few resorted to legal challenges. In September 1904, Thomas Joyce, who had been employed by Berwind-White for the previous five years, sued the company in a Cambria County court after he had left the Windber area. Claiming that he had never signed a permission form for store deductions, he sued for the cost of such deductions. Although the court granted that he had never signed the pertinent form, the deductions had been going on for five years, which, in its judgment, amounted to an implied contract. In any event, higher courts had ruled that such deductions were legal. Another form of protest—simply to leave the area and company—was practical mainly for single men.[56]

More commonly, Windber-area miners and their families protested the system by trying to shop elsewhere secretly. Many people interviewed told of how they had tried to do so. They also explained that it became virtually impossible to succeed in these endeavors, especially in outlying settlements, where only one road led into and out of the camp. In all such places, the Eureka store was situated at the entrance to the settlement, and spotters in the company's employ watched there for residents who returned with purchases made elsewhere. Train conductors even reported miners and their families who returned from Johnstown with purchases or who were seen walking with packages on railroad tracks. Even if a person succeeded in getting home with goods, he had to hope that neighbors or others would not report him.[57]

One couple told what happened in Mines 30, 35, 36, 37, and 40 when the company spotters noted an independent purchase. If someone had purchased a pig to roast and some flour, for instance, that person received identical items—a pig and flour—the next day from the company's delivery wagon. These items had already been charged to the family's deductible purchases for the next pay period. That such incidents were commonplace was confirmed by an independent grocer who reported that his delivery wagon was regularly followed into these settlements by a company store vehicle whose driver took note of purchases and purchasers until miners there stopped buying from the independent.[58]

Other forms of compulsion were involved in enforcing company-store patronage. Miners interviewed reported that foremen had told them at work that they were not buying enough at the company store and that they had better increase purchases if they expected to keep their jobs. Many people explained that farmers and other peddlers were prohibited from selling their produce in outlying settlements and borough limits. Borough ordinances allowed for licensing procedures but required burgess approval, and council minutes indicate that licensing was routinely refused if not prohibited. One woman who lived on a company-owned farm on the outskirts of town in the late 1920s even related that she had been visited by an important store official because she had bought what he considered an excessive number of canning jars. He had come to investigate and make certain that she was not selling food, thereby competing with the store.[59]

Miners and their families were not the only victims of the company-store system. Business people, potential competitors, and other consumers were also hurt by it. While the company enforced patronage of its enterprises primarily through the required credit and deductions, along with threats and intimidations routinely imposed on the miners, its stores also gained preeminence through a restriction of competition. Windber was never a geographically isolated community. It was situated not far from Johnstown and linked by railroad to the larger city, and from there to the main Pennsylvania line and major cities. Thus, Eureka's dominance in the community cannot be explained by the town's remoteness from civilization or by the absence of potential competitors. Rather, its successes resulted from its monopoly status, which derived from artificial constraints placed on miners and on competition.

Independent business people of various nationalities in Windber felt compelled to organize against the company store-system in the midst of the UMWA-Hendrickson unionization campaign. In February 1904, as part of an effort to reestablish absolute control over the miners, Berwind-White had launched an offensive against mining families who were not shopping at the Eureka Stores. The company's new offensive not only hampered the freedom of its workers but also threatened the livelihoods of local businessmen.

In February 1904, George Smith, manager of the company-store system, accompanied drivers on their rounds to the many boardinghouses in the community. He queried boarding "bosses," most often ordinary miners themselves, about their purchasing habits and then informed them that they had to buy all their food and goods from the Eureka Supply Company. If they did not, he told them, Berwind-White would discharge all their boarders from their jobs. Angry boardinghouse heads then told independent grocers and other business people about the new company decree.[60]

Hendrickson responded to the company's intimidating initiative by writing an article, "Company Store the Root of Great Evil," in which he printed the form newly employed miners were required to sign to authorize Eureka store deductions. He noted how unfair to other businesses this authorization was, printed existing laws on company stores, and charged that the company was not complying with the Store Order Law of 1901 then in force, which required coal companies to pay taxes on all such store deductions. In his view, the new order simply meant that the company's tyranny was being extended from miners to merchants.[61]

For their part, grocers responded to the new company order by meeting together and forming an independent grocers' association. Founding members invited other merchants to join the organization and hired two lawyers to represent their interests. Merchants and attorneys, perhaps naively, soon stated that they did not believe the Berwinds knew what was happening in Windber. Consequently, the grocers instructed their lawyers to visit company officials in Philadelphia to inform them of Windber events, but not so naively told the attorneys to proceed to appropriate state authorities if no satisfaction were forthcoming. Meanwhile, the renamed business organization, the Merchants' Association of Windber, continued to meet even as the company denigrated the movement and claimed that jealous merchants were meeting for reasons other than the compulsory store order system.[62]

The climax of the independent merchants' short-lived movement came in May 1904, after legal recourse had apparently been exhausted. The organization placed an important ad in the *Journal*. This ad contained a letter the businessmen had received from Thomas Fisher, general superintendent of Berwind-White. In what was an obvious untruth, Fisher claimed that there were no links of any sort between the managements of Berwind-White and Eureka.[63]

In placing this ad, the independent merchants played what amounted to their last card in their struggle against the company-store system. They had quoted Fisher's letter verbatim in a futile attempt to convince the miners that they had the right to shop wherever they pleased. Miners knew otherwise. That small independents had to resort to an authoritative company source to demand their right to a share of the miners' patronage showed how little power they held on their own. The fundamental weakness of their position in the community was revealed, and without a successful labor movement they could not hope to succeed.

Afterward, the businessmen's association continued in existence, but officers changed frequently, and before long, according to the *Windber Era*, its attention had shifted to early closings and to asking for Berwind-White's help in getting rid of credit deadbeats. In the ultimate irony, George Smith, Eureka Store manager, then assumed the chairmanship of the businessmen's group.[64]

Working people necessarily struggled on many fronts in the Windber area in the early century. Nativism permeated the environment, and nativism of a limited sort reinforced company dominance and the town's existing major class/ethnic hierarchy. The miners' union proved to be the only viable institution actively confronting nativism in the region by integrating American-born and foreign-born miners into an independent class-based organization that simultaneously respected immigrant cultures and challenged authoritarian company rule. To be sure, segments of the intermediary classes, including progressive reformers, journalists, and independent businessmen of all nationalities, had related interests and concomitant struggles of their own. But efforts to establish a YMCA that bridged the classes, a newspaper that was not subservient to Berwind-White, and a progressive business association that promoted industrial diversity and greater competition ultimately failed. The weak position of these classes in a company-town setting made even minimal changes in the status quo impossible for them to achieve on their own. The abolition of the company's monopoly and the enactment of reforms were therefore contingent on the success of the organized labor movement.

At the same time, the failure to open up the town, to diversify industry, and to break the company's monopoly set the conditions in which the miners' ongoing struggles for unionization would take place. Organizing would have been entirely different in a context in which competition and constitutional rights had prevailed. Miners would have had more room to maneuver; sympathetic merchants and politicians would have enjoyed more independence as well as more profits; and a company prone to oppressive behavior would have been less able to trample, routinely, on basic rights. As the events of 1903 and 1904 indicated, the autocratic setting of Windber ensured that organizing would remain difficult. Union miners of diverse nationalities would indeed have some friends among other classes, but ultimately it was they themselves who would have to lead the struggle to establish democratic rights in the workplace and the town.

8

The Strike of 1906

Shortly after daybreak on Monday, April 2, 1906, a thousand union miners from neighboring places marched into Windber to demand that Windber join them in a district-wide strike and stay out until a contract and a new wage scale had been signed. Immediately 3,000 of 5,000 local miners joined the group, and together they paraded through the main streets of town. Led by national UMWA organizer Joseph Genter, the gala turnout, comprised of a multitude of diverse nationalities, surprised friends and foes of organized labor alike. Windber miners vowed that they would stop work the next day, and before long, true to their word, the mines were essentially idle.[1]

In joining the strike in 1906, Windber's miners were responding to a peculiar conjuncture of events, grievances, and UMWA activities. Berwind-White's autocratic control had left a long trail of perennial grievances, and previous organizing campaigns had made their mark, but it was recent events and assertive actions by the district and national union organizations that convinced the men that there was a reasonable chance for local success at this particular time. The timely intervention by District 2's unionized miners on April 2 moved the men to act because it showed them in a concrete way that they would not be isolated in their effort to organize against their powerful employer, that they would have necessary outside help, and that new immigrants as well as English-speaking miners were welcome in the UMWA. Having weighed all the circumstances, pragmatic foreign-born and American-born miners concluded that a favorable time had come to try to gain greater freedom over their lives by becoming part of the mainstream American labor movement.

Meanwhile, shock and disbelief reverberated throughout the region. Coal companies, certain labor officials, and others had long proffered the stereotypical view that new immigrants were docile people willing to endure any degree of exploitation for the sake of a steady job. Largely because of Berwind-White's own self-serving propaganda, its mines were erroneously considered to be a nonunion stronghold immune to unionization and labor strife. Thus, the parade, the walkout, and the subsequent bitter strike shattered existing stereotypes and revealed the falseness of the company's claims that its largely new-immigrant work force was a satisfied lot and not interested in the labor movement. Even district and national UMWA officers were momentarily taken aback by the unexpected walkout of thousands of nonunion and mostly foreign-born miners. The masses of newly organized workers presented problems as well as opportunities and forced additional, heavy responsibilities on them and on the sparse resources of their treasuries.

The bulk of Windber miners stayed out in solidarity with the region's union miners until July, when a controversial district settlement was reached. Meanwhile, they endured company violence, nativist attacks, evictions, and armed suppression by the state police. In the end, through no fault of their own, they were unable to secure the right to organize—their central aim—or any of the other gains they had sought. But the strike was not without important consequences. Not only was it one of the most significant happenings in the town's history, but by looking at the landmark strike we can see what possibilities and obstacles new immigrant miners often faced in industrial America when they tried to secure justice and democratize their workplaces and communities.

Background

Windber miners were joining thousands of bituminous miners who went on strike throughout the nation on April 1, 1906. Challenging ethnic stereotypes of new immigrant docility, they were attempting to become part of a broad-based national and district-wide labor movement that was on the move. Rank-and-file union miners were restless in 1906. Months of negotiations between operators and miners had failed to produce an agreement through the interstate joint conference system, which had been reestablished in 1897 and 1898 to put labor-capital relations on a rational basis after years of strife. In 1904, national UMWA officers and the miners' scale committee had accepted employer arguments about depressed economic conditions and nonunion competition from West Virginia and agreed to a two-year contract that contained a general wage reduction of 5.55 percent. At the time, there was considerable opposition to the decision from within the union. Once the depression ended, the masses of miners made it known to UMWA leaders that they would not accept new wage reductions or a continuation of the then-current wage scale in the new 1906 contract. They demanded an advance.[2]

Meanwhile, coal operators continued to insist that, despite the end of the depression, they could not pay higher wages, even a restoration of the higher 1903 scale. In many regions, the two-year wage reduction had strengthened the hand of large operators, so that they saw the 1906 negotiations not only as a means of adjusting wage differences but also as a golden opportunity to destroy the miners' organization.[3]

The stage was therefore set in 1906 for a test of strength between organized labor and organized operators. Both sides anticipated a lengthy work stoppage. Companies formed new combinations and stockpiled coal. Within District 2 itself, coal operators founded the Bituminous Coal Operators' Association of Central Pennsylvania to coordinate strike policy for the disparate mining concerns in the region.

The new association's chief goal was to win the open shop. On March 17, 1906, George E. Scott, the organization's secretary, sent a letter on behalf of the executive committee to ask that all its members sign a written guarantee that they would not sign a wage-scale agreement with the union unless the agreement also included the open shop. The open shop was the one principle that all association members were honor bound to demand, and that was the one demand that union miners were honor bound to resist to the utmost to preserve their organization.[4] An informant wrote John Mitchell on April 2, 1906, that the Bituminous Association "had figured this year as a favorable chance to open the warfare" and had already contracted with the Baldwin Detective Agency to have agents everywhere in the district.[5] Once again, the issue was not so much which wage scale would prevail but whether the union would be able to survive.

Scott quickly invited the Berwinds to join the new Bituminous Association and was greatly disappointed when they refused. He was even more upset when Berwind-White acted without consulting him and when company officials subsequently blamed the ongoing organizing drive in Windber on the association. Scott had expected the Berwinds to be sympathetic to the operators' combination and the open-shop drive as a matter of right.[6] But the truth was that Berwind-White did not need the operators' association. Unlike many of its members, the New York–Philadelphia-based corporation already had the open shop and was therefore more concerned about maintaining the status quo than changing it. As a major national corporation with important interlocking interests, including the Pennsylvania Railroad and J. P. Morgan, it also had plenty of resources and allies with which to fight the labor movement.

Moreover, the company traditionally placed its own corporate interests above capitalist-class solidarity, and bituminous mining was a fiercely competitive, if not cutthroat, business. Interstate Commerce Commission hearings held in 1906 established that the Pennsylvania Railroad had systematically discriminated in various ways in favor of Berwind-White and against its competitors. One of the world's leading coal producers saw no advantage in cooperating with other coal companies in a movement that, if successful, would have allowed competitors to cut costs and be better competitors, and it continued to jealously guard its own coal supplies, cars, markets, and profits against competition, even after it faced a serious threat from below.[7]

As coal operators were making arrangements for a probable work stoppage, the UMWA tried to build up its strike fund for an anticipated lengthy and costly struggle. Rank-and-file miners were initially encouraged by the fact that, unlike 1902, contracts in both the anthracite and the bituminous industries were due to expire on the same date, March 31, 1906. Although a general strike by miners in both fields was unlikely, it was theoretically possible. Indeed, precisely what strategy the national union should follow to win a desirable strike settlement quickly became a hotly contested issue within the union. The strategy pursued by the union would ultimately have a great impact on the local strike in Windber.

On February 1, 1906, miners at the UMWA national convention voted to reject the soft coal operators' proposal to renew the 1904 scale and demanded that no contract be signed in any bituminous district until all bituminous districts had negotiated settlements. In other words, union members endorsed a strike-settlement policy that was national in scope, not a piecemeal one in which individual districts were left free to sign separate or partial agreements. Proponents of the national approach argued that a united front of bituminous miners was necessary to bring about the greatest success in this field, especially among recalcitrant employers.[8]

Within a month, however, a conservative-oriented John Mitchell was working to reverse the convention's endorsed policy for the bituminous field. On February 24, 1906, President Theodore Roosevelt had intervened in the dispute and urged him and Francis L. Robbins, president of Pittsburgh Coal Company, both members of the National Civic Federation, to settle. The threat of federal government arbitration and the willingness of a portion of the operators to settle seem to have convinced Mitchell that a piecemeal, district-by-district approach to negotiations was preferable to a national one. At a special miners' convention that convened on March 30 to coordinate strike policy, Mitchell prevailed against opponents who wanted to adhere to the original strategy of pursuing a national settlement. With the successful reversal of policy, individual district organizations were authorized to pursue separate negotiations and contracts with the disparate operators in their regions.[9]

The overall wisdom of Mitchell's policy change is debatable. What is certain, however, is that newly organized bituminous miners could no longer realistically hope for a united national front against recalcitrant employers. While it would be one thing for the UMWA to deal with operators such as Robbins, who had made it known that he would settle for the 1903 scale if a disruption could be avoided and a separate agreement were reached, it would be quite another matter to deal with Berwind-White and other open-shop employers who were virulently opposed to unionization and collective bargaining in any form.

Windber-area miners were not quiescent while these strike preparations and negotiations were taking place. In March it became clear that union-minded people were actively working underground to organize on the local level. While Windber miners had many long-lasting grievances, the lack of checkweighmen and Berwind-White's weighing system emerged as crucial sources of discontent in 1906. A miner's pay was based on the number of tons of coal he loaded, but miners felt shortchanged and cheated. Loaded coal cars were routinely passed, unweighed, over scales, while company weighmasters wrote down arbitrary figures to credit individual miners for the amount of tonnage they had loaded. In unionized mines, by contrast, it was standard practice to place checkweighmen elected by the miners themselves at each tipple to ensure a company's honesty and fairness. Rank-and-file anger had begun to coalesce around the issue of honest weights, and support for the union around a Slovak miner named Paul Bills. A stream of Slovaks who had been active in the UMWA's struggles in the anthracite region had emigrated from there to the Windber area in the preceding years. With them, they carried the union message.[10]

Previous organizing efforts had also informed Windber miners of the existence of the eight-hour day, pay for deadwork, and other conditions in vogue at unionized mines. They themselves worked ten hours or more a day and received no remuneration for deadwork. Once the strike began, miners openly voiced other grievances. They complained that Berwind-White had favorite contractors who benefited at the expense of some of the people working under them. Loaders, who made up a numerous sector of the labor force, disliked having to haul their own cars long distances in the rooms of the mine. Added to the perennial grievance of the company-store system was one related to the new company hospital, which opened in 1906. Although miners themselves paid in full for the construction and maintenance of the hospital via company deductions taken from their paychecks, they had no say in its management or policies. It too remained under the exclusive control of Berwind-White. Moreover, any expression of criticism automatically resulted in the discharge of a miner who had dared to voice dissent.[11]

Grievances related to work undoubtedly served as a catalyst for Windber miners to mobilize in 1906. However, the range of their discontent and concerns could not be narrowly confined to the workplace, given the nature of the company's dominance. In any event, all grievances, at the workplace or in the community, were secondary to the right to organize. Miners saw that there was no hope for improving working conditions or resolving other issues unless they could get the company to recognize the right of its miners to have some say in matters affecting them. The union—the UMWA—offered Windber miners the means to achieve this goal. They had observed the union's movement at the district and national levels and judged that the time was fortuitous for organization. In 1906, as in 1922, they sought to become part of an active, broad-based labor movement.

Thus, when UMWA organizers came to Windber in March, they found area miners surprisingly receptive to the union's message. Joseph Genter of Spangler, Pennsylvania, led the local organizing campaign, which met with great success in the final weeks before the declaration of the strike. Union reports indicate that the national organization was forced to call a foreign-speaking organizer, Joe Petok, away from the Alabama fields to help enroll the hundreds of Windber-area immigrants who were joining the UMWA late in March.[12]

That Windber-area miners were mobilizing in March is also confirmed by the actions of Berwind-White officials, who began to take active steps to forestall organization. Given these actions, the company's subsequent claim that it was totally surprised by the walkout on April 2 is incredible. In March, the company began to discharge known and suspected union activists in large numbers. In an effort to get rid of these staunch unionists permanently, it made a secret arrangement through a shady agent to pay transportation costs to send these men away to other mines. Meanwhile, its hired guards beat up UMWA leader Genter, who sustained such serious lifelong injuries that, when he died in 1919, he was honored throughout the district as a martyr of the 1906 organizing campaign in Windber.[13]

Berwind-White prepared for a strike in other ways as well. Although it had its own hired guns, it asked the Windber borough council on March 23, 1906, to provide more police to meet expected trouble. In a hastily called special session on that same date, the council concurred by appointing and deputizing six additional people. The company also hired detectives from the Tanney Detective Agency in Pittsburgh for the ostensible purpose of protecting company property.[14]

In a last-minute attempt to undercut the growing union movement before the April 1 strike deadline, Superintendent A. J. Cook posted an early announcement of a wage increase on March 30, 1906, but as the *Somerset Herald* reported, "company officials were disappointed Tuesday morning when the men failed to return to work." News of the increase and the company's statement that all other "terms, conditions, and customs of employment will continue as in the past" did not satisfy the workers. The miners were demanding union recognition "in the

Windber field, where it never heretofore had a foothold."[15] Neither the additional police nor the wage increase averted the drive for union.

The Strike in Windber
Early Phase, April 2 to April 15

It was the district miners' walkout of April 2 and the parade in Windber, however, that brought the masses of local miners out on strike with district and national forces. The failure of the interstate joint conference to arrive at a settlement at the end of March had made a strike in District 2 inevitable. Central Pennsylvania operators claimed that they could not possibly meet with district officers to negotiate a settlement until April 3. Consequently, Patrick Gilday, president of UMWA District 2, told subdistrict officers to issue a strike order to all locals. Suspension was to begin on Saturday night, March 31, 1906, when the current contract expired, and was to last until an agreement was reached at a conference tentatively scheduled to open April 3 in Clearfield. Approximately 35,000 union miners were directly affected by the strike call and obeyed the order. A thousand of these came to Windber to bring out Berwind-White miners.[16]

Windber mines were at a standstill after April 2. Berwind-White's immediate reaction to the impressive parade was to send a representative to Somerset to seek an injunction in court to restrain union officers and anyone else from interfering with Berwind-White employees. The representative arrived too late to secure the injunction that night, but Judge Francis J. Kooser granted a temporary one the following day. It restrained "the strikers from interfering with men at work, or causing any trouble to befall them or their property, interfering in any way or manner with the property of the Berwind-White Coal Mining Company, calling the men at work 'scabs' or any other opprobrious epithets."[17]

On April 4, 1906, Judge Kooser also appointed six additional deputy constables "in anticipation of trouble in Windber due to the laboring agitation." In the next two days, Sheriff W. C. Begley served eviction notices on 167 miners and their families, including union leaders. Tenants had five days to vacate the company housing.[18]

Meanwhile, Windber-area miners were pioneers in some of the best traditions of the UMWA. Forced to meet in fields or woods because they could not secure a meeting hall, they nevertheless organized a union local. By early April, the miners had created a genuinely multilingual labor organization, democratically elected officers who reflected the ethnic diversity of the work force, and consolidated a membership that included all mine workers, regardless of specific skill or occupation. Thus, English-speaking Charles Shank became president, with John Zimblik as vice-president; Frank Lydick became recording secretary; and John Pongalki was treasurer. By April 6, Genter was able to claim that every miner and motorman in Windber was enrolled in the union at that time and that most spraggers, blacksmiths, carpenters, engineers, and firemen were members too. Men paid on a day-labor basis had joined the tonnage men.[19]

In these early days of the strike, the atmosphere in Windber was widely reported to be mirthful and peaceful. The miners were awaiting results at Clearfield. Meanwhile, 2,000 or more miners at a time attended meetings held in fields and heard speeches delivered in English, Polish, Slovak, Magyar, and Italian. The wife of the English-speaking national organizer Joseph Genter gained considerable fame for a rousing speech at one important meeting when she praised God that there were no more scabs left in Windber.[20]

New immigrant miners were the backbone of the strike, popular support for which cut across nationality lines. Later the Dillingham Commission reported that Magyars and Slovaks had been the most active participants in the strike. In any event, only 200 men, at most, out of 5,000, were reportedly working in Berwind-White's eleven mines during the early phase of the strike. Moreover, the *Johnstown Democrat* reported that Windber miners were avoiding hotels and bars in order to save money for the duration, and that women were hiding their husbands' lunch buckets to keep them from returning to work.[21]

Individual new immigrant miners also took action to ensure the strike's success. For example, on April 6, Martin Smolko, a Scalp Level miner, wrote a letter to the editor of the national Slovak newspaper, *Slovák v Amerike,* to inform his compatriots about the Windber strike. Ethnic newspapers generally covered the national coal strike in some depth because so many of their readership were participants in it or affected by it.

Smolko reported that Windber miners were, above all, seeking union recognition from the company because they wanted to end their enslavement and be respected as free people. After warning fellow Slovaks that Berwind agents were trying to recruit miners to break the strike, he appealed to his brothers to stay away from the area. Calling on them to stand up like brave Slovak countrymen should, he concluded by quoting a familiar labor slogan: "All for one and one for all."[22]

In the meantime, the union prepared to confront the problem of evicted families. The UMWA leased seven plots of land in town and 25 acres outside, secured tents, and claimed that it was ready to provide shelter for 1,000 people, if necessary. But there are contradictory reports on how many evictions actually took place once notices were served. One report sent to the *UMWJ* from Windber on April 13 stated that threatened evictions had been a bluff. Company officials who hoped the strike would soon end had apparently told many who had been served with notices to remain in their houses.[23]

Meanwhile, unspecified union officials made a controversial decision by arranging for evicted miners to go to the Ellsworth mines in western Pennsylvania to work there under the 1903 union scale. Genter disagreed with the policy and advised the men against departure. He told reporters that very few would leave because Windber miners wanted a resolution of the conflict with Berwind-White. He himself wanted the best and staunchest union members on the scene in Windber.[24]

From the outset, the new UMWA local's primary demand was that Berwind-White recognize it. The resolution of all other grievances hinged on the right to organize. In addition, Windber miners made four specific demands: (1) an eight-hour day; (2) the employment of a union checkweighman on each tipple, and company collection of the checkoff; (3) a readjustment of the wage scale; and (4) reinstatement of everyone who had been discharged for union membership or union activities. It seems that national organizers were less concerned than local miners about formal recognition of the union. If the checkweighmen, checkoff, and wage-scale demands were granted, they told reporters, these concessions would be tantamount to union recognition.[25]

Windber miners did not immediately present Berwind-White with their list of demands because they were awaiting results of the Clearfield conference and the arrival of district officials who were in session there. However, because District President Patrick Gilday and other leaders were delayed longer than expected, Gilday telegrammed the local union to form a committee to meet the company's officials. Thus, on April 11, seven local miners and three organizers went to Superintendent A. J. Cook's office to present the union's demands.[26]

The company refused to deal with nonemployees, and the three organizers immediately withdrew. Seven local miners then presented the Clearfield scale and the terms cited above. Company representatives said they would accept checkweighmen but never collect the check-off. They also proposed to restore the company's, not the union's, 1903 wage scale, which the miners rejected. At this point, President Charles Shank of the Windber local is reported to have said: "Well, boys, there's nothing doing here. We might as well get out." Afterward a company spokesman praised the company's generosity and told reporters: "We are coming to believe that there has been too much prosperity in the Windber field." The stage was set for a long and bitter struggle.[27]

Meanwhile, the existence or demise of the district union was also at stake. Although Pittsburgh operators led by Francis Robbins did come to terms with the union quickly, it took nearly four months to get final contracts and resolutions of the strike in all other districts. District 2 would be the last to secure a negotiated settlement.[28] For four months it was forced to fight a fierce war with independent operators and the Association of Bituminous Coal Operators of Central Pennsylvania, as well as with open-shop companies like Berwind-White.

From the outset of the strike threat—indeed, even before—Windber town officials, along with local and county law enforcement authorities, cooperated fully with the company against the miners' efforts to organize. From the first days of the strike, sheriff's deputies, private and public police, and a force of hired detectives guarded mines and patrolled Windber, Paint Borough, Scalp Level, and outlying settlements. Then too, a new reserve force of repression had recently become available. In early 1906 the state of Pennsylvania had organized a new state constabulary that was supposed to enforce law and order in a neutral manner. However, the real purpose of the Pennsylvania State Police became clear shortly after its formation. Armed and deployed in key mining centers throughout the state, instead of other places previously designated, it had begun to tackle its first assignment: the suppression of the coal strike. Miners rapidly compared these mounted troops, who were supposed to replace the Coal and Iron Police but did not, to the Cossacks who violently repressed workers in Russia. This new state constabulary stood ready to actively help the operators to end the strike. All that was needed for repressive intervention was some measure of violence, whatever its origins—or the pretext of violence.[29]

If Berwind-White officials had any hopes of provoking an incident to bring in the state troops to help them end the strike, they were sorely disappointed in this early phase of the strike. Only a few minor incidents or clashes had occurred. On April 5, six strikers had been arrested in Mine 30 for a variety of offenses, including stoning a deputy, carrying concealed weapons, threatening a working miner, and resisting arrest. But such arrests were few and infrequent. Union leaders repeatedly advised strikers to fight for their rights within the law because violence could only hurt their cause. It is important to note that, throughout the first half of April, the Windber strikers were characterized, even by antilabor newspapers, as temperate, peaceful, and quiet. No major incidents or acts of violence had occurred, and the strike was effective. Yet once the events of April 16 had taken place, many of the same newspapers portrayed Windber strikers as "lawless foreigners and rioters," "ignorant and misled," "drunks," and a danger to American civilization that justified any measure of repression.[30]

The Windber Massacre of April 16, 1906

On April 17, 1906, Windber strikers made the front page of the *New York Times*, the *Chicago Tribune*, and most other papers in the United States, and they remained in the national spotlight until a couple of days later, when the San Francisco earthquake wiped out coverage of all other events. The news reports concerned events in Windber on the previous day, and such headlines as "Miners Shot Down in Battle at Jail," "Three Shot Dead in Riot at Windber When Mob Tries to Storm the Jail," and "Three Rioters Shot and Killed at Windber Monday Evening—Riot and Lawlessness with a Vengeance" conveyed the image of a crazed, drunken, ignorant mob of lawless, armed foreigners attacking defenseless people and defying legal authorities, even though the same reports often claimed, incredulously, that the miners had decided that day to return to work. Versions of the inaccurate story varied but little, because the Associated Press had simply taken and widely disseminated Berwind-White's explanation of the day's happenings. Articles were accordingly riddled with nativist and antilabor stereotypes, biases, untruths, and hysterical calls for large-scale repression. To a receptive English-speaking population, thereafter, the notion of Windber strikers was synonymous with that of foreigners, violence, and labor unrest. What follows is a far more accurate account, based on all the available sources, of what happened on that eventful day.[31]

A holiday atmosphere of peace and calm apparently prevailed in Windber during the early part of the day of April 16, 1906, except for the Easter Monday custom in which Hungarians and other foreign women "ducked" or threw water on the men who had acted in a similar manner toward them the day before. But before the holiday was over, three foreign-born men, including a Pole from Galicia, a Carpatho-Russian from Zemplén, and a Slovak from Moravia, lay dead in the streets, having been shot and instantly killed by Tanney Agency detectives in the employ of Berwind-White. In addition, a 10-year-old boy lay mortally wounded, and Paul Bills, a Slovak strike leader, nearly so. At least seventeen others required hospital care because they had been shot through their arms and legs. No weapons of any sort were ever found on the dead. One man had fallen just as witnesses had seen him standing a moment before, with both hands in his pockets and a cigar between his teeth. The boy, a curious onlooker, had been shot in the groin. Three wives and ten children living in Europe and the United States mourned the death of a husband or father killed that night.

The evening massacre was the culmination of a series of events that had taken place over a period of several hours during the day. It is necessary to reconstruct these events in order to evaluate what later took place.

The strikers had planned and held a peaceful mass meeting late that afternoon in a grove of woods near Seventh Street. Having failed to secure a hall in town, the union had rented land and erected speakers' platforms for the occasion. Five speakers, including three Windber priests, addressed a large crowd of 2,500 or more. The first speaker was UMWA organizer Joseph Genter, who spoke in English; the second was the Rev. Michael Balogh, a Greek Catholic priest who spoke in Russian; the third was the Rev. Leo Stefl, a Slovak priest who spoke in Czech; the fourth was the Rev. James P. Saas, a Roman Catholic priest who spoke in Polish; and the fifth was Paul Pačuta, a union butcher who spoke in Carpatho-Russian and Slovak. All speakers were reported to have urged the miners to pursue their rights and the union cause in a peaceful manner within the boundaries of the law.

Sometime near the end of the meeting, Constable S. W. McMullen, followed at some distance by several deputies, arrived on the scene. Miners considered his arrival a direct provocation, and a committee told

McMullen that he could not attend the meeting and had to leave. The miners insisted that law enforcement officials had no legal right to be at a peaceful gathering on private property that they were then renting. This position was simply "tit for tat" for the company's policy of denying them the right to meet together in suitable town halls or on company property.

But McMullen would not leave, and while the miners discussed physically ejecting him, he pulled two revolvers from his pockets and fired some shots into the air. Angry people quickly surrounded him, and he fled with a considerable crowd in hot pursuit. Eventually the constable took refuge in the home of a council member, but some of the crowd had spotted him. A number of hotheaded individuals then invaded the home to search for him and destroyed some of the owner's property in the process. Meanwhile, McMullen escaped by hiding in the cellar.

A short while later, a posse of deputies arrested eleven or twelve selected men, whom they charged with crimes related to the willful destruction of household property in the incident described above. Bystanders who had witnessed the events at the council member's house protested that those arrested were not the ones who had taken part in the damage. A delegation of miners and others soon followed the deputies and arrested men to the borough jail in the center of town with the apparent purpose of bailing out the miners.

Meanwhile, four hired Tanney detectives armed with Winchester rifles and fixed bayonets stood guard at the entrance to the jail. A crowd assembled and began to argue with borough officials, such as Burgess Thomas Delehunt. Protesting that the arrested individuals were innocent,they bargained to get the men released. Reports indicate that the crowd was successful in securing the release of four men, but when the burgess adamantly refused to consider bail for the incarcerated remainder, the crowd responded with boos and hisses. At that moment, some unknown person threw a stone or brick at one of the jail's windows, and immediately thereafter the detectives fired point-blank into the crowd.

The company's hired guards fired as many as four separate volleys straight into the unarmed and retreating gathering. Numerous random shots also strayed into neighboring houses and into a group of pedestrians on their way to evening Mass in a nearby church. The senseless slaughter of April 16 caused Windber to join the ranks of Lattimer, Pana, Virden, and other places in the pantheon of martyrs to labor's cause.

Windber-area residents woke up the next morning to find the town and surrounding mining settlements occupied by Troop A of the Pennsylvania State Police. After the shootings, Sheriff W. C. Begley had contacted Governor Samuel Pennypacker and hastily wired John W. Borland, captain of Troop A of the new Pennsylvania State Police, stationed at Greensburg: "Unable to cope with riot at Windber, send Company at once, Answer."[32] During the night, large numbers of deputies, private police, detectives, and sheriff's men stood armed guard at company offices, company stores, and all mines while authorities spread wild rumors that miners were stockpiling dynamite, as anarchists were believed to do, and preparing to destroy company property.[33]

Impartial reports suggest that foreign-born miners were occupied in quite a contrary manner. In the great state of confusion that followed the shootings, they were trying to find and aid their wounded and to gather their dead and arrange for their burials. Meanwhile, Somerset County authorities ordered all bars closed and transferred the nine prisoners arrested that day and still in custody to the Somerset jail.[34]

Events at the jail thus led to the beginning of massive repression to break the strike in Windber and elsewhere. The state was unquestionably a willing ally of capital and an additional resource for companies like Berwind-White, but violence had been necessary to justify the armed intervention of the state. Governor Pennypacker issued a proclamation of support for the constabulary. Troop A arrived in Windber via special train at 4:00 A.M. on April 17.[35]

In his monthly report to his superior, Superintendent John C. Groome in Harrisburg, Captain John Borland, commander of Troop A, described his regiment's activities in Windber shortly after its arrival: "At 8 A.M., the detachment formed and mounted went down the main street of Windber and dispersed a mob of between seven and eight hundred strikers and told them that all who assembled in crowds of three or more would be arrested." He added: "Mr. Robert J. Sample, correspondent, for the 'Pittsburgh Dispatch,' told me that the strikers melted away just like taffey from in front of our detachment and that they knew that our men were there for business."[36]

Borland's report signaled the antilabor, anti-immigrant, antistrike sympathies of the new American-born police establishment, which was welcomed by Berwind-White's management in the midst of false and fantastic rumors: "The people [Berwind people] of Windber thought that it was a godsend to them to be relieved from the rioting and bloodshed that had existed there for weeks." He went on to explain, "The city of Windber is of about eight thousand population and a large portion is foreign element." The detachment spent its time assisting the sheriff by serving warrants and patroling the areas in which the miners lived, while it "fully held down all the uprisings." It was to remain indefinitely.[37]

On April 18, 1906, the funerals of the three immediate victims of the massacre, the largest in the history of the town, were held. Thousands of people who could not fit into the overflowing churches waited quietly and reverently outside for the funeral processions to form to go to the cemeteries. The *Johnstown Democrat* reported that almost every miner in the Windber field was present. Joseph Genter led the UMWA contingent of the processions, and the national and district organizations sent representatives to attend. Foreign-born marchers chanted in their native languages, accompanied by the Italian and Slovak bands who played appropriate funeral dirges. The bells of St. John's and St. Mary's pealed throughout the ceremonies. The mortally wounded youth, son of an insurance agent, died on April 18 and was buried later.[38]

After hearing only two witnesses, the coroner's hastily convened inquest into the death of the three foreigners concluded: "The deceased came to their death by gun shot wounds made by guns in the hands of deputies in the discharge of their duty in guarding the borough jail." No one disputed the cause of their deaths, but the UMWA sought to hold the guards accountable for the unnecessary deaths and pointed out that the Tanney detectives who did the shooting had been hired by Berwind-White, not by the public. In mid-May, Annie Popovitch, wife of one of the victims, brought formal suit against seven deputies, who were charged with murder.[39]

Although bail for the deputies was set at $5,000 each, they were released almost immediately after they had been arrested. The Somerset Trust Company, acting on behalf of Berwind-White, furnished the bail. The deputies were never taken to trial, because a grand jury composed of county businessmen, farmers, and other nonminers and nonunionists returned a judgment that the case was "not a true bill." Given the contemporary atmosphere of nativism and widespread prejudice against organized labor, along with the hysterical reports about the "foreigners' riot" in Windber, the jury's hasty decision, made before miners and other witnesses had an opportunity to present evidence, was not surprising.[40]

After April 16, Windber's foreign-born population and the union were on the defensive in making known their side of the Windber story. The national union and the *UMWJ* found the English-language press accounts of the events of April 16 to be garbled, biased, and suspect from the outset. The *Journal* therefore awaited results of an investigation and refused to print the Associated Press version, which it later condemned as "gross perversions of facts."[41] Meanwhile, the national executive committee of the UMWA met to discuss the Windber situation and to promise its support to Windber strikers.

It was easier for the foreign-born to disseminate the strikers' version of events in ethnic newspapers than in the nativist press. Father Michael Balogh, the Greek Catholic priest who had spoken at the miners' rally on April 16 and who had buried two of the three foreign victims, accordingly wrote a letter to the *Amerikansky Russky Viestnik* to correct the paper's coverage and warn readers against believing the English-language papers about the bloodshed at Windber.[42]

Balogh, an eyewitness to what he openly called a massacre, placed the blame for the bloodshed squarely on Berwind-White. He described the day's events at some length, affirmed the peaceful nature of the afternoon meeting and of the crowd assembled at the jail, and deplored the willful murder of innocent people by company guards. In the course of his narrative, he claimed that the company had been trying for some time to create a violent incident to start an uprising. Then, after the bloodshed, it had tried to twist everything and make it appear that strikers and unionists were the cause of the violence when the opposite was the case.[43]

Of all the local English-language newspapers, the *Johnstown Democrat* was ultimately the fairest in its treatment of the Windber strikers, although it too contained an initial inaccurate story of the events of the day. After further investigation, it concluded that even if all strikers deserved to be shot as a matter of course, as some were then claiming, there was no excuse for shooting innocent passersby or retreating crowds, or for endangering people in nearby houses, as happened. In its final analysis, the paper doubted that there was any adequate justification for the violence.[44]

Soon after April 16, some papers printed rumors that Windber priests had not really supported the strike and that the Pennsylvania state constabulary had been called into Windber not to protect the company but to protect Father James Saas and his parish house because he had advised the strikers to go back to work.[45] While there is contradictory evidence about Saas's loyalty to the miners and the union, the support of the other priests is unquestionable. In any event, Saas was one of the three resident Windber priests who had spoken at the rally on April 16 and who later issued a joint statement to reiterate their support for the strike. In their statement, the priests claimed that miners had not threatened Saas or church property and affirmed that the strikers had been peaceful throughout the duration of the strike. "The people have been very orderly at the meetings, and nothing would have happened had not the deputy appeared on the ground. The appearance was practically the cause of the trouble that followed. This is the general opinion."[46]

The Windber Strike: Final Phase

Only after the massacre of April 16 did Berwind-White finally admit to the public that it had a strike on its hands. Its hopes that mass-scale repression combined with a nativist backlash would rapidly break the strike and result in an imminent return to work were disappointed. As a result, the company turned once again to the state for aid. On April 24, 1906, Judge Francis J. Kooser granted a permanent injunction against 146 strikers and union officials in Windber, and the sheriff's deputies began to carry out eviction orders. Early in May, Judge Francis O'Connor granted similar injunctions restraining forty-one miners in Mine 37 in Cambria County. Sporadic arrests of strikers were made daily, but on May 4 twenty miners, including the president and secretary of the Windber local, were surprised at night and hauled off to jail in Somerset. Charles Shank, union president, was formally charged with arson. Company deputies claimed they had seen him in the vicinity of a suspicious fire that had destroyed a vacated company house. As in other cases when strikers were arrested, bail was set at very high, even prohibitive, amounts—$1,000 each in this case. The union lawyers argued that the bail asked was excessive, and the UMWA struggled to obtain funds to get the men released.[47]

Meanwhile, Windber miners and their families lived in a state of terror under the repressive force of the state constabulary. People who were children at the time have recalled that troops sometimes shot at strikers without provocation and that they successfully prevented gatherings of

ŠTÁTNI VRAŽEDLNICI POD RÚŠKOM ZAKONA ŠTÁTU PENNSYL-VANIE.

Pennsylvania State Police, during Windber strike of 1906. Immigrants compared Pennsylvania's repressive state laws to those of Czarist Russia. The caption of this drawing from *Slovák v Amerike* reads in English: "Brave murderers under the Russian law of the state of Pennsylvania."

people. Whenever three or more people were seen standing together on a street, mounted police regularly rode between them and forcibly dispersed them. One woman reported that a non-English speaking uncle who had no knowledge of the local situation had come from Italy during this time. Among his first memories of America was an unwarranted incarceration that he did not understand. While walking down a street in Windber shortly after his arrival, police picked him up as a "suspicious" character, charged him with carrying a concealed weapon, and jailed him indefinitely.[48]

Although Windber was generally reported to be quiet throughout the subsequent duration of the strike, occasional clashes between troopers and miners, or union miners and nonunion miners, did take place. In

Depiction of the Pennsylvania State Police's nativist view of foreigners, c. 1915.

May the state police arrested fifteen people in Windber "for rioting," and throughout June they continued to patrol, execute eviction orders, and assist in arrests.[49]

At the end of June, Captain Borland reported to Superintendent Groome with great satisfaction: "Sheriff Begley told me personally that my detachment of twenty men was worth a regiment of deputies and that he would rather have my men than the National Guard of Pennsylvania." Berwind-White was also pleased: "The number of strikers was estimated at nearly four thousand and the Superintendent, of the Berwind-White Coal Company, was so well pleased with the way the situation was handled that he used every possible influence to have my men retained."[50]

Formal charges were brought against members of the constabulary for a variety of crimes incurred during the occupation, but all were unsuccessfully prosecuted. The county's nativist middle classes and antilabor sectors consistently heralded the performance of the state police, which stayed in town en masse until mid-July, when the district settlement had been effected and the Windber strike lost. A token force stayed for one full year. One member of the constabulary subsequently left the force, married a Windber woman, and became a prominent member of the local police force establishment.[51]

Bayonets may have awed Windber strikers, as the *New York Times* reported, but they did not produce coal or end the strike. More miners than ever were out after the massacre of April 16 than were before, when idleness was already nearly complete. But times were hard. Much of the financial support subsequently given to the Windber local had to be spent on legal costs for defense.

District relief funds were exhausted after the first week of the suspension, and UMWA representatives were therefore forced to visit sympathetic merchants every week to implore them to continue to extend credit to strikers on promise of future payment by the union. Independent aid was especially needed in Windber, because the company store now denied strikers credit, and store officials had gone door-to-door to collect passbooks. The Windber local was also at a disadvantage with merchants because of its recent formation and lack of funds and because the district UMWA had no prior history there. Then too, the strike hurt businesses drastically because so many were dependent on the $100,000 of wages paid semi-monthly to employees of the one major industry. Nonetheless, some merchants did grant strikers credit, and sympathetic homeowners, miners or others, took in evicted families.[52]

Meanwhile, many foreign-born miners responded creatively to their situation, thereby once again defying stereotypes of ethnic passivity or deference. In fact, precisely because they were immigrants and miners, they had options that other workers sometimes lacked. Many new immigrants thus chose this particular time to return to Europe or move to other regions in the United States. The temporary nature of their expected sojourn in America, their general lack of home ownership and civil rights, and the heavy preponderance of males in a local population with a severe gender imbalance made migration a more viable option than if large numbers of dependent women and children had been involved. Rather than become strikebreakers, these new immigrants adopted a strategy for survival that did not entail hurting the strike. The *Johnstown Democrat* reported that seventy-nine miners left Windber on one single day early in May and that they were part of a large ongoing exodus that began after the massacre. Parish histories also indicate that Windber's foreign population was greatly dispersed in 1906. The building of the Slovak and Italian Roman Catholic churches was seriously delayed by the departures, lack of funds, and overall disruption of community life.[53]

On June 6, newspapers again carried false headlines that the Windber strike was over and that the men had gone back to work. Union officials pointed out that such rumors were for "sentimental" effect only, that the miners were not fooled by them, and that the miners were as firm as ever in their support of the strike. By that time, half of the prestrike work force was reported to have left the town, and although some coal was being produced throughout June, tonnage never reached half of normal production. During these early years when the mines were constantly expanding, production normally kept pace, but in 1906 overall tonnage dropped. In the strike year, it was 22 percent less than it had been in 1905. The strike was costly to the company as well as to the miners.[54]

Meanwhile, the Bituminous Association met in Philadelphia on June 9, 1906. It claimed to have reliable information that District 2 was out of funds. Feeling that victory was in its grasp, George E. Scott commented: "In view of this, all the Operators in our District have to do for the next week or two is to stand firm; the battle is won."[55]

But on June 28, 1906, two large enterprises, the Buffalo and Susquehanna Company and the Morris Run Company, settled with the UMWA. Other companies quickly followed suit. A final contract between the operators and District 2 was negotiated on July 13 and ratified by a miners' convention shortly afterward.[56]

Newly organized miners, including Berwind-White miners in the Windber area, were not included in the controversial contract. Under the circumstances, Windber miners, who had not been defeated by armed repression, evictions, or nativist attacks, were once again isolated. They saw no hope that, alone, they could win their fight against Berwind-White and its powerful allies. In mid-July the miners who had remained in town began to return to work in large numbers.

Conservative union officials justified exclusion of the recently organized miners as necessary in order to salvage long-established locals from the operators' open-shop offensive, and they argued that the new contract was the best they could get. The operators had not won the open shop, but they had forced District 2's previously unionized miners to accept modifications in the collection of the checkoff that would seriously hamper union finances. Operators and miners considered eliminating the checkoff as a preliminary step toward achieving the open shop.[57]

As Windber miners returned to work, the national and district UMWAs tried to salvage what they could from the bad local situation. In the end, District 2 paid more than $3,000 in legal fees and court costs to defend Windber strikers, and they borrowed money to pay their debts to Windber-area merchants. After the strike's end, officials did what they could to help the many miners discharged and not immediately rehired by Berwind-White.[58]

Court cases resulting from the Windber strike were not finally settled until January 1907. In May 1906, UMWA attorneys had successfully gotten the entire panel of jurors for the Somerset County court cases quashed, but they had failed in their broader goal of obtaining jurors who were peers of the striking miners. Jury lists had systematically excluded unionists and miners.[59]

On September 20, 1906, a jury acquitted six men, including the officers of the Windber local, of charges related to the riot case of April 16. The same jury found thirteen others guilty of crimes in connection with the "riot." The district attorney asked the court to dismiss the arson charge against President Shank because of lack of evidence. In January the last cases were heard. At that time, fourteen foreigners were fined $1.00 each and the costs of the prosecution, which amounted to $137 apiece. Some of the miners were able to pay, but most were not and went to jail.[60]

The strike had been costly. On the eve of the strike the national union had had about $362,000 in its treasury. By July it was approximately $38,000 in debt. District 2 had expended its entire finances during the strike and owed a debt of $110,000 at its end. Neither the national nor the district organizations, however, regretted the money spent in District 2. National Secretary-Treasurer Wilson reported to the national convention in 1907 that all the funds of the International had been used up in the last weeks of the strike in District 2 in Pennsylvania and in District 6 in Ohio, which were the last to settle. He concluded: "This policy was absolutely necessary because it cannot be gainsaid that if the miners of either of those districts had been defeated, it would ultimately have been disastrous to all the miners of the United States." Gilday, too, stressed the life-and-death nature of the struggle and the reluctant acceptance of a less than desirable contract: "With the sinews of war exhausted and our men demanding more relief, I am free to state that we snatched victory out of the jaws of defeat."[61]

Spies and Factionalism

Many miners, besides Windber-area miners, were unhappy with the policies and results of the 1906 strike. Radical critics pointed to a lack of leadership on the national level in organizing nonunion mines and indicted the piecemeal approach to negotiations that had negated the original national strategy endorsed by the UMWA convention in February. A future District 2 president, John Brophy, was among those who condemned John Mitchell for the failed policy that had resulted in leaving Windber-area miners and other newly organized strike participants out in the cold.[62] From 1906 on, if not before, factionalism on the national and district levels, along with new struggles for control of the union's leadership, accompanied disagreements over policy.

The fate of the Windber strike figured prominently in these factional disputes. In 1906, National Vice-President T. L. Lewis, a critic of Mitchell, so distrusted the national president that he subsequently quizzed organizers who left Windber to find out why they had left when they left, and who had ordered them to other places. He was looking for evidence that his boss had consciously betrayed Windber miners. Mitchell was vulnerable to such charges, not only because of the conservative nature of his policies but also because of his willingness to deal with spies and shady characters. One of the most curious aspects of the Windber strike involved his transactions with a man whom John Brophy later described as "a sinister figure."[63]

Brophy was referring to Alfred Reed Hamilton, founder and publisher of *Coal Trade Bulletin* and a man who had many associations with coal operators, union leaders, and Pittsburgh newspapers. By 1906, Hamilton owned numerous diverse enterprises in the Windber area, and he occasionally identified himself as a consultant for Berwind-White. In fact, he was an unscrupulous character who gave information to selected operators and union officials about the opposition in return for favors. There were rumors that he was behind District 5 President Patrick Dolan's break with Mitchell at the 1906 convention, and he seemed to have been a central figure in other subsequent questionable or dirty dealings affecting the union. One of his shady transactions involved Windber directly and seriously compromised the integrity of the UMWA president.[64]

During the 1906 Indianapolis miners' convention, Hamilton apparently approached John Mitchell and proposed an unrecorded verbal deal that the union leader accepted at the time. But correspondence between the two men afterward indicates that each had a quite different understanding of the nature and terms of their agreement. The Pittsburgh publisher was willing to give Mitchell minor privileged information about coal companies, especially those located in the Pittsburgh and Central Pennsylvania regions, and also promised to use his inside contacts to help get a settlement in District 2 at the Clearfield conference. Indeed, Hamilton did send multiple telegrams and letters to Mitchell and did inform on coal operators. Of course, he expected something in return, and that was the point at which he and the union president came into conflict.

In return for his services in 1906, A. R. Hamilton unabashedly expected the UMWA to leave Berwind-White mines and the Irwin field alone. The national and district union organizations were supposed to desert the Windber local and pull out of the area entirely. As a consultant for Berwind-White, the informant could personally guarantee that Windber miners would get a raise. He later took credit for stalling evictions and arranging transportation for discharged workers. His primary goal from the outset was to protect Berwind-White from unionization and to keep the mines operating.

Hamilton besieged Mitchell with letters and telegrams throughout April, May, and June. Most were about events in Windber, and he implored, demanded, pleaded, and insisted that the UMWA leave Windber alone. On April 5, 1906, in a letter to Harry N. Taylor, an Illinois operator and childhood friend of Mitchell, Hamilton wrote: "As you know, I told John that I wanted these Windber interests protected, after they paid the advance, during the term of the present scale as I would not have to watch them constantly." On April 6 he telegrammed Mitchell: "All I want at this time is to get the Windber miners at work at advanced prices." On April 9 he again wrote the union president: "I was greatly pleased to learn that the Berwind situation would be worked out satisfactorily this week." Similar messages were sent throughout the duration of the strike.[65]

Hamilton bitterly complained to Mitchell that there was no need for organizers to be in Windber at all. He personally hated Joseph Genter and William B. Wilson, and did all he could, often successfully, to create dissension within union ranks. He bragged that he had spread rumors that Genter was receiving money from the Bituminous Association, and he was probably the source for many of the false stories printed in regional newspapers.[66]

As for Mitchell, on April 4, 1906, he confirmed that he had made an agreement with Hamilton and complained to his friend Taylor: "My arrangement with Hamilton evidently misunderstood by him. It was that men should be withdrawn from field over first of month." On April 9, he again wrote Taylor about the deal: "I am willing, as a matter of expediency and for the benefit of the people I represent, to enter into an arrangement with him [Hamilton] such as that entered into in Indianapolis, but my understanding is quite clear that it was not intended that the Berwind-White and the Irwin fields should be allowed to run unmolested." For him, the coal company's consultant was a considerable nuisance as well as an informant, and at one point the UMWA leader angrily wrote his Illinois friend: "If I had thought for one moment that he intended that I should abandon my efforts there, I should certainly not have considered his proposition at all."[67]

We will never know what small favors Mitchell may have granted Hamilton, but we do know that the union president did keep organizers in Windber throughout the course of the strike, which the national did support. However, it is impossible to say whether the national could have done more than it did, given the many other demands on it.

What is clear is that Mitchell's willingness to make a questionable secret deal with this agent compromised his integrity, fostered intraunion factionalism, and aroused fears of corruption and the suspicions of critics like T. L. Lewis, who favored a national, not partial, settlement of the strike. For the first time, Windber miners became a bitter source of contention and a potential tool for use in subsequent union power struggles. They deserved better than that. During similar events in the protracted strike of 1922, some Windber miners recalled a union "sell-out" in 1906.

At the national UMWA convention in February 1906, Patrick Dolan, president of Pittsburgh's District 5, angered many miners when he opposed a strike and supported acceptance of the operators' demands for renewal of the lower 1904 wage-scale. Dolan, who was widely viewed as a traitor, later claimed that a strike would not succeed because the thousands of nonunion [immigrant] miners who mined coal in the region encompassing Windber would never support a union. Many Berwind-White officials allegedly held the same belief.[68]

Yet, Windber miners defied such ethnic stereotypes. By going out on a strike for union in April 1906, new immigrants, including those who were on temporary work stints in the United States, proved through their behavior that they were not docile or poor union material. Long before World War I, before the personal decisions of many southern and eastern Europeans to remain permanently in the country, these working people acted as if they had choices to make about their immediate living and working conditions. They were actors on the historical scene, not just acted upon, and they concluded that Berwind-White's proffers of steady work and hastily arranged wage raises were insufficient guarantees of the social justice and greater freedom that they sought. Only unionization could bring about the desired results.

During the strike itself, miners of the many resident nationalities, American-born and foreign-born, supported the Windber union. Slovaks and Magyars figured prominently in the strike, but because they were the most numerous ethnic populations, that is not surprising. Italians might have been less active, as the Dillingham Commission reported, but the best evidence suggests that no single nationality played the role of strike-breaker. The UMWA's traditional policy of uniting culturally diverse workers on an industrial basis in a coalition for the common goal of democratic unionism had worked.

While new immigrants were successfully destroying one ethnic stereotype, however, Berwind-White and a nativist media were seeking to replace it with an equally erroneous, albeit contradictory, one. The coal company found a receptive American middle-class audience when it evoked nativist sentiments to support suppression of foreign-born miners and the strike. New immigrants were perceived no longer as docile creatures but as ruthless, violent radicals bent on destroying American civilization. Although the foreign-born had some English-speaking friends, they and their allies could not withstand the powerful combination of corporate influence, antilabor opposition, and nativism. Yellow journalism with a hysterical antiforeign bias prevailed in news coverage of the Windber strike from the day of the massacre on.

Nativism also strengthened the antiunion forces who wanted the coal strike of 1906 suppressed legally or violently. The state thus became a willing and eager ally of Berwind-White and other coal companies.

Antilabor and nativist state laws, biased jury selection and court procedures, and an exclusively American-born police establishment protected the interests of the coal industry's owners. The new Pennsylvania State Police were not neutral upholders of law and order but strikebreakers. Their use as such prompted District 2 miners to meet with Grangers and other dissatisfied people in Central Pennsylvania to try to form a labor party in time for the November elections. The coal company's having to resort to the use of state police to try to repress the coal strike might also have reflected the weakness of local popular support for the corporation, as Herbert Gutman has suggested elsewhere, but miners who wanted to win the battle for unionization in the future could not afford to dismiss the heavy repressive role the government played in aiding business against labor.[69]

Berwind-White came out of the strike with its commitment to the open shop intact. Once state troops had withdrawn, it continued to rely on old repressive measures, including massive discharges, to maintain that policy. Company control of the workplace and the community had, however, been threatened. Repression, timely pay increases, and public relations campaigns had not prevented the strike, and although Berwind-White continued to categorize national groups according to inherent racial traits after the strike, it could no longer assume—if it ever did—that its new immigrant work force was naturally averse to the organized labor movement. The company's subsequent statements about the strike placed the blame for it on a "fake labor leader."[70] Yet the labor leader around whom local discontent had coalesced was an immigrant Slovak! The strike was evidence that the company's previous policies to divide and conquer the various nationalities had not succeeded. In future years, Berwind-White would continue old divisive practices but would also turn to new ways to ensure isolation and division among its disparate work force.

Before the strike, new immigrants in Windber had expressed interest in forming national parishes and other institutions that were desired by their respective ethnic constituencies. Greater democracy and representation, rather than ethnic exclusivity, motivated these efforts, which were not designed to preclude interethnic or intraclass cooperation. After the strike, however, Berwind-White saw a golden opportunity to use its power and resources to foster the development of segregated ethnic enclaves that would keep its workers divided and nonunionized, and not of the autonomous ethnic communities envisioned by the immigrants. As it had traditionally done, it would rely on a handful of loyal ethnic leaders who would help keep the new enclaves separated and narrowly confined within acceptable limits that ensured an open shop. The controversy over the role of Father Saas in the 1906 strike suggested that, in the future as in the past, new immigrant miners would have to contend with ethnic company collaborators in their midst.

It is easy to fault John Mitchell and other national union officers for conservative policy choices made in 1906, for the subsequent factionalism, and for questionable and possibly even corrupt dealings. In fairness, however, they too faced a formidable unequal contest with capital and a hostile state and had heavy responsibilities throughout the United States. While no one can know for certain whether a national strategy would have failed or succeeded in 1906, it is clear from Windber's case that the national union's strategy for organizing the unorganized and incorporating them into the union movement was inadequate to the need, as John Brophy suggested. Unionization and strikes are not necessarily synonymous, but in Windber's case, given Berwind-White's resources and adamant opposition to unionization and the state's alliance with capital, it seems inconceivable that Windber mines could have ever successfully been organized in this era without a protracted, bitter, costly strike linked to district and national movements.

The conservative District 2 President Patrick Gilday seemed to denote the hope and tragedy embodied in the Windber strike effort in his report to the 1907 district convention. Windber miners had joined the strike despite a wage increase, and the hard-pressed district had been unable to help the Windber local as much as it would have liked to or should have. With an appreciation of the Windber miners' militancy and pro-union sentiments and without any illusions about the sacrifices and costs necessarily entailed in any serious struggle against Berwind-White, he warned the convention delegates: "If any movement is again made at Windber the national will have to be consulted and be prepared to support the men from the first week of the suspension."[71] The experience of 1906 had led him to conclude that the only hope for success lay in a healthy treasury and an all-out commitment on the part of the entire organization—local, district, and national.

That Windber miners were pragmatists who appreciated their own strength but also recognized the need for outside help in organizing Berwind-White mines is clear from their rational decision about *when* to strike and when *not* to strike. With the national and district organizations actively on the move in 1906, Windber miners seized their one available option, organization, when it seemed possible that they might succeed. They put up a good fight but, through no fault of their own, lost the war.

Windber miners would have to regroup and consider their options carefully before making another assault on their open-shop employer. For the time being, they would make the best of their situations, form or reunite families, and attempt to build their version of ethnic communities. The next decision to act would not be taken lightly, given their experiences in 1906 with corporate power, state repression, nativism, and a controversial union policy that had left them out of the finished settlement.

9

Rising Expectations

World War I is often considered a turning point in American history because it produced major social, political, and economic changes in American life. Such changes deeply affected the lives of Windber miners and their families.

Changes in immigration, immigration policy, and the attitudes of the American-born toward new immigrants could not but affect a large foreign-born population and their descendants. The outbreak of war in August 1914 led to the cessation or sharp curtailment of immigration, which in turn led to a labor shortage even as the war itself fueled demands for greater and greater quantities of coal. After the United States entered the war in April 1917, new demands for national uniformity, including "100% Americanism," along with the ongoing campaign for immigration restriction, fueled nativist sentiments that would escalate during the postwar Red Scare and climax in 1924 with the final passage of the Johnson acts severely restricting immigration from southern and eastern Europe. At the same time, the war and postwar conditions in Europe led many immigrants who once saw themselves as temporary migrants to reevaluate their position and remain in the United States permanently.

Because this was a "total" war, it required the mobilization of the nation's civilian population as well as the military, and it demanded organization of the nation's vast array of fragmented economic and industrial resources. America's participation thus led to genuine innovations and to changes that were continuations of ongoing trends. Significant developments included increased powers for the Presidency, the imposition of the military draft, the suppression of civil liberties, an unprecedented centralization of the economy, and an unusual partnership between government and business.

Before the war, the bituminous industry had long been plagued with problems of overproduction, regional differences, cutthroat competition, chronic instability, lack of any national organization, and labor unrest. In a war economy, coal—the source of nearly three-fourths of the nation's energy—was simply too valuable a commodity to leave the industry to chance or unregulated market forces. In May 1917 the Wilson administration's Council of National Defense therefore selected a committee, headed by Francis S. Peabody, a coal operator from Chicago, to mobilize coal production for the war.

The next person named to the new Committee on Coal Production (CCP), second in importance only to Chairman Peabody, was E. J. Berwind, whom *Coal Age* stated was "known throughout the length and breadth of America's coal fields" and who probably "has more important foreign connections than any other American coal operator." He became the coal industry's representative for Central Pennsylvania and for certain sectors of West Virginia.[1]

The composition of the new board quickly drew protests. Not only were the vast majority of committee members antiunion coal operators, but the only avowed prolabor representative was a conservative John Mitchell, then chairman of the New York Industrial Commission. When the soft coal operators on the CCP set maximum prices that were widely considered to be "price-gouging," Congress passed the Lever Act, which authorized President Wilson to set coal prices. Wilson promptly lowered the maximum price per ton by one-third. Self-regulation by the industry had failed and prompted intervention. By August 1917, labor protests induced the government to create a successor agency, the Fuel

Administration, with increased labor representation. UMWA President John White resigned his office in order to become a member of it.[2]

In the end, the government established a veritable partnership with the coal operators, but it also granted labor certain rights and set limits within which both operators and labor had to operate. In order to stimulate production, avoid labor unrest, and coordinate mediation of disputes, the government eventually created the National War Labor Board, which set a national precedent by recognizing the right of labor to organize and bargain collectively but which did not mandate employer recognition of unions.

Both capital and labor were affected by the wartime relationships with government. With massive profits, greater access to government, and exemptions from antitrust laws, capital was the big winner. Nevertheless, the labor movement—encouraged by the policies recognizing labor's right to organize and bargain collectively, and by the fact that the Secretary of Labor was William B. Wilson, a former miner and former District 2 president—revived during the war and, despite persecution of radical organizations, reached unprecedented heights in 1919, when workers in disparate industries engaged in the biggest strike wave in American history.

Throughout these eventful years, Windber miners and their families could not be immune from the great political, economic, and social changes taking place all around them. As we shall see, no single phrase can adequately describe the many ways in which the war and the postwar period affected these disparate working people, but one theme—the notion of rising expectations—does seem to have generally characterized their lives. The disruptions of "normal" life and turmoil of the era had led many to question the status quo, including the legitimacy of autocratic company domination, but also to believe that changing their lives for the better was becoming a practical possibility because of changed circumstances.

In other words, the war and the postwar years were not a quiescent period in the struggle of Windber miners for a more equitable and just life. As we shall see, these people, deeply affected by events, exhibited through their words and actions an underlying discontent that was not fully manifest until the strike of 1922. They were certainly not the "ignorant" foreigners that E. J. Berwind described before the Industrial Relations Commission in 1915, nor were they the contented lot of strike-proof foreigners cited by Berwind officials whenever the company sought to secure questionable contracts to supply coal for the Interborough Subway System of New York City.[3] Indeed they were keen observers of their environment and active if sometimes cautious participants in the making of their lives.

Throughout this time an unparalleled level of labor activity was taking place in the coal industry at all levels—national, regional, and local. This chapter seeks to show that such activity occurred even in Windber. In this nonunion town, miners were joined by small-business people and others who hoped to mobilize labor to create a more equitable society and bring political and economic democracy to the workplace and community.

The War's Immediate Impact and Early War Years

Windber's mining population, which contained representatives of all the various European nations who went to war in August 1914, naturally took great interest in that war from the outset. The war itself abruptly halted the flow of southern and eastern European immigrants to the United States, but that flow had been disrupted earlier. Throughout 1913 the *Era* had periodically complained that Berwind-White was suffering from a shortage of labor because of European affairs. In January 1913, Austria had called up growing numbers of men, including some from Windber, to serve in the Balkan wars. In February 1913 steamship agent Michael Shimko, working on behalf of the company, stated that his usual efforts to encourage immigration to Windber had been unsuccessful because the war was keeping potential miners at home in Europe. When World War I did actually start, the local newspaper reported that many Windber men would have returned home to their native countries to fight if they could have found transportation. Some were eventually able to do so, but, the newspaper concluded with some relief, those who had to remain expressed no ill-will toward other nationalities, engaged in no ethnic or racial clashes, and had reconciled themselves to peacefully but "breathlessly awaiting the outcome of the conflict."[4]

Initially, the war heightened the impact of a severe economic depression that already existed in the nation's coal industry. In the Windber area, the immediate effect was to cut the time that local mines worked in half. Throughout the century, Berwind-White had supplied much of the commercial transatlantic steamship trade with needed coal, and the disruption that resulted from the war meant that it had lost this important traditional market for the time being.[5]

Other markets were not immediately forthcoming. A surreptitious attempt by the company to supply coal to German warships through the Hamburg-American Line in August 1914 quickly ended when a federal investigation of the steamship line targeted illegal trading activities and the company's possible violations of the Neutrality Acts.[6]

The impact of the depressed market in local coal meant that Windber mines continued to work sparsely throughout 1914, 1915, and 1916. Meanwhile, the *Era* tried to raise morale through editorials that proclaimed "Let Boost Be Your Slogan" or "Windber Considers Depression Temporary." On December 14, 1916, the newspaper noted that the town was no longer in the stage of "reckless youth" and now needed to make a transition from a dirty mining hamlet to an enlightened city. Somewhat ruefully and nostalgically, it looked back to the prewar years as a golden era in which business had prospered, and it eagerly anticipated the arrival of peace, which it hoped would bring about a return to prosperity.[7]

During these early years of the war, outmigration from Windber increased. Men were reportedly leaving in large numbers for more lucrative jobs in munitions factories. In January 1917 the *Era* regretted that whole families were leaving the area because young men could not find suitable work there. The paper then took what was for it an extraordinary new step that would have further changed the status quo. It began to advocate the formation of a Board of Trade that would bring new industries into town. It also complained bitterly about food prices and the rising cost of living.[8]

Throughout the war, Windber's immigrant populations continued the uneven process of building their own institutions and consolidating their communities. In October 1914, Bishop Eugene Garvey, numerous fraternal societies, and Roman Catholics of all nationalities joined together to celebrate the formal dedication of the new St. John Cantius Church, which was still a territorial parish at the time. At the same time, St. Mary Greek Catholics were busy completing a new church building; Hungarian Roman Catholics were persisting in their efforts to establish a church of their own, even though progress was slow; and

Italians from diverse provinces were celebrating their national heritage with mammoth Columbus Day festivities.[9]

Fraternal organizations continued to perform routine duties and undertake new projects. By early 1915, despite Berwind-White's harassments, the members of six Slovak and Russian societies did succeed in erecting an independent common hall, only to find themselves engaged in an ongoing struggle with company officials and some of their own members over the issue of the Hall Association's independence from company control. Once the United States entered the war in April 1917, fraternals generously supported the war effort and their respective national causes in Europe. It is interesting to note that the *Era* criticized only one nationality—the Germans—for being slackers during the war, and that the newspaper was referring to Somerset County farmers, long-term American-born residents of German descent, not miners.[10]

Americans too continued the rich associational life they had pursued in prewar years. The Windber chapter of the nativist POSA remained active, frequently presenting American flags to schools, and temperance forces grew. Talk about "Americanization" became more frequent in certain quarters during the war, but the term was rarely or only vaguely defined. The *Era*, which occasionally employed the expression to mean some vague, higher form of civilization, most often used it to connote formal citizenship in the context of its reports on citizenship hearings.[11] Berwind-White's mines and the town's prosperity had been built on foreign-born labor. The company had sought to "curry favor" with new immigrants and had based its antiunion strategy on a divide-and-conquer policy that involved dependent ethnic enclaves and compliant ethnic leaders. In this context, middle-class movements to "Americanize" Windber's immigrants, no matter how the term was defined, did not go very far.

At the same time, however, more and more immigrants, by choice or by circumstance, were deciding to make the United States a permanent home rather than a temporary work residence. In November 1916 the *Era* concluded that "the men who have come to Windber to work have also come here to live" and cited, as evidence, figures that showed postal savings deposits were higher per capita in Windber than in any other place in Pennsylvania. Once the United States had entered the war, some of these foreign-born miners waived exemptions and served in the U.S. Army. But most immigrants retained interest in their homelands as well. In 1917 thirty-four members of Nest 97, Polish Falcons, Windber, volunteered to serve in the Polish Falcon Army in France in order to fight the common enemy and further the cause of Polish independence.[12]

Early on, the European war spawned new political organizations in Windber and an interest in new political issues. In January 1915 twenty-two charter members formally organized a Socialist club in town, and in July the Socialist Party of Somerset County offered a full slate for the fall elections. On the slate of seventeen candidates were nine people from Windber, including two American lumbermen, a Swedish carpenter, a German tailor, and an Italian storekeeper.[13]

It is significant that small businessmen and nonminers, American-born and naturalized citizens, were the nucleus of this new political party that had voted in a county convention to endorse women's suffrage and express sympathy for antiwar European socialists who were suffering from government persecution. Certainly there were individual socialists, Wobblies, and antiwar people among the Windber miners, but most immigrants were not eligible to vote. Moreover, as Berwind-White employees they were subject to authoritarian controls that precluded the right of any free assembly that might threaten that control. Businessmen and nonminers had an independence that miners lacked, and they had access to their own places for meetings. This partial mobilization by nonmining classes for broad political issues related to the war suggests that they and other such people might not long remain inactive in important local affairs. Meanwhile, the early war years also stimulated the wives of company officials, prominent businessmen, and town council members to organize debates about women's suffrage and form local organizations to support it.[14]

The Revival of Organized Labor

Of all the wartime changes that affected Windber miners and their families, none was ultimately more important than the revival of organized labor in the nation and in District 2. This revival began before the United States entered the war in April 1917. By 1916, national union leaders were beginning to assert that "the time is ripe" for a new labor offensive. The ending of the industrial depression of 1914 and 1915, a labor shortage, and the European conflict's creation of a new demand for munitions, coal, and other products had changed conditions to favor the growth of the labor movement.

The *UMWJ* was one labor vehicle that noted the new favorable climate and called for action. It challenged miners to act when it editorialized that "not in a generation can we expect an opportunity to come again such as now presents itself. Do not let it pass you by."[15]

Rank-and-file miners, who often proved difficult to control, and UMWA's District 2 did take action. It was they who led organized labor's revival in Central Pennsylvania. Delegates from District 2 came to the national UMWA convention in January 1916 armed with a resolution that called for a vigorous campaign to organize the nonunion sections of Pennsylvania, which included Somerset County. Their efforts were successful. The convention unanimously adopted the resolution.[16]

District 2's rebuilding efforts were greatly strengthened two months later. In March 1916 the national union succeeded in reviving the interstate joint conference system and winning major benefits in both the anthracite and bituminous fields. Nonunion miners were impressed by the union's ability to gain pay raises while securing the eight-hour day in anthracite and run-of-mine provisions in bituminous mines, without a strike or work stoppage.[17]

District 2 not only shared in these gains but also negotiated a new contract in April that restored checkoff provisions that had been lost in the 1906 strike. Not long afterward, organizers and others from various sections of Somerset and Cambria counties began to note that they were succeeding in reestablishing locals destroyed in 1906. District organizations, which had been defunct for many years, were also rebuilt.[18]

In response to District 2's organizing drive in 1916, nonunion miners in Holsopple, Hooversville, and other Somerset towns near Windber collectively organized and went on strike for union recognition, which mine operators continued to deny them. Meanwhile, miners from established locals repeatedly pressed District 2 to organize Somerset County, including Windber, in order to protect their own organizations. From 1916 until the strike of 1922, organizing Somerset County was an important item on the agenda of the district organization, and it repeatedly sought help from the national in various organizing efforts it undertook.

In December 1916, District 2 President James Purcell wrote John P. White, president of the national UMWA, to describe recent efforts. He explained his request for aid: "We have spent a large amount of money in the Somerset and other fields in organizing new Local Unions and maintaining strikes. We have spent in the neighborhood of $70,000 or $80,000, between strikes and paying newly appointed men to carry on the necessary work."[19]

Meanwhile, Windber's nonunion miners were indirect beneficiaries of the labor movement's advances. In order to be competitive and avert possible unionization, Berwind-White set wages and working hours that routinely reflected the strength or weakness of the regional union. As E. J. Berwind stated to the Industrial Relations Commission: "Though we will not work with the unions, we take their scale largely as the result of supply and demand. Changes in the rate are generally brought about by trouble occurring in some district or another." The company thus raised its wages when union gains were made, and in January 1917, for the first time in Windber's history, it reduced nominal working hours from ten hours to nine hours.[20]

Organizing in the region continued to take place. One Somerset County miner vividly captured the sense of optimism that prevailed early in 1917: "Times look good for the organization and scabs are scarce at present."[21]

The Government, Berwind-White, and District 2

The drive to mobilize labor continued after the United States entered the war in April 1917, but under new circumstances. The federal government began to assume greater control over the nation's economy. During the spring and summer of 1917, the Council of National Defense's new operator-dominated agency, the Committee on Coal Production, with Francis Peabody as Chairman and E. J. Berwind a member, attempted to coordinate production and mediate disputes. Conflicts of interest were inevitable.

Strikes at Berwind-White operations in Clearfield, Pennsylvania, and in St. Michael, a town located near Windber, came to the attention of the CCP in the spring and summer of 1917. The Maryland Coal Company that operated the St. Michael mine was a Berwind subsidiary whose officers were the same as those who managed the parent concern. District 2 UMWA records indicate that the strikes at these two places were among many that were taking place in the region at the time. The central issue in these various disputes was union recognition, which neither the CCP nor the successor agency, the Fuel Administration, were willing to require of operators.

Berwind-White responded to the U.S. government's efforts to mediate the strikes occurring at its respective operations with a customary denial of their existence. In a letter to Francis S. Peabody, CCP chairman, on July 30, 1917, Thomas Fisher, Berwind-White's general manager, claimed that no strikes were taking place, or had occurred, at any of its mines.[22]

Fisher explained away any difficulties in Clearfield by asserting that the company's miners there were surrounded by speculators whose coal concerns paid more and that the men had "quite naturally drifted away to avail themselves of this higher wage." As for St. Michael, he noted, "The mines never shut down." He even denied that Berwind-White officials managed the subsidiary company, although he made it clear that he served as its official representative.[23]

Despite Fisher's claim that no strike had taken place at St. Michael's, he did admit that "a partial suspension occurred there on May 17th." According to him, a brief minor disruption had been "fomented by a group of unnaturalized Austrians in the face of full-scale rates and working conditions." For Berwind-White and other nonunion companies, Austrians—the contemporary alien enemy—preceded "Bolsheviks" as a favorite scapegoat for any domestic strikes or union agitation.[24]

In any event, the company did not want these troublemakers back, and Fisher was glad that District 2 President John Brophy had found other jobs for them, because, he informed Peabody, "You will appreciate that nothing will be gained in production by reshipping these few men back to St. Michael." In June and July, Brophy did indeed have his hands full in coping with mass evictions at St. Michael. But because miners were in great demand then, he had been able to obtain jobs for many of the strikers in union mines.[25]

Meanwhile, in his correspondence to the CCP on August 1, 1917, another regional coal operator who used the example of Berwind-White to justify his company's own antiunion policies did admit that the union was making serious inroads, and even included a majority of his own men. Charles S. Ling, general manager of the Ideal Coal Company, commented: "We have been conscious of it [the organizing drive] and have been annoyed more or less by it, our men having made more demands and more exorbitant demands than ever before." In words that echoed Fisher's, he blamed the current strikes on unpatriotic union agitators and Austrians. Men had chosen to work at his company because, in his words, "they are disgusted with the United Mine Workers in other localities," but cowardly workers were now afraid to refuse to join the union.[26]

Ling's overall view of miners and foreigners was as unflattering as E. J. Berwind's had been in his testimony before the Industrial Relations Commission:

> As soon as intimidation by threats has lost its appearance of horror to our men they will be at work. The fact is that the nucleus of this organization consists of only the Southern European who is not intelligent, who is simply led by the man with a stronger will and is easily influenced and in things which he knows to be morally wrong. These men, or a majority of them, now are alien enemies. They are the ones who are responsible for the trouble in this locality at this time. We trust that there will be legislation, treaty or other arrangements effected whereby mine workers as well as other labor if exempted from military service shall be required to work six days a week and if subject to draft be drafted. If alien enemies they should be compelled to work steadily or be interned.
>
> We assure you, Mr. Peabody, that if Mr. Brophy will leave us alone that our mines will work and that our production will be increased.[27]

Ling's solution to labor unrest was forced conscription and detention of "alien enemies."

District 2 organizers had to confront the hostile worldviews of the Berwinds, the Fishers, the Lings, and other intransigent antiunion operators and managers. In early July 1917, District UMWA President John Brophy sent several of his men to Windber in connection with the ongoing strike at St. Michael's. Their task was to assess the local situation, but because of company harassment, the organizers quickly decided it was in their best interest to make Johnstown, not Windber, their headquarters. As usual, the union found that its major problem was not ethnic diversity or the disinterest of local foreign-born or American miners in the union, but the concrete difficulties involved in making contact with miners in a closed town that did not respect basic constitutional rights.

In making contact with their respective ethnic constituencies, English-speaking organizers were often at a disadvantage in Windber because they were more conspicuous than their foreign-speaking counterparts to English-speaking company guards and officials. On July 5, 1917, one such organizer, Pat Egan, complained to Brophy: "The English speaking men are very hard to do anything with as the spotters are onto me and follow me wherever I go." He added, "When I go to a restaurant they even sit down beside me but Fred and George [foreign-speaking organizers] have met with better success." He went on to report, "It is almost impossible for one English speaking man to do anything as they are watching me like a hawk." However, instead of despairing, Egan suggested diversionary tactics: "If I had another English speaking man who could talk to the men while the spotters were watching me, I think that we could get them in a short time."[28]

Meanwhile, foreign-speaking organizers were meeting with greater success in making contact with the more populous new immigrant populations in Windber. Not only were these ethnic union men generally less visible and well known to company guards, but they often had the advantage of reliable contacts within Windber's various immigrant communities as well. As Egan himself reported, Fred and George were finding new immigrants—without whom no initiative could succeed—receptive to the union message. About the work among ethnic miners, he commented: "John has not met many of [the] Hungarians but the Slavish and Italians . . . are satisfied to organize." The response by the foreign-born in Windber in July led him to conclude his assessment of Windber's potential for organizing by writing his boss Brophy: "I think that Windber can be got."[29]

However much Brophy may have wanted to organize Windber, he could do very little there or elsewhere with the district's limited resources and staff. In June and July he made repeated requests to the national for help, but none was forthcoming, and by August 1917 he seems to have dismissed the idea of organizing Windber for the time being.[30]

On August 8 Brophy wrote Robert Foster of UMWA District 20 to thank him for information he had requested on Berwind-White's West Virginia mines: "So far this Company has refused to recognize or deal with the United Mine Workers in this District, and is also using every effort to prevent the organizing of its workers at Win[d]ber." He added: "The only hope of getting this Company to recognize our Organization will be when the miners at Win[d]ber declare their intentions to be organized." He hinted that an earlier opportunity to organize there may have been lost: "At present it is very hard to reach the men, who, at this time, are receiving fairly good wages and have not the same incentive to take the stand that they had some 12 months ago." Berwind-White had responded to major union successes in the district by increasing repression and by matching some union gains, including a reduction of nominal working hours from nine to eight in June.[31]

Getting Windber "Free for Democracy"

Throughout the remainder of the war, the region's rank-and-file miners and District 2 continued to press national officers for a commitment to organize the nonunion fields of Somerset County. In April 1918, Windber miners themselves joined in the call. Greatly influenced by the mobilization of the labor movement, the spiraling cost of living, and their own wartime sacrifices, they began to demand that the union come in and help them to organize against an autocratic employer. In 1917, despite Berwind-White's attempts to avert unrest, outside organizers had found considerable latent support for unionization in Windber. Now, a short time later, local miners were taking the initiative and openly asking for help.

On April 1, 1918, one of these miners, Edward J. Robinson, wrote to ask the district president for advice on how to begin to form a union. He did not know how to proceed. His letter suggested that something was happening in Windber: "I have bin asked by a number of miners at this place to organize a union for them. But I am a miner myself and I would like to organize the Windber Field so Please answer this by return mail and give me all the details you can. Advise me at once."[32]

Brophy, who was out of town, did not immediately reply. Robinson persisted. He wrote again on April 15: "I wrote you two week ago about organize the Windber Field and I have not got any reply. Yet I hope you will take this matter up at once so as I can go to work and get things in line." Meanwhile, he had written a similar letter to District 2 Board Member James Feeley in Dunlo, who forwarded that letter to Brophy.[33]

Brophy responded to Robinson's inquiries with a request for a meeting. These communications were the first of many between the union president and the miner, who met in Nanty-Glo to discuss the Windber situation for the first time on April 21, 1918.[34]

Meanwhile, Brophy had again approached the executive board of the International about a campaign to organize Somerset County. On April 15, 1918, before the scheduled meeting with Robinson, Vice-President John L. Lewis informed him that the board had denied his request. In his letter to Brophy, Lewis quoted the board's conclusion: "In view of the unsettled condition of the industry and governmental affairs, we suggest that definite action be deferred until a more favorable time, and authorize the resident officers to meet changing conditions in such manner as may be deemed best in the premises."[35]

Lewis added that the board had to consider the problem of organization "from the standpoint of national interest" and that "it was the unanimous judgment that it would be folly to pursue a campaign of guerrilla warfare against only one section of the great interests of the Consolidated Coal corporation." He then turned to Berwind-White specifically: "Likewise, it was realized how firmly entrenched the Berwind-White interests are in their various nonunion operations." While he hoped that the two big concerns would soon be successfully unionized, "it was the consensus of opinion that any such campaign must be entered into with due deliberation and systematic preparation."[36]

On May 13, 1918, Robinson informed Brophy that he had been discharged and blacklisted. Board Member Feeley had unwittingly been responsible: "As Mr. Feeley has got me in wrong as these mail carriers have noticed his self addressed letters and they have bin took to these company's police. I have bin discharged from the company and they have me on the blacklist at each of their mines."[37] When it came to company-town autocracy, even some prominent District 2 officials were, it seems, naive.

Afterward, Robinson found a job in a small country mine, but the Berwind company's police continued to follow him everywhere. He furnished the district president with this description of routine intimidation:

> The Company has two of their first class men watching me and I just leave them go. They are at my heals all the time and they even set along side my door when I come home at night. . . . I never let on I see them, and if I stop to talk to any body, they stand on the other side of the one I am talking to. So you can see I have got to be very quiet for about 2 weeks and then they will see I am not trying to do

> anything. That is the way they do with every body they think is in favor of Union.[38]

Despite his current plight, Robinson remained optimistic. He continued to maintain hope that Windber miners would soon be organized, and he indicated that support for the union was widespread among at least one of the big new immigrant nationalities residing there. "Mr. Brophy, this is one of the best times to help me," he informed the district president, and then made a specific request: "If you can only send a couple of good Italians in this town to help, . . . ones that will work in the mines. . . . If you can do this, have them go to No. 41 or 42 mine as their is about 2,000 of them in Windber and they are all waiting untill some one gets it started."[39]

In order to avoid another detection by mail carriers and company police, Robinson concluded his letter by asking that Brophy and the union send any future correspondence meant for him directly to another party. That person was John Hritz, secretary of the Slovak National Hall Association, which had had many difficulties of its own with Berwind-White.[40] The Hritz address indicates that Robinson had important reliable contacts within the Slovak community and strongly suggests that Slovak miners were prominent among those whom Robinson claimed kept pressing him to get Windber organized. In any event, District 2 officials sent subsequent correspondence in unmarked envelopes to the sympathetic officer of the Slovak Club.

On May 21, Brophy reacted compassionately to Robinson's discharge and expressed pleasure that sentiment for organization was growing in Windber. He also stated that he expected to meet officers of the national in a few days and would discuss Windber with them then. He held out hope, saying, "It may be that later they [the national officers] will see their way clear to undertake the task of joining with us in a movement to reach the men of Windber with the message of Unionism."[41] But as on previous occasions, the national officers remained unconvinced.

By mid-June, Robinson was peeved and angered by the union's inaction. He wrote to express the frustration he and others felt:

> I will call your attention once again to the situation of this place as the people wants to no what is the matter that you people is not doing your part in getting this place Unionized because I am unable to do anything because they are at my heels all the time.

> All we want is to get organized, and then if it will go good with us, then we can make them come across and recognize us. The people are getting real out of umer because it has not bin organize a month ago. And they say if you do not go after this at once while the people are in sutch good hopes, they say the next time this is to be organized that they will turn and get the Companys to organize aginst you. So please do get this place on the list with the rest.[42]

Robinson then added a postscript because he said he had run into eight or ten men on his way to mail the letter, and he wanted to convey their sentiments. At the time the union was cooperating with government officials and the Central Pennsylvania Coal Producers' Association, an operators' organization, in patriotic speeches and rallies designed to boost coal production for the war. As if to impress on Brophy the seriousness of the Windber men, he told the union leader that the men he had just seen said that "they will give their Signature to work *every* day and even work *one* or 2 extra shifts per week if I can get this place organized."[43]

But Brophy could not win the national officers over to a commitment, nor could the delegates from dissatisfied district locals who met in special convention in Portage on August 15, 1918. To their demand that the national and district unions initiate a campaign to organize Somerset and Cambria counties within the next thirty days, President Hayes and the executive board of the International advised the miners to avoid precipitous actions for fear of harming the union's efforts to get a wage increase from the Fuel Administration.[44]

During the fall of 1918, the government, operators, and union considered the war the top priority, and each pressed the miners for increased coal production without interruptions. Windber miners responded, but in October, much to the dismay of *Coal Age,* they and others in the region abruptly ceased work for a short time. They had reacted to a false rumor that the war had ended and had gone off to celebrate.[45]

The war, which President Wilson proclaimed was being fought to make the world "safe for democracy," came to an end on November 11, 1918. During the conflict, labor leaders had taken up Wilson's democratic justification for American participation and applied the ideology to the home front. Thus, the war for democracy and self-determination was simultaneously taking place on two fronts. For organized labor, beating the Kaiser and autocracy abroad was one arena of the struggle. Ending corporate autocracy and establishing political and industrial democracy at home was the other front.[46]

The armistice immediately aroused the hopes of Robinson for imminent action, and he hastily dashed off a letter to Brophy. The document is important in that it reveals how powerful the democratic ideology had become on the grass-roots level and how working people sometimes adapted it to meet their particular circumstances. Robinson wrote:

> I am requested to write you about the situation in this nabor-hood.
>
> The People think that you aught to be able to get Windber Free for Democracy. They say that Windber augh to be freed and I think a little talk and a week's vacation all through Dist. # 2 will force Berwin Wite to sign our Armistice. They turned out the street lites and they even turned of the house lites when they got word of the 2nd peace news and hat the bells and whistles stop. So you can see how the People of this place feels about it. So I hope you to answer our calling.[47]

The Citizens' Association and Miner Initiatives

It soon became clear that Robinson and miners were not the only dissatisfied Windber-area residents, nor the only people whose expectations for a better and more democratic life had been raised during the war. From November 1918 through March 1919, a cadre of independent American-born and foreign-born businessmen organized and led a community-based civic movement that aimed at breaking Berwind-White's stranglehold on political and economic power in the town.

To achieve this goal, a new-founded Citizens' Association turned to two possible sources of assistance. First, it called on the federal government, which during the war had assumed a greater role in mediating labor and other disputes, to intervene against a corporation it considered un-American because of the coal company's blatant disregard of the rights of American citizens. Second, it urged UMWA District 2 President John Brophy to conduct an all-out organizing campaign in Windber. Members believed that unionization and recognition of the rights of labor were essential to bringing about genuine political freedom and economic fairness for all classes living in the company town.

The independent business movement originated at an unlikely moment, in fact, in the midst of a deadly influenza epidemic that was sweeping the nation. From October 1918 to January 1919, Central Pennsylvania and Windber were especially hard-hit. *Coal Age* reported that tonnage was down 80 to 90 percent in mid-December. UMWA District 2, which had a modest death-benefit system for union members, paid out an unprecedented $25,000 in benefits for the month of December alone. The expenditure for a normal month was $3,000. Windber residents later recalled that "people dropped like flies" during the epidemic and that the town was itself quarantined for a time. The Windber borough council spent more than $1,600 in three months for the services of quarantine guards.[48]

The deadliness of the flu epidemic in Windber is undeniable, as was the need to take extreme measures to contain it. Yet the quarantine imposed during November by Berwind-White, not some neutral authority, seems to have been the final straw that, in combination with a long series of perennial grievances, precipitated the formation of the Windber Citizens' Association in November and December 1918. Small businessmen apparently felt much aggrieved at the treatment they received. Their subsequent complaints resembled those of union organizers and others who could not bypass quarantine regulations during the smallpox outbreak in 1899, even though company officials and others could.

The Citizens' Association subsequently charged that the quarantine system had been extremely autocratic and unfair, that it had proved "to the citizens of Windber, as well as to the farmers who trade in town, what it means to have a despotic corporation in the saddle." In approving 1,500 travel permits, a local official had functioned as a dictator "with a dual personality." He had denied all permits to "those not connected with the big interests," but had given them away readily to favored company employees "for the mere asking."[49]

Windber attorney F. A. Millott spoke to a small group of business people who had gathered in the Lyric Theater on November 22, 1918, to organize an independent body of taxpaying citizens. According to the speaker, the arbitrary administration of the quarantine demanded action, but the more general problem was the political "system" with which citizens had to deal. "This corporation which controls the elective and appointive offices of this Borough has no right to impose its policies on the tax payers," he explained. "This corporation imposes its paid employees on you as the officials of this community and its employees are dictated to in its policies of appointments." The lawyer then added that the company itself dictated borough policies and trampled on the rights their ancestors had fought for and that were now denied them. He insisted: "This corporation has no right to dictate to its employees or impose its policies or deny the right to form an organization."[50]

To support his argument for an organization, Millott then introduced ideas that are sometimes termed radical but that were actually widespread or part of the American mainstream during the immediate postwar period. These ideas included federal control of basic industries and federal encouragement by law of labor's right to organize. Both ideas had gained much popular support because of wartime needs and the government's recent successful experiences in managing the railroads, organizing the economy, and mediating labor disputes.

Millott told his audience: "I believe in Federal control and supervision. While I do not have the Berwind-White Company in mind, I believe coal companies should be controlled by the Government, just as they control the railroad companies."[51] He continued, "There is a Federal law encouraging the organization of employees for the purpose of securing their rights." He gave two regional examples: "For example: The Altoona shops of the Penna. Railroad Company, employing some of the most skilled mechanics, all intelligent men, were never able to obtain their rights until Federal supervision opened the way. Today they are organized." In the second instance, "There is under way now, I understand, in Johnstown a move to, by the steel mill men, take advantage of this Federal act to obtain the same recognition and rights as secured by the Altoona shop men."[52] Millott emphasized that independent organization was the goal and that everything could not be accomplished in one night. "No doubt many, very many, more men would have attended this meeting, but we have to believe that fear of antagonizing this corporation has held a large number of men from attending," he concluded.[53]

The new Citizens' Association continued to meet weekly on Tuesday nights, and by mid-December a local newspaper reported that it had 400 members and was growing.[54] By then, however, Berwind-White had begun to take action against employees who attended the meetings. It discharged and blacklisted them.

On December 20, 1918, Secretary of Labor William B. Wilson received a petition from the Windber association in which the petitioners described the coal company's response to its existence:

> We have been holding public meetings at the Lyric Theatre and on the outside we have nearly all of the Berwind-White Coal Co. officials congregated on both sides of the entrance with the Burgess and Coal and Iron Police watching to see who goes in and they have discharged some of their employees for attending these meetings and they have brought pressure to bear on some of the business men and they had to resign or move. The laboring men in Windber are treated that way. What we want is a square deal for everybody and that a man can vote as he pleases and join a society or attend a meeting if he wants to as long as he and the society do what is right. We have not hurt anybody nor any property and we do not want to. It is only our rights that we are after. We do not think it is right for men to be fired just because they go to a meeting of citizens.[55]

The petition ended with a request for help or an investigation. It was signed by the associations's officers: President, J. W. Kelly, merchant; Vice-President, H. M. Habeeb, merchant; Secretary, E. M. Berkhimer; and William H. Bolopur, barber.[56]

On December 28, 1918, Robinson again wrote to Brophy, but this time because businessmen—not miners—had asked him to write in order to get Windber miners organized. His letter indicated that small-business people recognized that breaking the company's monopoly of borough affairs could not be achieved without unionization of the miners. Robinson wrote:

> The Business Men have called on me to bring this matter [organizing Windber] up once again as 90% of the Business men have all come this way and more so. The Citizens are coming this way more so each day. I have bin offerd a *hall* 3 month free of chare so it will give us a very good start. We have bin doing some fine stunts in gitting things to come this way. We have formed an organization, namely, The Business Men and Citizens Association. This is to set the B. W. Coal Co. and others to set back on elections and we can get the men in the Borough offices that will stick to us and give us justice. So

> Please take this matter up at once and advise me at the very earlys time as I am in good faith that you may take action at once and I am at liberty to give my hand at your call.[57]

Brophy appreciated that something important was happening in Windber. On January 1, 1919, he wrote Robinson: "I am pleased to learn that among many of the business men of Windber there is favorable sentiment toward the United Mine Workers and a growing recognition of the fact that our Organization will mean greater political freedom in that community." He then asked Robinson to send him information on the ethnic composition of Windber miners. Robinson accurately conveyed the diversity of the mining population, but his estimates of the numbers involved were far too high. Nonetheless, he put Brophy in touch with numerous merchants and leaders of the association.[58]

On January 13, 1919, the Citizens' Association sent Secretary Wilson a new petition, signed by eighteen members, including the officers, other leading merchants, and a variety of skilled carpenters, railroad workers, car-shop workers, and an assistant foreman at the mines. The petition described the company's ongoing intimidations and discharges of employees for their attendance at the organization's public meetings, as was their right "under the laws of this country." It alleged that the corporation's threats then extended to the independent businesses who later employed such men.[59] In various ways, the petition revealed that the association's conception of American citizenship was quite different from that of the company's. At the time, the town's only newspaper, a company organ, was actively engaging in red-baiting against the civic group.

The petitioners reported: "We have no organ with which to make reply, therefore, to such infamous falsehoods as that we are 'bolshivikis' and are adherents of the red flag. We assert that there are no more loyal citizens of our country than can be found in Windber and among those who have subscribed to the principles of our association."[60]

What the men wanted, they added, was an investigation by the government "in the hope that such an investigation and public hearing will assist the laboring men here in securing their full rights." They described the following situation:

> There is here no antagonism to capital as such, but if conditions such as exist here are permitted to continue we fear that it will lead

> to such an attitude on the part of the working men. The business men of this town who are independent and free to voice their opinions, together with the men who labor, are practically a unit in opposition to the system employed by the said coal company in coercing and intimidating the poor men whose livlihood is their daily wages. In the coal company's system of following decent citizens with their hired policemen, their acts at election time in distributing marked sample ballots to their employees, and their browbeating attitude towards their laborers, we believe that they are enemies of society and of good government.[61]

In their view, the coal corporation was the real enemy of American society and of the American government.

On January 15, 1919, J. W. Kelly and F. A. Millott met in Washington, D.C., with Hugh L. Kerwin, assistant to the secretary of labor. The result of the meeting was that Kerwin assured the men that the Labor Department would send a Commissioner of Conciliation to Windber to investigate the situation, especially the formal charge that Berwind-White employees were being systematically discharged for having attended the association's meetings.[62]

Kerwin ordered Frederick G. Davis, a Commissioner of Conciliation, to Windber. On Tuesday, January 28, Davis met with association members who presented him with affidavits from discharged employees, including Verden Thompson, David J. Powell, and others. That evening the civic group met, and Millott, one of the speakers, reiterated a basic theme: "Any corporation that will follow a man to a public meeting and then discharge him for attending is a corporation that is an enemy to our government." In response to the company's ongoing charge that the group, composed of respectable middle-class American-born and foreign-born businessmen, were "Bolsheviki," he countered that the Berwinds were "Hohenzollerns."[63]

Davis completed his investigation, and in a confidential report to the Department of Labor dated February 18, 1919, he substantiated the civic group's charges, and concluded that the Civic Association was correct in stating that Berwind-White officials "acted together in all matters pertaining to local politics." At the organization's first meeting, Attorney F. A. Millott had said something that led the company to believe the businessmen were interested in a unionization effort there, hence the discharges and intimidations. He understood that "this company is operating under non-union conditions and their history is that they stop at nothing to prevent unionism at their mine." In his judgment, the entire situation was highly unfavorable to organized labor because of the harassments of the Coal and Iron Police and the actions of a handful of ethnic and club leaders who helped to limit grievances and control votes.[64]

The Labor Department's investigation revealed the promise and limitations inherent in this sort of action and protest. Davis was sympathetic to the dissident businessmen, miners, and organized labor movement, but the purpose of his inquiry was limited. Having ascertained some basic facts in a few short days, he considered the affair ended once he had obtained a promise from Superintendent Simpson to reinstate the employees who had been discharged for attending the meetings. He also found that Simpson "was very pronounced in his decision to combat any effort to unionize their workers."[65]

Beyond reinstatement, the government offered no other aid and no guarantees that businessmen or miners could even continue to meet without undue intimidation. Undoubtedly the association had hoped for more. The report was never even made public, although John Brophy later requested and obtained a copy of it.[66]

The civic group continued to press its case. On February 9, 1919, a committee of Windber businessmen met with Brophy in Altoona. They told him they had formed the association to obtain "a square deal for all citizens opposed to the policy of autocracy, as practiced by the company." One committee member was George Habeeb, a Syrian department store owner, who had supplied the hall for the businessmen's meetings. He had refused to renew an arrangement, a common practice, whereby Berwind-White paid the owners of halls $10 a month to determine who could meet in them.[67]

The Citizens' Association had printed a pamphlet, "An Outline of the Purposes and Principles of the Windber Citizens' Association of Windber, Pa.," which the Windber men gave to the union president. The document declared their members' intention to regain rights "as vouchsafed to them under the Constitution and laws of our country." It also showed that the democratic ideology of the Great War had made an impact on their consciousness: "If our starry banner is to carry to oppressed peoples across the seas the message and guarantee of freedom of oppression, it ought to mean that we at home, from one boundary of our country to the other, also enjoy those blessings to the full."[68]

Although the pamphlet discussed all sorts of concrete grievances, including the company-store system, the association's main concerns were civil liberties and political freedom. It offered the following description of contemporary voting procedures:

> The voter of Windber must pass through the famed Indian gauntlet before arriving at the polls. A long line of company officials, private police, bosses, both major and minor, and their hired help, must be passed. And lest the memory of the voter be short, a marked sample ballot is neatly folded and pressed into the palm of his hand or gently slipped into his pocket when he ascends the steps to vote.[69]

The publication further contended that the company had discharged employees who voted in precincts where a company candidate lost an election, and it declared: "It is not right that any class of men or business should control the borough and use it as a mere department of their operations." As a result of these abuses, the civic group adopted seven goals, among which were its intentions (1) to work to elect independent citizens to office and (2) to promote the education of the foreign population so that they could take part in community affairs and vote.[70]

By adopting a program designed to secure real attainment of basic constitutional rights such as freedom of speech and freedom of assembly, along with political freedom and the right of labor to organize, the businessmen were making a direct challenge to Berwind-White's rule. Their promotion of "the principles of right conduct between employer and employee to the end that each may enjoy all the rights to which he is justly entitled" conveyed their sense of justice, but it also indicated that they recognized that, without UMWA organization, chances for success in securing the rest of their program were slim or nonexistent.[71]

Meanwhile, miners from Windber and other places were besieging Brophy with requests for action. J. H. Davis, a Windber miner, wrote Brophy on February 8, 1919, and made an urgent plea:

> Having at last located you through William Green at headquarters, I am asking you for God sake and For the Sake of Suffering Humanity [to] send some one in this region to help stay the iron hand of the miners' oppressers. For these people in and around Windber are trying their Best to organize with no one with credentials to lead them

> and they are a going to tear loose some of these days with no charter, no leader, no representation with the national organisnation. And you know what that will mean another black eye for the U. M. W. of A. in this field.[72]

Davis said that he had been in the area for about two years, enumerated various grievances existing in Windber mines, and claimed to have been in strikes in West Virginia, where he had seen Mother Jones. He also said he was working in a scab mine only so he could get close to some Slovaks, whom he hoped to win to the union. He urged immediate action:

> The miners here stuck to the flag till the war was won and now by [God] the flag must stay with them. We have all kindy of miners here, Italians, Slavish, Hg., Poli. But the Italians are there with the big mit. Well Friend you must send some of your men over here but send them on the git for these dam thugs here are hell. . . . So get busy for the sake of Humanity, get busy, get busy.[73]

At the same time, on February 11, 1919, the members of UMWA Local 64, Holsopple, made an urgent appeal to fellow unionists to organize Somerset County in order to safeguard their own interests. Once again, District 2 locals responded to their initiative by besieging District 2 President John Brophy with resolutions asking that the district and national organizations send organizers into Somerset County as quickly as possible.[74]

In February and March 1919, Brophy, the District 2 executive board, and district locals energetically tried to win the support of the national organization for a drive to organize Windber and Somerset County. In terms of popular support, at least, this period in early 1919 seems to have been an optimal moment to inaugurate an organizing campaign. The rank-and-file were insurgent and enthusiastic and wanted to retain the 2,000 or more new union members and the various organizations founded since 1916. Windber-area miners and businessmen seemed to be poised for action too.

It was in this context, on March 9, 1919, that one of the largest fires in Windber's history devoured a major section of downtown Windber. The fire of unknown origin had begun in the Lyric Theater, and it destroyed the hall in which the Citizens' Association met as well as some of the businesses of the top leaders of the organization.[75]

Time was running out for the independent business group's civic movement and for joint action by it and the union. Brophy again argued the case for organizing Windber with Hayes and Lewis on March 21, 1919. In a confidential letter written to another union officer on the following day, Brophy reported the results of that meeting. He told Robert Harlin, president of the UMWA's District 10, in Washington state, that Lewis had expressed grave doubts about the entire project because of the powerful interlocking nonunion and financial interests involved there. Moreover, Lewis believed there was some reason to hope that Consolidation Coal Company would accept a union in its Maryland operations at contract time, and he feared antagonizing its officials.[76]

Brophy did not share Lewis's optimism about the Consolidation Company. The conservative course and policy recommended by the vice-president meant inaction in the Pennsylvania field. The result of the meeting was that "President Hayes apparently acquiesced to a policy of this kind which Mr. Lewis summarized as being the only plan that could meet with success without the loss of life or expenditure of large sums of money."[77]

The failure of the national UMWA to formulate a plan or to commit itself to a well-financed coordinated drive to organize Somerset County at this time doomed the Citizens' Association in Windber as well as many of the district's new locals. It is impossible to say what, if any, other considerations had influenced Lewis in the decision not to organize. It is worth noting, however, that by then he had developed a highly controversial mysterious relationship with Alfred Hamilton, the same man who had entered into a questionable deal with John Mitchell involving Windber in 1906.[78]

In 1921, two years later, a Windber-area miner and his wife used the example of the Citizens' Association to warn a young activist, Powers Hapgood, not to say anything about the union when he was in town because he would be run out of Windber, or perhaps even killed, if company officials learned he was collecting information for John Brophy. They told him what fate had befallen the association, and Hapgood quoted the woman in an unpublished portion of his diary:

> [She said:] "There was that big fire a couple of years ago that started in the theatre where the Citizens' Association held its meetings. It took down almost a block and Habeeb's store with it. Mr. Habeeb

opposed the company and he got burnt out and moved. We can't prove it was the company that done it, but we think it did all right."[79]

In January 1919, *Coal Age* had reported that there was considerable friction at the Windber mines and that the friction had led to the formation of the Citizens' Association. No other news about these events was forthcoming until June, when the operators' journal briefly noted that Habeeb and another Windber merchant had purchased some land near Meyersdale, where they presumably relocated.[80]

Postwar Militancy and the Red Scare

Contemporary grass-roots initiatives to organize Windber had failed for the time being, but Windber-area miners and citizens could not ignore the unprecedented labor unrest occurring throughout the nation and the region in 1919. During this one year alone, 4 million or more workers in key industries took part in major strikes that sparked the Red Scare and other hostile, antilabor reactions. Steel workers and coal miners were among those who went on strike in the nation and in the immediate Johnstown region in the autumn of 1919.

Such strikes were fueled by the economic situation. Working people continued to suffer from the ongoing rise in the cost of living. A survey by the National Industrial Conference Board in 1919 indicated that the cost of major food products alone, the most important item in a worker's budget, had risen 85 percent during the period from July 1914 to July 1919. Meanwhile, real wages had declined. Overall, wages were worth 14 percent less in 1919 than they had been in 1914.[81]

Many found their postwar plight intolerable. Former miners, for example, had been implored and induced by the U.S. government to serve their country by leaving lucrative jobs in the munitions industry and returning to the mines in the summer and fall of 1918 to bolster coal production. Several months later, when the war had ended and the coal market had declined, they frequently found themselves unemployed, underemployed, and ignored by that same government.[82]

After the war, the expectations of working people remained high, and rank-and-file miners were representative of a new militancy. In September 1919, at the UMWA's international convention held in Cleveland, organized miners enthusiastically endorsed a large wage increase, but they also supported more radical goals, including a six-hour day, nationalization of the coal mines and railroads, and democratic workers' control of such industries. Operators had been hoping for a return to the status quo ante, with wages and working conditions resembling those of 1913. Meanwhile, even as the convention met, steel workers launched the Great Steel Strike. Thus, it was in the midst of much industrial unrest that the postwar struggle between capital and labor in the coal industry was coming to a head.[83]

The showdown was being waged over two different interpretations of a term in the wartime bituminous coal contract, the Washington Agreement, which had been negotiated under the government's auspices and signed on October 6, 1917. One of the contract's provisions was that it was binding "during the continuation of the war and not to exceed two years from April 1, 1918." The dispute centered on the dating of the war's conclusion. The operators contended that the war was not over because the Treaty of Versailles had not been ratified, that therefore the contract was still in force. With the armistice signed, the hostilities ended, and their economic situation steadily deteriorating, the miners wanted redress and declared the war officially over effective October 31, 1919. The contract no longer applied, as of November 1. A strike was imminent.[84]

In the heady days of October 1919, District 2 held its biennial convention in Johnstown, where the Great Steel Strike was in progress. Long before the coal strike of 1922, considerable friction had developed between the national UMWA's acting vice-president, John L. Lewis, and John Brophy over the issue of organizing the unorganized, especially those in Somerset and Cambria counties. The dispute involved more than miners. Early in the year, District 2 had aided a drive to organize iron workers at Cambria Steel Company, and in September miners employed by the company had gone out on strike to support the steel workers, even though the national UMWA had not authorized either action. Throughout the steel strike, District 2 did what it could to aid the Johnstown effort, and with far greater enthusiasm than the national organization. For example, Joe Foster, District 2's Hungarian-speaking organizer and a Windber native, traveled throughout the district to help set up cooperative stores that would aid the strikers.[85]

The October District 2 convention enthusiastically endorsed the 60 percent wage-increase demand of the UMWA's recent Cleveland convention, approved the proposals for the six-hour day and the nationalization of the mines, called for a labor party, and offered aid to the Johnstown strikers. "A rousing vote adopted the resolution declaring that the entire Pennsylvania mining field should be organized by a vigorous campaign," the *Johnstown Democrat* reported, and President Brophy singled out Windber as one of the district's main targets.[86]

The convention also implicitly recognized the importance and difficulty of organizing the company town when it adopted a resolution submitted by the Portage local. The resolution recommended that the charters of all locals be draped for thirty days as a gesture of respect for the memory of Joseph Ginter [Genter], who had died in Iowa in May. Ginter [Genter] was remembered as

> a gallant fighter for the right of the miners to organize, a pioneer of the movement in Central Pennsylvania, a man who suffered much in the cause of labor, whose efforts to organize the miners in Windber, in 1906, resulted in his being waylaid and brutally beaten by hired thugs, causing broken health from which he never fully recovered.[87]

On November 1, 1919, in the midst of the ongoing steel strike, the nation's unionized coal miners went out on a massive strike of their own to support a wage increase and the radical demands of the UMWA's recent Cleveland convention. The strike was effective in District 2's organized mines, and even some nonunion mines joined in the stoppage. Dominick Gilotte, District 2's Italian-speaking organizer, stated that no more than sixty miners were at work in Berwind-White's St. Michael's operations. The *Johnstown Democrat* reported: "It was further stated that 60 per cent of the Windber miners took the day off and that there was much activity in the big coal town last night, when company officials had reason to believe that Saturday's layoff was more than the customary celebration of a church holiday."[88]

Berwind-White reacted to the possibility of a strike at its Windber-area operations by suppressing civil liberties in more than the usual ways. The Windber borough council, firmly under the company's control again after the demise of the Citizens' Association, passed a new sweeping ordinance that unabashedly denied American citizens the constitutional right of assembly anywhere, at any time, for any reason, in Windber.[89] Only the burgess, a Berwind-White office employee, could approve and grant petitioners a specially written, stringently limited, conditional permit that he could revoke at any moment. On November 5 the *Johnstown Democrat* reported: "Burgess B. B. Barefoot of Windber yesterday issued a proclamation stating that conditions existing elsewhere are 'causing a feeling of unrest among the employees of this vicinity,' and calling upon all persons residing in the borough 'to proceed about their daily vocations quietly and peaceably; not to gather upon the streets and highways of the borough and vicinity; not to injure any property, public or private, and not to interfere with other persons lawfully at work.' "[90]

In the end, the showdown between the coal operators and the UMWA was premature, muted by the intervention of the government, which had first tried to prevent the strike, then sought to suppress it, and finally determined to resolve it. Rank-and-file unionists widely violated the injunction the government had secured against the strike on November 8, 1919, even though Acting President John L. Lewis revoked the strike order and urged compliance in a famous statement that concluded: "We cannot fight the government."[91]

The strikers remained out for several more weeks and returned to work only after the government had granted an immediate pay raise of 14 percent and agreed to establish a representative commission to investigate conditions in the coal industry before recommending a final Bituminous Commission award. Meanwhile, on December 12, 1919, District 2 operators, unhappy with the settlement and possibly bolstered by the collapse of the steel strike in Johnstown, issued a formal statement of protest, concluding: "It is merely a postponement of the showdown which, in our opinion, is bound to come."[92]

Shortly before April 1, 1920, President Wilson announced his support of the majority report of the Bituminous Commission, which had granted the miners a 20 percent increase for tonnage rates and a 27 percent increase for day labor rates. A strike was averted when representatives of the miners and operators met in joint conference in New York and approved the award on March 31, 1920, although neither side was happy with its terms. Because the new contract, based on government policy, was not to expire until March 31, 1922, the postponed showdown, a classic confrontation between labor and capital, was rescheduled. From 1920 to 1922, both operators and miners maneuvered for position. That the momentum had shifted—that employers were increasingly on the offensive and workers on the defensive—is obvious from the terms in which the nature of the oncoming fight was expressed. Capital spoke of reestablishing the "open shop" or instituting the "American Plan," while labor proclaimed a policy of "No wage reductions!"[93]

Windber-area miners, their families, and other citizens were among the working people who experienced a general rise in expectations during the war and in the postwar period. The long-standing grievances they had against Berwind-White and the company-town system persisted after 1906 but existed in a new context after 1916. If for no other reason than that Windber was a major coal producer, its population could not be immune to the UMWA's new initiatives and District 2's mobilization, to wartime social conditions and cost-of-living increases, to the federal government's increased powers and its precedent-setting policies endorsing labor's right to organize and bargain collectively, to the democratic ideology that simultaneously stemmed from the government's prosecution of the war and from the labor movement's offensive to extend democracy at home, to the Red Scare, and to the unprecedented wave of postwar labor unrest that was occurring all around them.

Windber-area miners and business people responded to these events by trying to get Windber "Free for Democracy," as Robinson had put it. They made repeated requests that the UMWA come in and organize Windber, and they solicited help from the federal government against a corporation that they considered to be autocratic and un-American. Their grass-roots initiatives, overlooked in district and national accounts, indicate that latent if not always overt support for unionization was widespread among the town's foreign-born and American-born mining populations as well as among many small-business people of various ethnic origins. This sentiment for organization was a powerful undercurrent that was more fully expressed in action when miners went out on strike with the district and national union in April 1922. Robinson was perhaps quite close to the truth when he noted that these workers seemed merely to be waiting "untill some one gets it started."[94]

In 1920 and 1921, as in the preceding years, Berwind-White miners, District 2 locals, and District 2 itself continued to make passionate appeals to the national to initiate an organizing campaign in Somerset County, but those appeals were denied.[95] John L. Lewis and the executive board of the International reiterated their interest in a full-scale campaign there but always postponed the drive to a more favorable time in the indefinite future. Responsible individuals could reasonably argue either that the UMWA unwisely missed a golden opportunity to organize these nonunion fields, or that it wisely avoided a serious, probable costly defeat, given the power and resources of its corporate opposition.

Meanwhile, the question of organizing the unorganized in Cambria and Somerset counties was becoming more and more intertwined with union politics, rivalries, and personalities. On the whole, Brophy, who was interested in formulating a concrete program to implement democratic nationalization of the mines, and in forming a political labor party, was far more radical than Lewis.[96] Eventually he would emerge as Lewis's major rival and the representative of an alternative to conservative policies. Nonetheless, his pushing for a joint organizing drive in the period from 1916 to 1922 seems simply to have been a responsible and human response to initiatives that originated "from below."

Berwind-White's antiunion policies were not substantially changed by the events of the war and the postwar periods. Old prejudices about the docility of southern and eastern Europeans were retained, as were old forms of repression. The company continued to recruit new immigrants and to rely on ethnic enclaves and key ethnic leaders to keep the nationalities separated. "Americanization," even of the open-shop variety, remained the antithesis of its policies.

In July 1921, District 2 President Brophy granted Powers Hapgood permission to work as a nonunion miner in order to investigate the conditions that prevailed in nonunion fields for the Bureau of Industrial Research. The young activist began his tour in the midst of a severe industrial depression at a time when the labor movement was increasingly on the defensive. Nonetheless, this tour brought him into close contact with the specific grievances, conditions, and general tenor of nonunion life.

Hapgood began his travels by working a stint at Scalp Level's Mine 40, and in the conclusion of his *In Non-Union Mines: The Diary of a Coal Digger*, published in April 1922, he not only quoted a Berwind-White miner from Scalp Level but made an unwittingly prophetic statement:

> Grievances of one kind or another always seem to exist [in nonunion fields]. The greatest is the feeling among the miners that there is no organization to enable them to have an equal chance with the operators in the settling of differences. When the discontented young miner at Scalp Level said: "The company's got a man where it wants him here and the man can either take it or leave," he expressed

the feeling of hopelessness that most unorganized miners feel in respect to getting their rights. That feeling of hopelessness gives way to intense determination when a drive for union organization begins, and this intense determination, if it is met with force by the operators, leads to the long strikes, the bloody conflicts, and the suffering that have existed in the dreary mining camps of Colorado and Alabama and which are going on now in the mountains and valleys of West Virginia. Some day these Pennsylvania miners will begin to join organizations.[97]

10

The Strike of 1922

The hopes of Windber's nonunion miners for a better, more just, more democratic life, which had been raised during the war and repeatedly frustrated in the postwar years, were visibly aroused once again during the spring of 1922 when the UMWA became a militant defender of labor's recent gains. As the ongoing working-class support for various concepts and proposals—including nationalization of the miners and railroads, workers' control and industrial democracy, and the formation of a labor party—indicated, miners, other workers, radicals, and progressives had not yet given up on the possibility of achieving some sort of imminent reconstruction of the existing social and political order.

Still, wage workers were confronting great economic hardships. In particular, the depression of 1920–1921 and the pressure, even on union operators, to decrease miners' wages because there were "too many mines and too many miners," and the success of nonunion operations in capturing a greater share of a diminished coal market, were widely felt throughout all coal fields. In this context, the UMWA—the nation's largest trade union, with a membership of over 500,000 in 1921–1922—seemed to offer mine workers the best opportunity for maintaining decent wages and for meeting broader working-class aspirations.[1]

Meanwhile, the massive strike wave of 1916–1922, unprecedented in American history, was reaching its denouement. A working-class mass movement that had begun in the context of plentiful jobs and a government somewhat sympathetic to organized labor now culminated in defensive strikes in coal, railroads, and other industries in 1922 in the context of depression, open-shop employer offensives, and a government hostile to labor. As David Montgomery has argued, the United States' participation in World War I had accelerated developments that paved "the way for a decisive confrontation between the working class and the state in the years 1919–1922."[2]

Given the depth and extent of postwar dissatisfaction in Windber, it should not be surprising that Windber's foreign-born and American-born nonunion miners made an autonomous decision to join the nation's organized bituminous miners on strike in April 1922. Nevertheless, their action seems to have stunned Berwind-White officials, Washington politicians, and the mass media, who wrongly assumed that this company's new immigrant work force was strike-proof. Meanwhile, delighted union officials could only comment that the walkout was "surprising but it ain't surprising."[3]

Probably all were amazed at the duration and dimensions of the strike. For almost seventeen months, massive numbers of ordinary working people made enormous sacrifices to try to gain a union, which they saw as the only way they could assume democratic control over their workplaces and community. With District 2's help, Windber-area miners and their families stayed out for one whole year beyond the date when John L. Lewis and the International had negotiated a national strike settlement that left them and other newly organized miners out of the final contract. Ultimately the grass-roots strike was lost—but through no fault of the strikers themselves, who had won it on the local level.

The Somerset County strike, the subject of Heber Blankenhorn's classic *Strike for Union*, was of great local, regional, and national significance.[4] In terms of local impact and long-lasting memory, this strike was perhaps the single most important event in Windber's first century. Elderly people interviewed in the mid-1980s cited it and the strike of

1906 as the community's top two landmark historical events, followed by the floods of 1936 and 1977. The class warfare, outmigrations, wholesale evictions, harassments of private gunmen and the state police, life in tent colonies, and the disruptions within community institutions, affected the lives of everyone who lived in the area in March 1922.

By coming out on strike in April 1922, tens of thousands of Somerset County's nonunion miners, including those in the Windber area, made an enormous contribution to John L. Lewis's success in negotiating a contract for previously organized fields in August 1922. But their exclusion from the settlement also exposed the inadequacies of the International UMWA's strategy for organizing the unorganized. Thus, the Windber strike helped spawn the formation of a radical movement, first from within the UMWA and then from without, to provide an alternative to John L. Lewis's conservative policies. District 2's John Brophy offered a genuine alternative both to corporate power and to "business unionism" through "The Miners' Program" of nationalization, a labor party, and a shorter work day and week—a program endorsed by the national union in 1919.

During this eventful era, Windber miners pioneered in new methods of struggle, including picketing of the multinational coal corporation's offices in New York City. Because of their actions, the strike itself became an issue in New York City progressive politics. As we shall see, the miners succeeded in convincing Mayor John Hylan to send a committee of prominent New Yorkers to investigate living and working conditions in Windber because the city bought coal for the subway from E. J. Berwind, who was president of the Interborough.

The prolonged nature of the strike revealed the depth of the miners' commitment to unionism, but it also showed the tremendous power capital could wield against organized labor, and the importance of the role played by federal, state, and local governments in labor disputes. Berwind-White had amassed huge economic resources and corresponding political clout. It had at its disposal armed gunmen, the state police, nativist prejudice, evictions, injunctions, and other antilabor laws, in what one historian has termed labor's "lean years" in the 1920s.[5] In addition, the unequal contest enabled the coal company to fuel intraunion battles that left victims, including Windber miners and their families, in their wake.

It is probably not accidental that many of the individuals—Lewis, Brophy, Hapgood, Blankenhorn—who played key roles in the Somerset

County strike of 1922 went on to make important contributions to the American labor movement in the 1930s. Presumably, each had learned something from his previous experiences. By the time of the New Deal, bitter realities had even led Lewis to appreciate the wisdom of his chief rival, John Brophy. In 1933 and again later in the decade, the UMWA president called on the former District 2 president to help organize the unorganized in mass unionization drives for the UMWA and then for the Congress of Industrial Organizations (CIO). Hapgood joined in too, while Blankenhorn advised the new National Labor Relations Board and served as chief investigator for the LaFollette committee on civil liberties in industry.[6]

As we look at the narrative of the local strike and consider its wide-ranging impact, three issues deserve special attention because of their broad historiographical significance. First, there is the persistent issue of ethnic stereotypes. Despite the work of Victor Greene on the anthracite region, and a number of other excellent studies on immigration and labor, the view that southern and eastern European immigrants were poor union material remains dominant in many circles.

This old ethnocentric view, perpetuated by coal operators, mill owners, and others, primarily on racial and ethnic grounds, needs to be continually challenged and critically examined. In 1906, Windber miners took actions that supported Greene's conclusions that a multinational working class was not an insurmountable obstacle to unionization and that new immigrants were often more responsive to organized labor than their American counterparts.[7] Such early actions in Windber and elsewhere should have but did not jettison another erroneous belief, cited by David Brody and others, that new immigrants became interested in unions only after they had decided to remain permanently in the United States sometime during World War I.[8] Obviously, a broad range of particular circumstances at given moments in time—not one single causal factor—determined whether or not workers would unionize or strike. That Windber's foreign-born miners may have decided not to return to Europe is not sufficient reason to explain why they supported unionization and went out on strike in 1922. They expected to return to Europe and had done likewise in 1906.

Second, the issue of why nonunion new immigrant workers acted to unionize *when* they did, in general, needs to be examined. The Windber case refutes the old notion, originated by social scientists Clark Kerr and Abraham Siegel, that new immigrant miners went on strike because they were "isolated masses."[9] The reverse is true. Windber miners went on strike in 1922, as in 1906, because they were in touch with the mainstream American labor movement and wanted to be part of it; they did not strike because they were "isolated" from the dominant American middle-class society. Their grievances were perpetual and deeply felt throughout the nonunion period, but at key moments they mobilized for union because they had hope. As their expectations for change rose during the war and the postwar period, they had approached the district and international unions time and time again for a commitment to help them organize. Now, finally, with the district and national UMWA organizations on the move, they understood that they had a reasonable chance to succeed against their autocratic employer. They did not have to be isolated from their fellow workers.

Third, issues related to immigrant communities and "Americanization" need to be addressed. The Windber strike indicates that the successful mobilization of the American labor movement was not contingent on the imposition of cultural conformity or the destruction of ethnic communities, as Thomas Göbel and others have suggested.[10] That the strike occurred at all indicated that Berwind-White's reliance on ethnic stalwarts and ethnic enclaves to keep the workers compliant and divided had failed. It was in moments of such strife that the class divisions that existed within the immigrant communities surfaced most fully. At the same time, World War I, Red Scare deportations, and Congressional measures to restrict immigration were forcing new immigrants to consider in a serious way what it meant to be an American, or at least a noncitizen in an American setting.

Conjuncture of April 1922

Heber Blankenhorn, who chronicled the events of the strike in Somerset County, stated: "I don't believe anybody will ever know exactly where the Somerset strike started." In the first week of April, town after town "exploded" in spontaneous outbursts in which miners stayed home from work. Windber was one such town, and its example was contagious. Neighboring towns, such as Jerome and Boswell, followed suit soon after. At the same time, a similar walkout was occurring in the Connellsville coke region, which had not experienced such an upheaval since 1891.

Berwind-White and Frick miners were strengthened in their resolve when each learned that the other had come out also.[11]

From the standpoint of Windber-area miners, it was they themselves who had taken the initiative on the local level to organize. Union-minded workers had been working in secret for some time for organization. In March 1922 they once again made an effort to get District 2 and the International involved, and some visited union locals in neighboring places. On March 30, 1922, William Parks, recording secretary of Local Union No. 2233 in Beaverdale, Pennsylvania, wrote John Brophy: "I have been authorized by our L. U. to request you, to get in touch with the International Union, and have organizers sent into the Win[d]ber field." He explained the reason for the request: "There has been a few men here from Windber, who states, that at this time, it would be possible to do something in that Field. Hoping you will take this matter up as soon as possible."[12] After the strike was under way, a *New York Herald* reporter wrote: "Miners of Windber say that they decided on the strike themselves and that the union knew nothing about it until Windber asked that men be sent to organize it."[13]

By contrast, John Brophy liked to think that, under his leadership, District 2 had aroused Windber-area miners from the outside and led them out on strike. His response to the Parks letter is therefore illuminating: "The matter of reaching non-union fields and having the non-union workers join with the union in the strike on April 1st has been attended to in this district and adjacent territory, including Win[d]ber." However, he was not especially optimistic: "Just what the response to this appeal will be remains to be seen."[14]

John Brophy had recently returned from an International UMWA Policy Committee meeting that had coordinated plans for a nationwide suspension of mine workers in case no contract had been reached by March 31, 1922. Near the end of their deliberations, in what many historians have considered an afterthought, committee members approved a decision to issue a call to all nonunion miners to join the union members on strike.[15]

Yet the invitation to the unorganized must be considered more than a simple afterthought, if only because nonunion miners were widely reported to be of strategic importance to the strike's outcome. For months before the strike, coal operators, newspapers, and prominent officials of the U.S. government had openly proclaimed that nonunion coal production would meet the nation's needs in the event of a strike and that nonunion fields held the key to the union's defeat. Already, cutthroat competition between the nonunion and union sectors of the overproducing "sick" bituminous industry was occurring. In the past year's campaign against wage cuts, the *UMWJ*, the union's official organ, had repeatedly and vigorously urged nonunion miners to organize and thereby do their part to maintain existing wage rates.[16]

When Brophy returned from the Policy Committee meeting on March 24, he immediately proposed a plan to get nonunion miners in Somerset County out on strike with the union men. In an oral history and in his memoirs, he described the somber response he received from district officers and organizers. Union men had dubbed Somerset County "Siberia," as it was understood in the days of czarist Russia, and the seriousness of the undertaking was not lost on the men. Windber miners had to break *out* of Siberia to get outside help. District organizers had to break *into* Siberia to provide help.[17] A Windber strike leader described the situation the organizers faced:

> These organizers didn't dare come in here with their own car. You could not come in on the streetcar, the Johnstown Traction Company, because if they got on the streetcar in Johnstown, the Traction Company conductor would stop in Moxham and go to the dispatcher. He would call the General Office that a stranger is on the streetcar. They all knew these people. . . .
>
> When the car arrived in Windber, the Berwind's Coal and Iron Police was there. As soon as they [the organizers] got off, they wanted to know who they were, where they was from, what their business is. And when they found out, they told them: "If you know what's good for you, you'll get back on the streetcar and get back out of here." They wasn't allowed in here. They had to get out.[18]

But Brophy was not deterred. He set up a strike headquarters in Cresson, Pennsylvania, and arranged for a union printer to print special strike cards. The cards appealed to the nonunion men to strike, quit work, join the fight, and end the fear that dominated their lives and ruled their company towns, where democratic rights were nonexistent. He also ordered mass quantities of *Penn-Central News*, a district union newspaper, for distribution, because it too carried the strike appeal. Organizers were told to take the newspapers and the cards, which fit neatly into their pockets, and to enter the county from all four directions at once. The question was not which organizers could avoid detection but how far each could get before he was arrested or deported from the county. If arrested, the men were instructed to make as much fuss as possible in order to attract attention and publicize their purpose.[19]

On March 31 the invasion of the county began. That evening, Arthur Taylor wrote Brophy from the jail in Boswell and informed him, "Cowan's in Jail." Board member David Cowan and two assistants had entered Windber earlier that day. They had bought up that day's issue of the *Windber Era* and substituted the *Penn-Central News* in its place. They rented a hotel room and began to distribute the strike cards on the town's main streets. At about 2:30 P.M. they were arrested and taken to the municipal jail without formal charge. Several hours later, Burgess Barefoot held a hearing in which he charged that the men were suspicious characters and guilty of distributing literature without a permit. Cowan protested that he was an authorized agent of the union newspaper, but in order to be released the men had to return to their hotel room and turn over their literature. At 8:00 P.M., three Coal and Iron Police—Morris, Wilkinson, and Collins—escorted the organizers out of the county on a streetcar. During the course of the day, two or three other organizers had also been arrested in Windber.[20]

Cowan later testified in a Somerset County courtroom packed with miners at a hearing requested by Berwind-White and other coal companies for the purpose of securing a permanent injunction against the union. For the companies, the issue was unlawful conspiracy; for the union it was civil liberties. Percy Allen Rose, a Berwind-White attorney, assumed that it was a crime for Cowan to have entered Windber. Rose asked, "You were in the borough of Windber March 31?" Cowan quickly responded, "Yes, and they put me in jail for it." Rose then asked if he had distributed any literature, which he also considered a crime. Cowan laughed and said, "No, they caught me too soon."[21]

In April 1922, self-generated local grass-roots initiatives consciously linked up with a planned district-wide effort that was itself designed to implement the stated policy of the national organization. Autonomous movements therefore came together in the walkout that surprised so many. The strike was not as "spontaneous" as it must have appeared. Blankenhorn himself stressed that a visible presence on the part of the union was what made Somerset County miners commit themselves to the strike. Apparently almost any sign—the presence of an organizer or word of the attempted distribution of literature—was a sufficient spur to those local men who had wanted organization all along. But detailed court testimony also indicated that many of the miners who led the work stoppage had not known about the appeal to nonunion miners and had not seen the literature distributed on March 31.[22]

The successful convergence of these autonomous movements was critical to the success in District 2 in getting nonunion miners to join and then stay out on strike. In most nonunion situations, one of these movements without the other would have doomed an effort by either. The walkout simultaneously showed the depths of the miners' discontent and their awareness of contemporary regional and national affairs. It is also useful to remember that all nonunion regions in the nation did not come out on strike in April 1922, and that all district presidents did not act to implement the last-minute decision of the Policy Committee to call on the nonunion miners to join the strike. Blankenhorn noted that nonunion miners came out en masse in only three regions of the country, all located in Pennsylvania. In each case, district and local UMWA officials had taken specific actions and issued an appeal. Where this was not done, as in Kentucky, no nonunion walkout occurred.[23]

The strike movement was uneven but contagious. St. Michael miners came out on strike on April 1. When they reached their workplaces that morning, they found the union's strike cards, which the fire boss had carefully distributed as he made his morning rounds to check for dangerous gas. On April 4, Windber men attended a special meeting of St. Michael strikers in Nanty-Glo and again told union officers that Windber was ready.[24]

Evidence suggests that spraggers and motormen were the catalysts of the immediate Windber strike. Miners who held these positions enjoyed a certain "miner's freedom" as they moved about the mines from workplace to workplace. In the course of their regular jobs, they came in contact with workers of various nationalities and learned how to communicate with them. A Windber strike leader who had been arrested for distributing the union newspaper described how the local had secretly organized. He himself was a Slovak spragger.

> The haulage men of Berwind-White—I mean mine motormen and spraggers—were in contact with all the miners in the mine because they always ordered their cars, the amount of empty cars that they wanted or the loads that they pulled out. The haulage men got

> together in 1922 to get it organized to come out on strike, and we avoided informers. [We] made sure they didn't contact them so they didn't know anything about it.[25]

On Wednesday evening, April 5, approximately 300 men, including many spraggers and motormen, met together in a company-owned park. The state police, which had been sent to the coal region in preparation for April 1, chased them away. The men regrouped on a public road and formed several committees to act as messengers to go find union representatives in other towns.[26]

New immigrants used their kinship and ethnic ties in the process. A Hungarian miner, Steve Foster, thus went to the union town of Nanty-Glo on Thursday to find his brother, the Hungarian-speaking organizer Joseph Foster, and board member William Welsh. Other men went to DuBois to locate District 2 Vice-President James Mark. David Cowan was approached by Windber miners at a ball game in Portage. Jim Gibson went to South Fork. Others went to Cresson to see T. D. Stiles, editor of the *Penn-Central News.*[27]

The meeting on April 5 was notable for other reasons. While it was in progress, Berwind-White's Superintendent Booker arrived and spoke to the workers to try to avert a walkout. He asked them what they wanted but was not prepared for their response. Jim Gibson, subsequently president of Local 4207, spoke up and said, "We want organizers." Booker then promised that if the men remained at work, he would rescind the 40 percent wage cut the company had instituted on April 1 but not disclosed until April 3 or 4.[28]

That the men's grievances went beyond specific wage rates became clear from their response. They told the superintendent that they wanted their next wage scale in the form of a union contract instead of arbitrary posted notices or verbal promises. Activists later testified that news of the "secret" Berwind-White wage cut, which had coincided with their appeals to the workers to rise up and organize, functioned as a live firecracker under the men but denied that the wage cut alone had tied up the mines. Meanwhile, according to Blankenhorn, other nonunion companies in Somerset County were furious that Berwind-White officers in New York and Philadelphia had chosen that particular moment to reduce wages.[29]

The overtures of the Windber committees evoked positive action on the part of the union. On Thursday afternoon, April 6, Stiles and

Cowan came to a field in adjacent Cambria County to speak to 2,500 workers assembled there. The field was owned by the Samuel Hoffman family, oldtime farming residents who had fought their own battles with Berwind-White and remained uniquely independent. As the organizers approached, the miners chanted: "No work tomorrow!" They immediately elected nine committees of three miners each to take down names and collect dues. On that day, they organized three separate locals and elected officers. Local 4207, the largest, included the miners of Mines 31, 35, and 36 as well as those of Lochrie Mine 41. Local 5229 covered Scalp Level, which included Berwind-White Mines 30, 37, and 40, located in Cambria County. Local 5231, the smallest, was composed of Berwind-White miners at Mine 42.[30]

The process of closing down the mines was uneven but thorough. Miners in Mine 36 quit work on Thursday morning, April 6. Motormen Steve Foster and James Murray, a popular ball player, drove the first motor of the day into the mine and derailed it. When the first shift of workers arrived at the blockage, the word "strike" quickly surfaced, and the men walked out.[31]

In Mine 30 in Cambria County, all of the 173 miners, loaders, scrapers, and drillers had joined the strike by April 9. A majority of these did not work after April 5. Mine 35 lagged a bit by comparison. Here the last day of work for most spraggers, motormen, and other company men was either April 8 or April 9, but a minority, including spragger Martin Madigan, Local 4207's treasurer, were not on the payroll after April 5, the date of the miners' meeting. In any event, employment records in the company's general office indicate that it was extremely rare for workers not to have joined the strike. Most work records note an interruption in employee work histories, dating from April 1, 1922.[32]

On Sunday, April 9, union officials held an even larger meeting. John Brophy, International board member Charles Ghizzoni, Tom Stiles (editor of *Penn-Central News*), John Brezezina (a Polish organizer who had once worked in Windber), Joe Foster, and George Bassett spoke to a crowd of 4,000. On that day alone, $2,000 in dues were collected, and dues were 50 cents a person. Foster was placed in permanent charge of the organizing effort in the Windber area.[33]

Once under way, the strike in Windber was effective. On April 20, *Coal Age* admitted that there was no production forthcoming from mines there and that the South Fork branch of the Pennsylvania Railroad was closed as a result. On May 10, 1922, the antiunion *Somerset Herald* reported: "The mining industry of Somerset County is at present facing the greatest crisis it has been called upon to meet since the development of this region started thirty years ago."[34] At an injunction hearing on May 23, Thomas Fisher testified that the stoppage had a disastrous impact on his company's business and that Berwind-White was purchasing coal for its steamship contracts from Belgium, France, and Great Britain. He regretted that Windber's 3,500 miners were losing business to others and no longer producing the daily average of 250,000 tons.[35]

The retroactive wage cut was undoubtedly a powerful mobilizing force, but not the essence of the miners' struggle. As of April 1, daymen who had been earning $7.50 a day were reduced to $5.00 a day. Tonnage rates dropped from $1.28 to $1.01. Suspicion that Berwind-White weights were dishonest remained an embittering issue. There were other grievances concerning pay issues. In recent months, certain categories of nonproductive work that had been modestly remunerated were reclassified and put into the category of deadwork, where no pay was received. Hapgood had been impressed with the discontent over deadwork and short weights when he had worked in Berwind mines in 1921. All of these grievances contributed to the sentiment for union, which offered the only hope of genuine redress.[36]

Joseph Foster wrote a short statement, "Why Are the Windber Miners Out?" in which he expressed many of the miners' long-standing grievances. Of Hungarian descent, he had entered the mines in Windber with his father twenty-six years before, when he was eleven years old. He noted that he had felt cheated every single working day throughout these years and did not believe he had ever received proper credit for the coal he had loaded. Whenever the workers got a raise, which he surmised was related to union advances elsewhere, his pay never seemed to rise. He charged that company weights were simultaneously reduced when pay raises were advanced. He had been discharged several times for asking questions about the weights or suggesting that his motor needed repairs or telling an official who asked that he was voting for a judicial candidate disliked by the company. Hours and working conditions were terrible. He ended by quoting miners who attended one of the early meetings: "We are no longer slaves and we are done loading three ton for two. We will never return under a scab system. We want union to protect our rights."[37]

More than a year later, with the Windber miners still on strike, John Brophy submitted a brief to the U.S. Coal Commission, formed after the national settlement, on violations of civil liberties in Somerset County. He listed six demands of the nonunion men as follows:

1. For collective bargaining and the right to affiliate with the union.
2. For a fair wage.
3. For accurate weight of the coal they mine. (Experience teaches us that this can be secured only when the miners have a checkweighman.)
4. Adequate pay for "dead work."
5. A system by which grievances could be settled in a peaceful and conciliatory spirit by the mine committee representing the miners and a representative of the operator.
6. But above all they struck to secure their rights as free Americans against the state of fear, suspicion and espionage prevailing in nonunion towns. Against a small group of operators controlling life, liberty, and pursuit of happiness of large numbers of miners. To put an end to the absolute and feudal control of these coal operators.[38]

Windber-area miners had struck for union, as the title of Blankenhorn's book indicated. Without an organization, democratic representation, and a union contract, the workers were at the mercy of arbitrary company actions, such as "secret" wage cuts, unfair discharges, and blacklists. The right to organize and to engage in collective bargaining were at the heart of the struggle, but the struggle to gain the right to organize necessarily involved a struggle for constitutional rights, including free speech, free assembly, and the right to vote without intimidation.[39]

On Strike with the National

The period from the onset of the strike until the Cleveland settlement of the national strike on August 15 was one filled with great hope for Windber-area miners. For the time being, they were part of the mainstream of the organized labor movement, on strike with unionized miners throughout the nation and with various other formerly nonunion miners, no longer pariahs. Blankenhorn described this period as an era of relative naiveté, as well as genuine mass democratic aspirations and mobilization.[40] However, Berwind-White, other coal companies, the three local boroughs, and the state were not idly standing by while the miners organized.

Berwind-White used three powerful weapons—guns, courts, and evictions—to defend the open shop. From the first days of the strike and throughout it, company gunmen, guards, special deputies, and the state police maintained a pervasive presence in the Windber area. That the real purpose of the state police, who were commonly called "Cossacks," was to break the strike was widely assumed in labor circles. In Windber, the events of 1906 and the state constabulary's role in them were recalled.

Brophy and the union were subsequently able to support their contention that the state police were sent into the coal fields as strikebreakers by publishing a secret memo written by Lynn G. Adams, superintendent of the state police, to coal operators on March 18, 1922. In order to have information to dispatch state troops effectively, Adams had asked operators a series of eleven questions that showed his department was not concerned with enforcing the law and maintaining order impartially. One question asked was: "Have you an understanding with the Sheriffs of the Counties in which your operations are located?" Another was: "Can you supply me with the name and description of all known radicals living in the vicinity of your operations?"[41]

Berwind-White's General Manager Thomas Fisher replied to Adams on March 29, 1922. He reported that the company had 3,544 men at work in the Windber-area mines, and another 461 employed in Clearfield County. In the event of a strike, he expected the mines to operate with the old employees and did not intend to import strikebreakers. He and John Lochrie, operator of an affiliated coal company, had made a personal visit to the sheriff of Somerset County, and the company did indeed have arrangements with the sheriffs of Somerset, Cambria, and Clearfield counties. As for information on radicals, Fisher noted: "We have the information if required by you."[42]

Despite the strikebreaking role maintained by the state police in general, something unusual and unexpected happened in Windber. Sergeant George Freeman, commander of state troops at Windber, volunteered to protect the striking mine workers from the numerous gunmen whom Berwind-White was currently importing.

Throughout the strike, Freeman continued to amaze coal operators because he insisted on enforcing laws impartially. He stunned observers when he testified at an injunction hearing in May that miners had been peaceful, that they had a legal right to hold peaceful meetings when not trespassing, and that police and gunmen had no business attending such meetings or listening in on them. He checked the activities of company gunmen on various occasions, and even accompanied union representatives to witness affidavits that area miners and families subsequently filed to charge violations of their civil liberties. By summer, the operators were actively seeking his removal, and the union vigorously protested to keep him from being transferred away from Windber. It was Freeman's individual initiative, not state policy, that caused the state police in Windber to serve as a rather impartial arbiter of law and order in the 1922 strike instead of an overt repressor, as they had done in 1906.[43]

The company and the Windber borough council were quick to mobilize their own repressive forces once the serious possibility of the strike arose. Berwind-White, which traditionally used the Tanney Detective Agency in Pittsburgh, sent to outside agencies for men. On April 6, the day many miners struck, the Windber council met and authorized Burgess Barefoot to employ three special police officers and to purchase five automatic revolvers and ammunition. In addition, a rifle case was installed in the police station, and the Citizens' National Bank and the Windber Trust Company each loaned the borough a Winchester, while Berwind-White supplied five shotguns and seven rifles. In the first week of April, the burgess also posted "no trespassing" signs on the company's many properties.[44]

The Windber council continued to authorize the use of special police paid for by Berwind-White. On May 1 it approved the hiring of twenty-four extra policemen without pay and one with pay; on June 6, an additional thirty-three without pay and one with pay; on July 6, another twelve without pay and two with pay. When the strike approached its second year, the council again built up its forces. Blankenhorn noted that 122 deputies were on the sheriff's regular payroll but paid for by Berwind-White, which had more such men on the county's rolls than any other single company in Somerset County. Company employment records from April to July further confirm that its officials had hired and maintained at least 100 extra special police and guards for ongoing use in the Windber vicinity.[45]

There were many charges of petty harassment and intimidation by the company guards throughout the strike, but there was one event that mobilized the miners with potential serious consequences. Early in May, company guards had invaded the home of John Rykala near

Mine 36 on the pretext of looking for illegal liquor. The husband and wife claimed afterward that one of the guards had raped Mrs. Rykala. Neither the sheriff, who on other occasions had characterized the strikers as "ignorant foreigners," nor Berwind-White officials gave the charge any credence. Consequently, Brophy and the union provided legal counsel, investigated the case, and petitioned the governor. Union leaders had to convince a large crowd of Windber men not to lynch the accused. Eventually the case did get to court, but it was dismissed on the grounds of an imperfect identification by Mrs. Rykala. This incident seems to have incensed the miners more than any other single criminal action of the strike.[46]

Closely linked to police power as a means to combat the strike was the injunction and the full legal force of the state. On April 19, 1922, Berwind-White and about nineteen or twenty other coal companies filed bills of complaint in Somerset County Court seeking a preliminary injunction to stop the defendants from "doing or causing to be done any act or thing that will infringe upon the rights of the plaintiff to employ non-union labor." Among other things, the company's brief stated that the purpose of the UMWA was to "encourage strikes," that it had engaged in many lawless acts in the county and elsewhere, that it had "distributed large quantities of scurrilous, libelous and defamatory literature," and that "the defendants entered into an unlawful combination and conspiracy with one another . . . to destroy the cordial relations existing between the complainant and its employees."[47]

A series of widely publicized hearings followed in which the companies, led by Berwind-White, and the UMWA fought an instructive battle over the right to organize. The companies were unhappy with the temporary injunctions immediately granted because they were limited to enjoining the UMWA members from committing acts of violence or intimidation, but not from organizing nonunion miners through peaceful methods of persuasion. Brophy issued a warning to all striking miners to obey the injunction rigorously. The operators then sought a stricter permanent injunction against organizing per se.[48]

At the April 28 hearing, the companies charged that the union was engaged in an unlawful conspiracy and cited the infamous *Hitchman v. Mitchell* case, which UMWA lawyers said did not apply. Employees of District 2's nonunion companies had not signed "yellow dog" contracts and had not made verbal agreements to refrain from joining a union as a condition of employment. Judge John Berkey chose not to rule on the right-to-organize issue and scheduled final hearings on the injunction for May 22. However, he refused the operators' requests to issue an order to halt the meetings of strikers and ordered company officials, armed guards, and police to stay away from the miners' meetings. As before, organizing could continue peacefully and without intimidation.[49]

The hearings from May 22 through May 25 were more of the same. Brophy was asked pointed questions insinuating that he or the district organization had received money from competing operators to organize Somerset County mines. When Berwind-White's case came up in court, David Cowan, the Foster brothers, James Gibson, Martin Madigan, and others testified that the company's miners had voluntarily asked District 2 for help and that the union was not engaged in an unlawful conspiracy, as charged, to destroy cordial employer-employee relations, which the union maintained were nothing but "a sham of the thinnest veneer." According to Blankenhorn, the evidence presented was so convincing that afterward Berwind-White officials blamed the entire strike on the wage cut.[50]

But Berwind-White also switched tactics in court. For the first but far from the last time, the company claimed that the national union had not authorized the strike in the nonunion fields and did not support it. For the moment, the charge was apparently without effect. Brophy successfully refuted it by citing the convention's actions and the authority, composition, and decision-making powers granted the Policy Committee.[51]

The UMWA defendants also argued that they were not unlawfully interfering in the operator's business, which Thomas Fisher testified had been severely damaged. When Berwind-White produced six or seven witnesses who had quit striking and claimed intimidation, the union successfully refuted the charges one by one, and Sergeant Freeman testified that the miners had been law-abiding throughout the strike.[52]

On the final day of the hearing, the union went on the offensive and put the guard system on trial. William Gwyn and Sherman Finney, two miners from the hard-coal region, testified that they had answered an ad in a Wilkes-Barre newspaper that told of work for $5 to $14 a day where no labor trouble existed. They had been shipped in a special train with forty-three others to Windber, where they were lodged in company houses. They reported that some had refused to work when they learned that a strike was on and that the guards had offered to protect them. One guard, a man named Martin, told them he would furnish them with guns, and he advised the strikebreakers to shoot anyone who accosted them.[53]

Berkey, an elected judge, repeatedly postponed a final decision on the injunction with regard to the UMWA's right to organize nonunion fields. At this stage of the strike, he quoted the Pennsylvania Constitution and maintained that miners had a legal right to free assembly and speech. But while he allowed the status quo—peaceful organizing—to continue, he also began to make specific rulings that hampered organization. He limited the number of pickets, required meetings to be held within strict distances from company premises, and even regulated the voice level that Powers Hapgood and other organizers could use when talking on a public road.[54]

Evictions were a third powerful legal weapon employers had at their disposal to try to break the strike. Berwind-White had threatened to evict strikers by April 10, but negative publicity about the police and operators' actions had led to a temporary delay. In May, sheriffs and deputies began to execute large numbers of eviction orders.

Berwind-White and Consolidation Coal Company led the nonunion initiative. District 2 reported that Berwind officials had issued 196 eviction orders, as of May 6, 1922. In other places, evictions were a matter of public record, but not so in the Windber area, where private company guards did the evicting and kept no public records. At the same time, company officials either denied the evictions or admitted to very few. But by May 1923 the Windber police estimated that 600 had occurred in Windber proper. Strikers claimed that 2,000 had taken place in the company's area operations. Blankenhorn preferred the estimate of Joseph Foster, who believed that there had been about 765 evictions, including 46 at St. Michael's. Given that each company house contained families and boarders, even the lowest figures suggest that thousands of people were affected by the orders. Others who anticipated receiving such orders left before receiving formal notification.[55]

Men and women who were young children at the time of the strike recalled their horror at the scenes in which they witnessed families (sometimes their own) and all their earthly possessions being delivered into the streets by armed guards. Soon after the strike began, company managers and their assistants had made the rounds of all company houses and collected the storebooks. Credit, formerly required, was no longer available at the company stores. But strikers viewed the evictions as the ultimate act of inhumanity. No allowances were made for ill children, women about to give birth, or similar exceptional hardship cases.[56]

Union miners who owned their own homes and sympathetic businessmen responded to the evictions by taking in as many people as they could. One Italian woman who lived in a private home described her reaction to the eviction process and what she and many others did when they learned of the wholesale evictions which were taking place:

> So then when they went on strike, Berwind went there and threw everything out from these families. That was a shame. You don't do stuff like that. You stop them, the book [storebook]. You don't give them nothing. You don't give them the [electric] power, okay, the light, but don't throw their stuff out. These people were stranded. They were out there picking up their things. I told my husband, "Well," I says, "yonze go pick some families up." So they [husband and others] went out, too, and they brought me a family. She had a couple kids.[57]

All families could not be so accommodated, and the union found itself faced with a tremendous task in finding housing for thousands of people in the Windber area and in other parts of District 2, where evictions were also taking place. From May on, there were tent colonies of strikers and their families situated on the outskirts of Mines 40 and 37.

Meanwhile, the strike had taken the ethnic leaders who were the linchpins of the Berwind system by surprise. Once such ethnic foremen and contractors realized the depth of the strikers' feelings and the gravity of the situation, they were careful. For several months they maintained a low visibility and bided their time. They had not dared to preach defeat during the initial phase of the strike, for fear of losing whatever influence they might still have had over the workers. At least one long-term ethnic contractor chose the moment of the strike's outbreak to leave the mines permanently and open a small business.[58]

What had characterized the strike from its onset was the emergence of new ethnic leaders who openly supported the union and who sought to displace the influence of the old ethnic leaders who were loyal to the company. The new leaders arose from the ranks of the miners. To organize successfully, they had had to avoid known ethnic enemies and company spies. The Foster brothers and other local unionists who

A row of company houses with a tent colony of evicted families in the background, as seen from one approach to Windber, 1922.

emerged from the mining ranks had been inhabitants of their stratified ethnic communities. Conscious of their class position, they happily offered the foreign-born miners a viable alternative to Berwind's ethnic linchpins.

The emergence of the new ethnic union leaders, which appeared to be so spontaneous, was the result of years of exploitation in nonunion mines and towns. Blankenhorn described the phenomenon of mass volunteers, old unionists and new, of all nationalities, tramping from one place to another throughout the county in April and May to aid organizing efforts. Ordinary miners visited relatives in other towns, and people who had never in their lives given speeches were suddenly appearing before crowds of hundreds or thousands.[59] Without the conscious participation and commitment of such laboring men and women, the Windber strike could never have been sustained as long as it was, even with the help of district organizers and officers.

Moreover, the new ethnic leaders and speakers represented many different nationalities. Given the composition of the work force, inclusion of all major ethnic groups in the union was necessary for success. In a democratic manner, miners thus elected officers of the new Windber-area locals who, in general, reflected the ethnic composition of the region. Union meetings were always lengthy because of the multiple speeches in disparate languages. The class experience had made it possible for working people to transcend ethnic divisions, but enormous practical difficulties were involved. Above all, miners of all nationalities had to find some way of meeting and acting together in an autocratic setting.

Joe Zahurak, a Slovak who became president of the Windber local in the 1930s, and Alex Kada, a Hungarian friend, were two of the many ethnic union activists in 1922. On their own initiative, they used Kada's car to drive to different places to give speeches whenever they got word that some miners, especially those with large families, were thinking about returning to work. They urged the miners to stay out and offered encouragement. Between the two of them, they could speak Slovak, Hungarian, Russian, Polish, and English. Whenever possible, they got an Italian friend to join them.[60]

Because organizers could not get into town in their own cars or on the streetcar, Zahurak and Kada again acted. The local men had been forced to go into hiding when company guards told them to get out of town. But until Kada's car was blown up in the middle of the night, the two men took it on themselves to bring outside organizers to local union meetings. Zahurak explained: "We had to go to Beaverdale to get a Slovak organizer, Mr. Slivco. We had to go to Nanty-Glo to [get] the Hungarian-speaking organizer, which was Joe Foster. In Homer City there was John Ghizzoni and Charles Ghizzoni, the Italian." The miners met in ball parks or farmers' fields, and they wanted outside speakers who spoke their languages. Zahurak noted: "You had all them nationalities at that time—that generation. You had to use the different nationality speakers to interest them in staying out."[61]

Immigrant societies could not but have been affected by the disruptions of the strike. Fraternal societies, one of the twin pillars of the ethnic communities, continued to proliferate on the very eve of the strike. Members of a new organization, the Russian Culture Society, had inadvertently chosen this most inauspicious moment to apply for a charter. The result was that, on May 1, when the Windber council authorized the employment of twenty-four extra police "without pay," it also unanimously decided to oppose the society's application in court because "it would be a very bad thing for the town." The new Russian society thus became one of the first victims of the strike.[62]

From the strike's inception, foreign-born miners had rallied to support the strike. They quickly took steps to take charge of their own institutions and run them in their own class interest. They disciplined strikebreakers and ousted the ethnic leaders who supported Berwind-White. The Slovak societies serve as an example of what occurred within various immigrant communities. On April 23 the St. Thomas Chapter 304 of the National Slovak Society resolved to deny sick aid to members who worked during the strike. On May 7, Lodge 48 of the Pennsylvania Greek and Roman Catholic Union denied any sort of aid to scabs. On May 21, Local 292 of the First Catholic Slovak Union, or Jednota, passed a similar measure.[63]

The fraternals also voted to expel the members who worked during the strike and were not up-to-date in the dues payments. In July, Lodge 48 decided to throw out all working members who had not paid their dues for the last thirty days. On June 20 an outraged Kunrad Valent, a member of one of the Jednota lodges, appealed to the secretary of the national organization. He claimed that the Windber local had previously allowed a member to owe dues for three months before suspending him, but that in his case, "when owing two months dues they refused to take any more money from me, and told me they had scratched me off the books because I was working during the strike."[64] The national secretary upheld the local's action: "I do not know as what your local society may have against you: These strike and strikebreaking business do not concern us, but for non payment of dues they had right to expel you."[65]

During the strike, fraternal societies also contributed money directly to the UMWA or helped their own members who were on strike. In July 1922, St. Thomas Society 304 and the Slovak Political and Educational Club each contributed $100 to the UMWA. In June, Lodge 48 members chose to follow a policy they said had been adopted in many other local fraternals—they spent $100 to aid their own members who were on strike.[66]

Women and children also rallied to support the strike, which was a family affair as well as a community one. The strike naturally had an enormous impact on all members of the family, and active female support was of critical importance to the outcome. Without the help of foreign-born and American-born wives, sisters, mothers, and daughters, the men could not have remained out for three, let alone seventeen, months. The prevailing sexual division of labor had decreed that women bore the primary caretaking responsibility of home and children. Only by their own staunch commitment to a union could they have endeavored to see their families through the hunger, evictions, and hardships of tent-colony life, and the many deprivations that their children as well experienced.

In addition, women and children played a special role in disciplining strikebreakers. Newspaper reports indicate that they were often prominent in the ranks of the pickets who confronted strikebreakers on their way to or from work, although injunctions greatly restricted these activities. Despite fines and harassment, immigrant women and children continued to offer strikebreakers bread crumbs and pennies as symbols of shame.[67] This gender-based tactic was perhaps as effective as it was because it hit at male concepts of manhood that underlay gender roles in the family economy of the miners, as well as at notions of class solidarity.

It is also possible that company guards and police sometimes allowed women greater leeway than they allowed the men at such times, for political and other reasons. After all, Windber miners and the general public had been outraged over the rape incident previously described. On another occasion, Robert Baylor, who was driving strikebreakers through a Windber street, swerved and hit a striker's young child. When he tried to continue on his way, an angry crowd stopped his car. Fourteen guards then threw tear gas into a group of women and children who were standing in the vicinity. At least one man required medical care. The union gathered affidavits from witnesses to this and other incidents and later gave them to Gifford Pinchot, who became governor of Pennsylvania in 1923. One of the new governor's first actions was to begin a formal investigation of police activities in the coal strike.[68]

In any event, women were not immune from arrest. On June 7, 1922, seventeen Windber women were arrested and taken with their babies and young children to Johnstown, where they were rather quickly released. A coal and iron police officer had charged them with disorderly conduct for attacking miners en route to work. As the strike went on, however, Somerset County women who were brought to court on such charges were treated much more harshly than before. Judge Berkey began to set bond at $100 per person, but the fines did not stop the women, who continued to picket and shame strikebreakers.[69]

Meanwhile, as the national strike continued into May and June 1922, officials of the federal government repeatedly stated that the administration's policy was "hands off"—which was exactly what the coal operators wanted. But the Herrin massacre on June 22 and the railroad shopmen's strike on July 1 led President Harding to take some action. First, he summoned Lewis to the White House and tried to win concessions from him. Having failed in that, he then called miners and operators to a secret conference on July 1. Over the next few days, negotiations went on, but to no avail. The operators presumably insisted on a district-by-district, not interstate, settlement and on reduced wages—two demands rejected by the UMWA.[70]

It was in this context of secrecy and public conjecture that the fears of Windber and other newly organized miners became aroused. Since May, Berwind-White had been claiming that the national union did not support the Somerset County strike. Then too, Windber miners must have remembered that, as newly organized miners in similar circumstances, they had been shut out of the final settlement of the 1906 strike.

In any event, the miners voiced their concerns and asserted themselves into the politics of the situation. On July 4, 1922, the three Windber-area locals (and many others) sent a resolution to Harding, Lewis, and Brophy to demand that "we expect, and think it only just, and proper that we should be included in any settlement of the present coal strike, which is made, and that such a settlement should carry with it recognition of the United Mine Workers of America, in these newly organized coal fields, of Somerset County."[71]

On July 10, Harding tried to break the deadlock, but he only offered the miners resumption of work at the old wage scale until August 10. Then, a commission of three miners, three operators, and five presidential appointees would award a new scale and make a thorough investigation of the industry. Lewis asked for a week to consider the arbitration proposal, which the UMWA and many operators ultimately rejected.

On July 18, Harding wired the governors of twenty-eight states to invite coal operators and miners to resume work under the guarantee of armed protection to those willing to work. Governor William Sproul of Pennsylvania immediately sent 1,100 troops, including the cavalry and machine-gun battalions, into the bituminous coal counties. The presence of soldiers actually diminished coal production. Miners, not soldiers, dug coal.[72]

The fears of Windber miners that they might be overlooked or "sold out" in the conflict with the operators were temporarily assuaged. On July 24 the Windber-area locals sent Brophy a resolution that expressed confidence in both the national and the district organizations "for the manner in which you have handled the strike situation, up to the present time, and that you continue to handle it for our best interests. Until its final determination."[73]

By August, the railroad strike was beginning to affect coal supplies. With a stronger hand, Lewis thus invited coal operators to meet in the old interstate joint conference manner in Cleveland on August 7. Few operators attended, and the conference adjourned while attempts were made to induce more attendance.

The deadlock was broken on August 14, when T. H. Watkins of the Central Pennsylvania operators' association suddenly reversed his position and joined the conference. On August 15, through a combination of bluff, bluster, and hard, often secret, dealing, Lewis succeeded in signing up operators who accounted for the production of about 40 million tons of coal. Getting other companies to sign up proved to be difficult, but a breakthrough had been made. Lewis termed the Cleveland settlement, which retained the old, higher wage scale for the old unionized mines, a "great victory." Critics asked "great victory for whom?" Certainly not the Somerset County and Connellsville newly organized miners who had been arbitrarily left out of the settlement.[74]

Carrying On Alone

While the national settlement was still news, fifteen district leaders and Somerset County delegates met to consider forming an opposition ticket to field against Lewis and the national officers for the December election. Nominations were due August 22. They wired Brophy in Cleveland to discern his sentiments and received a negative response to their proposal. Brophy later denied that he ever received their message, which had been presumably intercepted by Lewis forces, who then faked the reply. Despite the bogus telegram, the delegates went ahead

UMWA miners decide to continue the strike at outdoor meeting, Bantley Place, Scalp Level, 1922.

and nominated Tom Stiles for president as well as other candidates for national offices. On August 16, Powers Hapgood wrote his parents that he did not want to mix in union politics but that he "always knew Lewis was unprogressive and a politician, but I did not realize before that his political ambitions led him to dishonesty. He must be defeated."[75]

Intraunion dissension was an inevitable by-product of the terms of the national settlement. Although the threatened challenge from District 2 to Lewis's reelection in mid-December 1922 from District 2 fizzled, the cooperation between the district and the International from the August settlement until then with regard to the fate of Windber and other miners who had been left out was tenuous at best.

Meanwhile, the Cleveland settlement gave Berwind-White important ammunition with which to discourage the strikers. On August 17 the company announced that it would not sign the Cleveland agreement and would continue to run nonunion. In its customary manner, it raised wages to the old rates to meet the competition of the union mines, and vaguely suggested that it might consider the employment of a checkweighman but would never collect the checkoff. Old ethnic bosses began to preach that the strike was unwinnable.[76]

District 2 officials debated what to do about the thousands of excluded nonunion miners who remained out on strike in their territory. On August 18 they pledged to support them, even though they were momentarily preoccupied with formulating policy, signing up the district's union operators, and concluding on August 23 a district-wide settlement, based on the Cleveland agreement, for unionized operations. Unlike the national organization, the district executive committee set an important precedent by refusing to sign up some of the mines of a company if it did not sign up all of its various mines. Meanwhile, Brophy appealed for financial aid from the International to continue the Somerset strike while the district board took immediate action of its own. It approved an assessment of $1.00 per working miner per pay period to support the ongoing strike effort.[77]

From August 26 to August 29, District 2 officials and seventy-two delegates from forty-two locals representing 25,000 miners who had been excluded from the national agreement met together in Somerset to decide on their future course of action. They simultaneously invited nonunion operators from the struck mines to a joint conference, but none came. The miners concluded their debates by resolving to pursue the strike until victory, and they thanked District 2 for its efforts on their behalf.[78]

Meanwhile, Foster, Hapgood, and other organizers in the field were kept busy explaining the International's betrayal and trying to persuade the miners to continue the fight with district help. Union leaders dissatisfied with the Lewis settlement were caught in a dilemma. Hapgood wrote his parents: "While we know that Lewis has gone back on us, we have to defend him before the men to encourage them because if they will only stick solid they can win without help from the international union."[79]

In September, the men overtly acknowledged the importance of women to the success of the ongoing strike. District 2 and UMWA locals called a mass meeting of women to explain the district's relief policy, which could provide minimal help at best. There was no point in trying

Customers and clerks at a Graham Avenue store established by the union to provide necessities and relief for the strikers.

to continue the strike if the women in mining families opposed doing so. They had to feed their families somehow. After listening to the union's explanations in various languages, however, the women declared their resolve to continue the strike.[80]

Relief payments and resolve were especially needed after the national strike settlement, when Berwind-White again evicted strikers en masse in a futile attempt to force an end to the ongoing local strike. The union began to move miners and families to union mines and towns whenever possible, so the men could work and support the strike. When winter approached, the UMWA rented almost any sort of structure or shelter it could find—chicken coops, barns—and constructed some barracks for the remainder. In October, New York reporters photographed the tent colonies, and New York City investigators interviewed the families who had not been relocated. The number of infant deaths due to exposure and poor living conditions will never be known, although individual instances are known. Any summary of the casualties of the strike must take these deaths into account.[81]

Ethnic unionists also made occasional requests for aid from outside fraternal societies. In September, District 2's Polish organizer, John Brzezina, took leaders of the Polish National Allied Societies on a tour of Somerset County, including visits to the various Polish societies, in hopes that the larger organizations would provide some financial assistance. Meanwhile, ethnic organizers maintained a watch over foreign-language newspapers and informed or protested to editors when they published misleading ads for "steady jobs and better prices than union mines" in Berwind-White or other nonunion mines in Somerset County. Unlike the Connellsville area, where coke operators imported black strike-breakers from the South, Berwind-White seems to have relied on recruitment of new immigrants from the anthracite region.[82]

On November 14, 1922, George Swetz, an officer of Jednota Lodge 196, wrote to the secretary of the national organization: "There has been a strike on here since April 1. The people are as upset as if it was the war and in the same way lodge 196 is upset. The leadership is moving as I wrote you last spring."[83] The analogy of the strike to the world war in Swetz's statement was no exaggeration, given that class war was indeed occurring.

Class warfare within the immigrant communities had eventually spread to the other bulwark of immigrant communities—the immigrant churches. As a result, three priests lost the support of their parishes; two left town in 1923, the other left a little later. At some indeterminate point during the strike, but probably after the national strike settlement, the Hungarian, Polish, and Italian priests infuriated miners by preaching from the pulpit that the men should return to work. Father James Saas, the object of some controversy in 1906 and the only priest present during the onset of both strikes, was especially detested. Joe Zahurak was one of those interviewed who described the situation:

> There was Father Saas in the Polish zone, the Polish church, and Father Lach was here in the Slovak, but Lach didn't take no part in it to preach in church what the Berwind-White wanted. But Father

> Saas announced in the '22 strike, told the people, go back because they will not recognize the union. Father Fojtan of the Hungarian Roman Catholic Church, also done the same thing. In his preaching in church he told the membership that they'd better go back to work 'cause they're not going to be recognized as union, that they're just going to be suffering for nothing. Up at the Italian church was Father Leone. He was up there. He also took part in it and ordered and asked them to go back.[84]

Zahurak then explained how the striking miners reacted to what could only be considered betrayal:

> When these priests announced in church, in their sermons to the people, and asked them to go back, Father Saas then next morning found a coal miner's tools—pick, shovel, tamping bar, and augers to drill the holes, a hatchet, a saw, and a sledgehammer. That was the entire tools for a coal miner. That was tied up and put on his front porch at the parish. "Father Saas, you can take these tools and go ahead! Go to the mines!" The same thing happened to Father Fojtan, the Hungarian priest. They gave him a set of tools and put them on his front porch over there, at the rectory over there. And Father Leone, he got in trouble over that. That's when he left up there then.
>
> Father Saas was got rid of because the people, most of them, got discouraged. Father Lach had brains enough. He sympathized with the people, Father Lach did. He sympathized with them and everything else. They're up against a tough battle. He didn't think the working man's going to be able to win, but he didn't insist that they'd better go back because it's in vain that they were striking. Now other churches, like the Protestant churches and the Evangelists or different American-speaking nationalities, weren't involved that much because they [American-speaking nationalities] weren't there. Italians, Slovak, Polish, Russian, and all these [were] here because that's a majority of the coal miners.[85]

It was true that foreigners constituted the bulk of the mining force and the strikers, but Protestant ministers also occasionally sympathized with the strikers and felt the wrath of the company as a result. When Elizabeth Houghton, a New York reporter from the *World*, came to

Windber in October, she talked to the Rev. Dr. G. W. Pender of the Methodist Episcopal Church, who was about to depart for a new parish in California, Pennsylvania. She quoted him on his reasons for leaving:

> I am not leaving because of company pressure, but relations have been far from pleasant since I spoke at union services on the rights of the working man. I was told in open meeting that if I did not stop expressing myself on labor matters I would not stay in my position another year. My best chorister had been bought away from me and two company officials who were on my board, resigned at the end of the year giving no reason, but the majority of the congregation have supported me.[86]

Pender also told the reporter that the company used un-American methods in suppressing any leaders who opposed its policies and cited the case of the Civic Society's recent rise and fall. Mr. Mickle, one of its leaders, had been discharged by the company and never rehired.[87]

The morale of the Windber strikers, which had been at its nadir after August 15, rose again in October because of a union initiative that carried their cause to the public and to government officials in New York City and Washington. Blankenhorn, Hapgood, and others from the Bureau of Industrial Research devised the plan. Miners' delegations from Berwind-White, Consolidation, and other Somerset County coal mines came to New York City in late September and petitioned Mayor John Hylan, the state fuel administrator, and the city's Board of Estimate. E. J. Berwind, the coal magnate, was in a vulnerable position because of recent investigations that showed he used his position as a director of the Interborough Rapid Transit system to obtain the city's lucrative coal contracts for Berwind-White. New York subways ran on Windber coal![88]

Arrangements had been made for representatives from Berwind-White mines—James Gibson, Joseph Phillips, Mrs. Harry Beal—to testify along with other Somerset County delegates in hearings before the Board of Estimate on September 26. The delegation charged that E. J. Berwind was a nonunion operator who refused to meet with them, cited many of their specific grievances, and generally made their case. "We are plain rank-and-file coal diggers, ready to go to work on the same terms as the 500,000 other members of the United Mine Workers have accepted," they emphasized.[89]

The miners made a connection between their lack of rights and poor working conditions, on the one hand, and the city's unnecessary coal shortage and poor quality of subway service, on the other. The Interborough had attempted to justify high coal prices and poor service in an ad that claimed that the coal strike had cost it an additional $1 million. The City of New York had been paying $7.65 for a ton of coal before the strike, but Berwind-White miners proved that they had been paid only $1.28 a ton then. Moreover, the miners claimed they had mined the coal under un-American conditions and argued that the same financial interests that took advantage of the New York public were trying to crush unionism in Pennsylvania through the use of armed guards, evictions, criminal actions. They concluded: "We are the victims of gigantic steel and transit corporations who are more determined to crush unionism than to give public service."[90]

The Windber-area delegates then called on the City of New York "to see that the miners for its public utilities are not the victims of slavery of an un-American life." They made two specific requests. First, they asked the city to press Berwind-White to produce its coal under conditions that were fair to labor. It could immediately do so by using its influence to get Berwind to confer with his employees. Second, they asked the city to send a committee to Somerset County to investigate conditions in Berwind-White mines and thereby verify their charges.[91]

E. J. Berwind formally replied to Mayor Hylan's request that he meet to confer with Windber-area employees by issuing a public statement in which he denied the delegation's charges, which he termed "baseless and malicious." He said his company had always met with bona fide employees but never with "outsiders." He also said that he had paid "the very highest wages" for twenty-five years and had granted his men "fair treatment." He concluded his statement, however, by agreeing that a meeting could be arranged.[92]

In a letter to Charles Craig, city comptroller, the delegation of Pennsylvania miners protested Berwind's statement. After citing prior refusals of meetings, wage scales, working conditions, and the company's union-crushing activities, they concluded that they had been "short-weighted, defrauded and oppressed," that they had "struck against slavish conditions and joined the union," and that they were merely asking for "the same contract that four-fifths of the country's coal operators have already signed."[93]

The Windber-area miners then publicized their decision to test Berwind's offer to meet with its bona fide employees. When long-term employees of the company went to his office and were rebuffed, State Fuel Administrator William H. Woodin intervened and got E. J. Berwind to promise that the company's vice-president, Harry A. Berwind, would confer with a committee in Windber, but not in New York, on October 2. Pursuing their policy of publicly testing the company's good faith, the miners then left New York, which Berwind had wanted, but vowed to return to picket if the promise were broken, as they fully expected it to be.[94]

On October 2, Harry Berwind broke the promise by failing to appear at the Windber meeting. Instead, two company officials, Thomas Fisher and E. J. Newbaker, passed off the miners' delegation to Superintendent Baylor, who abruptly informed them that former employees could apply for work only as individuals and not under a UMWA contract.[95]

E. J. Berwind's failure to live up to his promise resulted in James Marks and six Windber-area miners coming back to New York to picket Berwind-White offices on Broadway in mid-October. They were pioneers of a new tactic. Apparently the spectacle of Windber miners, donned with miners' hats and head lamps, signs in hand, picketing outside the Berwind residence and offices was unusual enough to attract considerable attention. As a result, the city government decided to send an investigative committee to Windber, newspapers increased their coverage of the ongoing strike and committed some reporters to the Windber field, and progressive circles organized fund-raising and speech-making campaigns to aid the Somerset strikers.[96]

Coal Age denigrated the miners' initiative as nothing more than "a little grandstand play by union officials to get some notoriety and possibly sympathy and help from outside sources."[97] But strike relief and publicity and investigations were heartily welcomed by the strikers, whose morale was immediately bolstered by the New York campaign. By contrast, a simultaneous effort by a Windber delegation to obtain a hearing with President Harding failed utterly.

Mayor Hylan responded to the miners' initiative by appointing an investigative committee headed by David Hirshfield, the city's commissioner of accounts. Four other prominent New York City officials—John Lehman, assistant corporation counsel; Thomas T. Moran, examiner of the Bureau of Investigations, Department of Finance; Amos T. Smith, mechanical engineer of the office of the secretary of the Board of

Windber miners picketing at Berwind-White Company offices on Broadway, September 1922.

Estimate and Appointment; and Mrs. Louis R. Welzmiller, deputy commissioner of public markets—served as members. The group spent five days in Windber in late October and early November holding hearings, investigating conditions, interviewing the police, visiting tent colonies, and touring a mine. Initially, Berwind-White offered to cooperate, on

Hall where the Hirshfield Committee held hearings, Windber, October 1922.

the condition that the hearings be secret. When the committee refused, company officials blasted the inquiry. The company had withdrawn its special force of armed guards but only for the duration of the visit of the investigators.[98]

The New York City visitors were appalled by the treatment they received in Windber. Not only were they followed by company spies wherever they went, but their mail and rooms in Johnstown were ransacked as well. A New York reporter, Elizabeth Houghton, who had written several important sympathetic articles about the strikers before the investigation, complained: "A reporter is about as welcome a sight here as a stick of dynamite on a front porch."[99]

Strikers and their wives testified in detail about their many long-term grievances and the additional hardships associated with the strike.

Hirshfield Committee entering a mine, October 1922.

Afterward, the committee returned to New York. The strikers faced a rough winter. In December, as Powers Hapgood was busily trying to stretch relief money to buy shoes for the children living in tent colonies, lawyers filed a tax report on the estate of the recently deceased Mrs. E. J. Berwind. She had not left her husband any money because, as she had specified in her will, "he does not need it." She did, however, leave him her vast array of jewels, which were worth nearly $500,000 and which included a diamond-studded dog collar worth more than $22,000.[100]

On January 2, 1923, the Hirshfield Committee made the results of its investigation public. "According to the tales of horror recited before the Committee, the living and working conditions of the miners employed in the Berwind-White Coal Mining Company's mines were worse than the conditions of the slaves prior to the Civil War" was the devastating conclusion it had reached.[101]

Tent colony of evicted Berwind-White families at Mine No. 40, October 1922.

The report itself cited evidence that the committee had collected and described from its experiences. The document included a brief history of the investigation, an outline of its methods, summaries of testimonies heard and related findings, and photographs that illustrated the conditions it deplored.

The Hirshfield Committee report was a scathing critique of the Berwind Company's practices in both Pennsylvania and New York City. The investigators calculated that the company had made a profit of $1,617,000 from Windber coal sold to the Interborough in 1921 and that it would make at least $2,500,000 in 1922. At the very same time that the company had cut miners' wages in Windber on April 1, 1922, to $1.01 per ton, it had raised the cost of coal to the transit system by $1.00 per ton. In fact, it was the New York public—not Berwind-White—that was actually financing the strikebreaking fight.[102]

Barefooted women and children in one of the tent colonies in Windber, October 1922.

As a result of the investigation, David Hirshfield called on the federal government "to take over the coal fields, utilize them for the benefit of the people and place it beyond the possibility of any man, or group of men, to restrict coal production or its distribution." He also declared that the transit should be operated for the benefit of the people and that coal should be bought only from those "who pay their employees a living wage and treat them like human beings."[103]

The Hirshfield Committee report led to no important steps to change the existing situation, even though its general conclusions were published in the nation's major newspapers. One of the great ironies was that *Coal Age*, which belittled the inquiry, reported its findings accurately, while the *UMWJ* did not do so until two years after the release of the report.[104]

Evicted Berwind-White families living in henhouses and cowsheds at Mine No. 38, Seanor, 1922.

By January 1923 there was significant intraunion dissension and conflict between the national UMWA organization and District 2. Brophy was also emerging as a serious potential rival to Lewis, who had won reelection to the presidency in December. None of these developments augured well for the fate of Windber-area miners.

The ongoing Somerset and Connellsville strikes and the burst of publicity resulting from the New York initiative did, however, force the Department of Labor to take an interest in mediating the conflict. During the fall of 1922, government officials conferred frequently with John L. Lewis and other International officers, but as late as October 18, 1922, a labor conciliator reported to Hugh Kerwin, assistant to the secretary of labor, James J. Davis, that the International executive board had not yet formulated a policy for Central Pennsylvania.[105]

In mid-November, Kerwin sent several men to investigate conditions in Somerset County. Information in December led him to conclude that about 40,000 miners were still out in the Somerset, Cambria, and Connellsville regions and that the struck companies would not recognize the union, the principal demand. He seemed to think that the results of the district and International union elections in mid-December might lead to the union's acceptance of the best the Labor Department could offer—a vague promise that the companies would not discriminate against strikers after the strike was over.[106]

E. J. Berwind refused to promise even this much. In response to an inquiry, he informed the secretary of labor on January 10, 1923, that his company would continue to run on an open-shop basis but that all other matters, including reinstatement of former strikers, would be left to the discretion of local mine officials.[107]

There was much for Windber miners to be demoralized about in January 1923. Events beyond the miners' control directly or indirectly affected their lives. The most devastating blow was the national UMWA's calling off the strike in the Connellsville coke region on January 18, despite local and independent reports that thousands were still out and that the strike remained effective.

On January 25, Hapgood informed his parents: "The men in Somerset County are much disturbed by the calling off of the Connellsville strike and there are vague rumors among them that the strike here is to be called off also."[108] Foster was just as upset. Moreover, in connection with its calling off the strike, the national organization announced that it was discontinuing an assessment for relief aid, as of January. Thereafter, if the Somerset strike were continued, the district would have to finance it alone. The organizers in the field and the striking miners eagerly awaited District 2 decisions.[109]

Other events simultaneously intensified the intraunion discord. Lewis might have interpreted his uncontested electoral victory in mid-December as a mandate to move against his critics. In any event, afterward, he began to red-bait progressives in the union from a new position of strength.

In January the three-member Nationalization Committee of the International UMWA, which included Brophy, came out with its proposals in a pamphlet entitled *How to Run Coal.* Lewis apparently considered its publication a usurpation of his authority. The result was that, on February 1, Lewis's ally, Ellis Searles, editor of the *UMWJ* and a man who opposed the union's nationalization program and refused to allow discussion on the subject in the union newspaper, made a bitter personal attack on the members of the committee, whom he termed "Greenwich village coal miners." One by one they resigned, but the *UMWJ*'s attacks on Brophy continued.[110]

Shortly after the Connellsville settlement, Hapgood and Foster—who were each responsible for coordinating the strike in one half of Somerset County—began to attend meetings of dissident progressive miners who organized an intraunion movement to oppose Lewis, the settlement, and union corruption. Eventually, Lewis called for the men's expulsion from the UMWA for what he called "dual unionism." By March, District 2 and Brophy had became the favorite targets of the national UMWA's red-baiting campaign. The attacks were so vicious that even *Coal Age* objected and stated that Brophy was Lewis's strongest rival and a scapegoat. Tongue in cheek perhaps, the operators' journal then suggested that Brophy's radicals used pens for weapons—unlike the Lewis people, who preferred guns.[111]

Meanwhile, District 2 leaders decided to hold a referendum to determine whether the miners working in the district's union mines were willing to continue the $1.00 assessment per pay period to support the strike. National organizers, who had not previously helped in the Somerset strike, came into the district to campaign against the referendum. District officials campaigned just as hard for the assessment. On February 21, Richard Gilbert, district secretary-treasurer, announced the results of the referendum. Of 32,402 miners, 24,206 (75 percent) had voted to continue the financial support.[112]

Throughout the winter, the union continued to move miners and families to union mines and towns. Dispersion of the population continued as individuals sometimes left for large cities to find work. Although some of the old workers began to return to the mines, most of the miners who were working in the Berwind-White mines as the second year of the strike approached were imported. Furthermore, Blankenhorn used figures he had obtained from mine inspectors' reports on the number and nationalities of Berwind-White employees, as of January 1, 1922, and January 1, 1923, to prove that Americans—not the foreign-born—were the primary strikebreakers. As late as the spring of 1923, the mines were operating with a work force that was still 35 percent below that of the prestrike period and considerably less efficient.[113]

Union officials acted to reinvigorate the strike when it entered a second year. The district issued new strike cards and literature in an attempt to pull out all regional strikebreakers on April 2, Easter Monday, 1923. The effort was at least partially successful, but no one knew how long the new strikers would remain out.[114]

District 2 called a special convention for April 17, 1923, to assess the situation. A delegate from the Windber local reported that 950 of its workers were still out on strike. Scalp Level had 700 in that position. The total number of strikers still residing in the county was estimated to be between 5,000 and 6,000. The sessions were otherwise stormy. Representatives from the national and the district fought over many issues as they blasted Lewis or Brophy in turn. In the end, however, Brophy got almost all he wanted from the rank-and-file, including a decision to continue the strike.[115]

Coal companies responded to the spring initiative by instituting yellow-dog contracts and by securing a new injunction, which if obeyed would have made picketing and strike relief impossible. Foster, Hapgood, and others simply violated it. But the long months of sacrifice were beginning to be felt. In June, District 2 cut the amount of strike relief to $4.00 a week per man because of slack work in the union mines, and on June 6, delegates from twenty-four Somerset County locals met in Johnstown to assess their plight.[116]

The nature of their lamentable predicament is clear from the types of resolutions they passed. One resolution asked Governor Pinchot to provide adequate protection for the striking miners of Jerome, who had recently been assaulted by hired guards and strikebreakers. A second resolution called on the International executive board of the UMWA to send financial relief and "to make a public statement that they are back of the Somerset strike morally and financially" in order to refute the recent statements of nonunion operators who said that they were not. A third resolution expressed appreciation to the union miners who had helped them, and simultaneously reiterated their unanimous determination to "continue the strike for a Union contract and recognition of the United Mine Workers of America."[117]

The strike persisted into July and August. Some last-minute appeals for aid from the national and a truce with Lewis failed. How to end it then became the problem. Finally, after almost seventeen months, the district considered it unwise to carry on and called on the delegates from twenty-two Somerset County locals, a remnant representation of the original strikers, to meet in Johnstown in mid-August. On August 14, only one day short of the one-year anniversary of the national settlement, the locals voted to end the strike. But they held their heads high in doing so.[118]

The formal and final resolution the delegates adopted revealed their grass-roots generosity and basic fair-mindedness. They ascribed their defeat to the "brutal tactics and tremendous financial strength of the coal companies, as well as to the weakmindedness, selfishness and un-Americanism of strikebreakers." They refused to blame the national organization, even as they praised the officers of District 2, "who left nothing undone that was in their power to do for our benefit." Despite the tragedy inherent in the national settlement, subsequent controversial decisions, and intraunion bickering, they were only temporarily abandoning the fight. They concluded: "We . . . reaffirm our belief in the principles of unionism and declare our intention of keeping alive our union, even though we are working on an open shop basis, and await the first opportunity of winning a contract and union recognition." In defeat they were more charitable than many of their contentious leaders had been in victory.[119]

The seventeen-month-long strike of Windber-area miners in 1922 and 1923 was a landmark event from which we can draw a number of important conclusions. First, the walkout by Windber's nonunion, mostly new immigrant miners, along with their prolonged efforts on behalf of the union despite severe hardships, is further evidence that, under certain circumstances, these foreign-born people were organizable and good union recruits. They were neither the company's strike-proof, docile work force nor incipient anarchists ready to rise up at any moment to challenge the existing system. These workers were intelligent, pragmatic people. Such obvious truths need to be restated, because despite a wealth of good labor and immigration studies, ethnocentric stereotypes about "foreign" peoples (or their motivations) remain important, if often underlying, strands in American labor historiography.

Second, Windber miners struck for union in 1922 because they saw a chance to end their isolation, gain democratic control over their workplaces and communities, and become part of a broad-based class movement. They seized the opportunity to join with fellow workers, not the middle classes. Both new immigrants and American-born miners had been attuned to the political and economic issues of the war and post–World War I years, when organized labor had raised the specter of change. Although Windber miners struck primarily to end local perennial injustices, their aspirations for freedom from an autocratic employer and the slavelike conditions described by the Hirshfield Committee were of more universal significance. It is not coincidental that these workers struck for union at a critical juncture in the history of the American labor movement and at the end of a strike wave during which important sectors of workers had mobilized as a social movement that had raised important issues like "workers' control."

Third, the immigrant communities within which Windber's various nationalities resided had not averted organization. Berwind-White's open-shop strategy, based on segregated, stratified ethnic enclaves and loyal ethnic leaders, had failed. Class-conscious ethnic unionists who lived within these communities understood that they had to end company domination wherever it existed. To succeed in organizing, they—and only they—knew how to avoid their own ethnic spies, discipline their own ethnic strikebreakers, seize control of their own ethnic institutions, challenge the leadership of the company's old ethnic bosses, solve many practical difficulties, and transcend ethnic boundaries to work together with miners of other nationalities in a class-based union. That these miners did succeed, even temporarily, in all of these objectives was a significant accomplishment that made the prolonged strike possible.

This discussion of immigrant communities should not obscure the fundamental fact that American and ethnic strikers confronted a class establishment that was overwhelmingly American-born. Many miners were not American citizens at the time, but they had a sense of justice, and by going on strike they were actively rejecting the open shop, which corporations throughout the nation were then proclaiming as the "American Plan." Meanwhile, the United Mine Workers was offering them a more inclusive, labor-oriented Americanization "from below."

Once the strike was over, the company found that it could not completely return to the status quo ante. Foreign-born miners were not so docile after all. Although it continued to rely on old autocratic means and old ethnic bosses to keep its diverse workers unorganized, it also opened a night school for its employees that was in part designed to curb dissent and to co-opt potential ethnic union leaders whom it invited to attend. Ethnic miners and strike activists did attend, but they were not co-opted. It was the same leaders who led the unionization drive in 1933 when permanent unionization was effected.[120]

Fourth, the Windber-area miners who received none of the benefits of the national strike settlement had nevertheless made an important contribution to the great victory (for the older-organized miners) that John L. Lewis proclaimed in August 1922. The nonunion walkout in Central and Western Pennsylvania had immensely strengthened his hand during the negotiations. Coal operators and government officials, who had counted on this region's nonunion coal production to defeat the strike, were disappointed. Indeed, production in local Berwind-White mines for the entire year of 1922 totaled only about half of what it had been in 1920 or 1921, and much of this annual tonnage had been produced in the three months before the strike.[121]

Historians have noted that the great victory Lewis proclaimed in 1922 was a Pyrrhic one, given subsequent developments, coal-industry problems, and the expansion of the open shop. But this insight was not so obvious in 1922. Nor was the subsequent ossification of the labor movement that David Montgomery described. For all the strike's failings, for the moment anyway, the UMWA stood firm. Its membership had successfully foiled an attempt to destroy the organization and impose drastic wage cuts.[122]

Fifth, the primary reason for the defeat of the Windber strike lay in the enormous economic resources and political powers that Berwind-White could wield against foreign-born and American-born miners who were trying to organize. Without similar resources and clout, there was no way that Windber strikers, alone or with the generous help of District 2, could win the protracted struggle against the combined might of the corporate interests—Berwind-White, the Consolidation Coal Company, Cambria Steel, J. P. Morgan, John D. Rockefeller, the Pennsylvania Railroad—who were represented in Somerset and Cambria counties. Moreover, the contrast between the relative friendliness toward labor exhibited in 1917–1918 by the Wilson administration, and the negative policies of state and federal governments in 1922–1923, underscored the importance of the state in all matters that affected the labor movement. Nevertheless, UMWA policies and dissensions, about which more will be said in the next chapter, must also bear some share of responsibility for the defeat of the Somerset County strike, which Windber miners had won on the local level.

Throughout the lengthy strike, Windber-area miners had remained loyal to the union. It is significant, however, that for most of them "the union" was synonymous with the District 2 organization, especially after

August 1922. In an ultimate sense, it made little difference to them whether union corruption, secret deals, or political and economic realities had been responsible for the terms of the national settlement and the national's lackluster support of their drive. They faced the tragedy of their exclusion and many other disappointments, but they persevered because they believed in the justice of their cause. Furthermore, the UMWA was the only alternative available to them. It provided hope and the organizational means with which to carry on the struggle against their own exploitation. They had taken a step toward economic and political democracy by joining the union even though, as Blankenhorn noted, the UMWA was an imperfect democracy, as were all other democratic institutions.[123]

11

The Long Depression and the New Deal

The demise of the union in 1923 enabled Berwind-White and other major open-shop employers to reestablish authoritarian rule, which lasted until the New Deal period. The companies' attempts to reassert control after the strike of 1922 took place in the context of a decade that historians have accurately described as the "tribal twenties" and labor's "lean years."[1] Long before the Great Depression of 1929, hunger, poverty, and depression marked the lives of the bituminous miners and their families, who lived in central and southwestern Pennsylvania. Berwind-White's privileged position in the coal industry—its links to the Pennsylvania Railroad, public utilities, the Navy, steamship companies, and J. P. Morgan—could not ultimately protect it from the ongoing crisis in the coal industry, the competition of other fuels, or the impact of the depression. By the mid-1920s, Windber had already peaked in terms of industrial preeminence and population growth, and Berwind-White was exploring possible lucrative mining investment opportunities to add to its holdings in West Virginia and Kentucky.[2]

This chapter examines how Windber miners and their families confronted the many serious problems of the 1920s, and how they acted to fulfill their long-term desire of gaining greater control over their lives by unionizing in 1933. The argument is that important material, social, political, and other changes that helped shape the struggles of the 1930s took place in the 1920s. Thus, Windber's immigrant miners and their families contributed to the successful revival of the United Mine Workers after the union's debacle in the 1920s and loyally supported the New Deal, which enabled them to make a radical break on the local level with many of the autocratic economic and political structures that had dominated their lives throughout the nonunion period. In the long run, however, their mobilization in the 1930s, and the reforms of the New Deal, could not resolve the many problems and inequalities inherent in a capitalist society.

Nativism Again

The efforts of Berwind-White and other coal companies to run their mines on an open-shop basis after the defeat of the strike of 1922–1923 were immediately strengthened by the resurgence of a semi-autonomous nativist movement in Pennsylvania and the nation. When the Ku Klux Klan expanded into the bituminous mining region of Pennsylvania during the summer of 1923, it rapidly assumed the leadership of nativist forces who fomented the bitter "cultural wars" of the 1920s.

The KKK was only the most recent manifestation of native-born hostility to the new immigrants, and only the latest regional organization to espouse nativist prejudice. Nativism had had powerful support from sectors of the American-born middle classes throughout Central Pennsylvania during the twentieth century. In 1906, Berwind-White and state authorities successfully exploited such sentiments to justify the state police's brutal repression of the Windber strike. These sentiments again came to the fore during World War I, when demands for immigration restriction and immigrant conformity to "100% Americanism" increased throughout the nation, and during the postwar Red Scare, which aimed at government suppression of aliens, radicals, and the unprecedented wave of strikes. By positing a widespread ongoing threat to the nation, leading businesses and conservative American-born elements of the middle classes led a broad-based movement to defend a particular narrow undemocratic conception of "Americanism" into the 1920s. In Central Pennsylvania, the defeat of the Johnstown steel strike of 1919 and the Somerset coal strike of 1922–1923, in which new immigrants had assumed prominent roles, strengthened the existing nativist and antilabor climate.

Nonetheless, it is difficult to imagine how the self-proclaimed goals of any fraternal organization could have been more offensive to the masses of Windber miners than those of the Ku Klux Klan were. The exclusively white, native-born, militant Protestant organization openly espoused a peculiar amalgam of anti-Catholic, antilabor, anti-immigrant policies, and it also used terroristic tactics and tried to provoke violence in order to advance its cause.[3]

By the 1920s the recently refounded Klan was tending to take its core notion of white supremacy and its promotion of traditional racial hatred toward blacks as givens. The organization never hesitated to foster old forms of racism against old targets, but during the Harding-Coolidge period it increasingly turned its attention to new targets. Thereafter, the KKK directed its primary energies against Roman Catholics, the labor movement, and new immigrants, all of whom it perceived as imminent threats to white supremacy and/or the Klan's singular conception of "Americanism."

For the KKK, Roman Catholicism was now defined as the nation's leading enemy. All Catholics were automatically assumed to be servile subjects of a foreign power, the Pope, and parochial schools were viewed as part of a papist conspiracy to overtake America. The Klan made many sensational false or misleading statements about Catholic beliefs and practices, but it was most known for having widely distributed a bogus "Knights of Columbus Oath" in which Catholics promised to "'hang, burn, boil, flay, and bury alive' all non-Catholics." Catholics were, by nature, un-American and dangerous, according to Imperial Wizard Hiram Evans, who told Klansmen: "The Klan is Protestantism personified" and "The very spirit of Americanism and Protestantism are the same."[4]

Another enemy was organized labor. The Klan sided with open-shop industrial employers against the broad-based labor movement. Its creed advocated a "closer relationship of capital and labor" and "the prevention of unwarranted strikes by foreign labor agitators," which included the coal and railroad strikes of 1922 and 1923. Evans explained the Klan's rationale: "With the free influx of red agitators, at liberty and eager to stir up strife among their fellow countrymen in the great industrial, manufacturing and mining centers of the United States, capital cannot feel secure. It must rise and face this very present danger."[5]

In its fight against organized labor, the Klan came into open conflict in various mining centers with the United Mine Workers, whose membership included some Klansmen and sympathizers, despite a union provision prohibiting joint membership. The permanent presence of African-Americans and Mexicans, many of whom had initially been imported into southwestern Pennsylvania during the strike of 1922 by particular coal operators, to break the strike, increased regional racial and ethnic tensions. In centers where American-born nativist and other racist miners predominated, union leaders had to struggle to enforce the UMW's stated nondiscriminatory, inclusive, class-based policies.[6]

The Klan's third new target was the foreign-born, primarily the new immigrants from southern and eastern Europe, whom Evans called "the scum of the Mediterranean and middle European countries." To the KKK, such people were inferior and unassimilable, Roman Catholics and Jews, totally lacking in the virtues of the Anglo-Saxon race and inherently incapable of becoming "true" Americans. They were responsible for the nation's labor troubles and for crime, especially bootlegging. Thus, the Klan vigorously supported immigration restriction and rejoiced at the passage of the Johnson Act in 1924, which it did not believe had gone far enough. It continued to advocate the passage of other bills that required the forced registration of aliens and easy deportation. The Imperial Wizard stated the Klan's position: "The Klan prescribes America for Americans." In its view, inequality was essential to "Americanism"; the nation's founders had meant only "white men" when they proclaimed that "all men are created equal." A clause in the Klan's creed said: "I am a native born American citizen and I believe my rights in this country are superior to those of foreigners."[7]

In the 1920s the Klan expanded from its Southern base to the North with great success. A provisional chapter of the KKK was founded in Windber in December 1923 around the same time that the organization was rapidly spreading throughout western Pennsylvania. On August 25, 1923, an armed clash between the KKK and Catholic residents of Carnegie, Pennsylvania, occurred in which one Klansman was killed. The KKK had selected Carnegie for the site of a mass demonstration because of its large Catholic population, which it hoped to intimidate.

Hiram Evans, who was in attendance at the violence, rejoiced at the news that the KKK now had a martyr and recommended that the tactic of inciting violence be repeated. He was widely quoted at the time as having said that the death of Thomas Abbott would bring 25,000 new recruits to the KKK in the Realm of Pennsylvania. He may have been right. One year later, the western portion of the Pennsylvania Klan had a peak membership of about 125,000. Klan publications had used the death of the martyred Abbott to defend "Americanism" from perceived attacks by "un-American" Catholics and foreigners.[8]

In any event, the timing of the Klan's founding in Windber suggests that its origins were related to the publicity about the Carnegie riot. The first evidence of its existence comes from *The Imperial Night-Hawk,* which reported on December 26, 1923, that the Provisional Klan of Windber had sent $88 for the fund established for Abbott's widow and children. But Windber might also have been affected by the geographical spread of a powerful and contagious social movement. Windber's native-born Protestants were undoubtedly aware that Altoona, Johnstown, Somerset, and many other neighboring towns had already established local klaverns.[9]

In Windber the Ku Klux Klan enjoyed the support of members of its former nativist bulwarks—the Patriotic Order Sons of America, the fire company, and certain Protestant denominations. Throughout its history the major dividing line in the community had been, simultaneously, a class/ethnic/religious one. It was the dominant American-born/Protestant middle classes—those who had migrated to Windber to manage the local mines, operate many of the businesses, run the town, and teach the school children—who originated numerous discriminatory practices against the new immigrants. The strike of 1922 had reinforced such segregationist and nativist tendencies. While the masses of foreign-born miners had acted independently in terms of their class interests, Americans had been the primary strikebreakers. It is not surprising, then, that, in the aftermath of the strikers' defeat, hostility toward the foreign-born and their majority religion surfaced.

As KKK activity rose in the immediate area, many of Berwind-White's American-born bosses were increasingly willing to tolerate the overt harassment of foreign-born miners by English-speaking miners at the workplace. With the UMWA out of the picture, the focus of derision now centered on religious beliefs. Nativists taunted Catholics and ridiculed them for worshiping a "cement Jesus."[10]

Meanwhile, the fire company proudly maintained its policy of prohibiting the foreign-born from joining its ranks, and in 1925, Reverend Cartwright of the First Presbyterian Church, a frequent speaker at KKK rallies, delivered the high school baccalaureate sermon. From 1924 to 1926, crosses were occasionally burned on hills surrounding the foreign enclaves. In late August 1925 some 6,000 Klansmen and families from Windber, Scalp Level, and other Pennsylvania towns gathered in a grove outside Windber to hold the Windber Klan's second annual picnic. The *Era,* which ignored the issue of persecutions and discriminations, treated the KKK as an important respectable institution that held big picnics supported by donations from local businessmen and churches. On occasion, white-clothed, masked Klansmen also attended the services of the United Brethren Church or other sympathetic Protestant churches in town and made donations to them.[11]

To defend themselves against the KKK and nativist harassment, immigrants turned to their traditional institutions. Shortly before Easter Sunday 1926, the immigrant churches and fraternal societies organized a massive Catholic Day parade and celebration to express pride in their religion and place of birth and to counter recent KKK activities in the region. One strike leader and lifelong resident of Windber described the event as the biggest he had ever seen in the town. Even the oldest people in the ethnic communities took part in the parade. Although the KKK burned crosses that same night on hills overlooking Windber and Scalp Level, the mobilization of the Catholic foreign-born masses had the desired impact. KKK activity in the area diminished afterward, and by 1927 and 1928 intra-KKK factional conflicts and other crises had doomed the state organization.[12]

Windber had escaped the bloodshed at Carnegie and Lilly, mining towns with large immigrant Catholic populations that the KKK had singled out for intimidation in 1923 and 1924.[13] While chance and immigrant mobilization probably played some role in this absence of full-scale confrontation in Windber, the old restraints against the most extreme forms of nativism might also have applied.

In the Windber setting, nativism itself was replete with contradictions. The prejudices of many resident American-born Protestant managers and foremen, along with those of resident businessmen, teachers, borough officials, farmers, and English-speaking miners, were undoubtedly stirred up by the culture wars of the 1920s at the height of the KKK's power, 1923–1926, but the movement was also undercut by practical realities and by its own successes. The ongoing need for labor, the demise of the UMWA locals during the strike in 1923, and the passage in 1924 of national legislation that severely restricted immigration from southern and eastern Europe allowed the nativist sectors of Windber's middle classes to feel more secure in their dominance over the masses of Catholic immigrant miners.

Berwind-White did not have to create the KKK to be a prime beneficiary of it. The Klan's nativist activities reinforced company rule and the existing class/ethnic/religious hierarchy that kept the masses of "Hunkies," "Polaks," and "spaghetti eaters" in their place.

The Aftermath of 1922

Windber miners did not abandon the dream of unionization after the defeat of the 1922–1923 strike, but between 1922 and 1933 they had little or no faith in International UMWA President John L. Lewis. They had been among the 80,000 newly organized miners, mostly new immigrants, who had been excluded from the Cleveland settlement, which Lewis had personally negotiated to end the national strike in August 1922. The settlement had been a severe blow to the local rank-and-file. It had given Berwind-White added ammunition for its charges that the International did not support the strike. For Windber miners, thereafter, allegiance to the union meant allegiance to the progressive leadership of District 2, which had assumed the burden of the strike and mobilized on behalf of the strikers.

It is worthwhile to look at the settlement of the strike and at Lewis's role in it, because what happened in Windber and Somerset County in 1922 became a serious matter of contention within the UMWA throughout the 1920s. The fate of the Windber strike subsequently became a prime argument for the region's dissatisfied progressive miners to found a number of organizations designed to challenge Lewis's conservative leadership from within or without the union.

With regard to the strike settlement in 1922, Lewis always claimed he had made the best deal possible at the time. If so, it shows how bankrupt the International UMWA's policies were when it came to organizing the unorganized and protecting the newly organized, who were basically left out in the cold except for district help. Moreover, critics immediately blasted Lewis for allowing companies to sign the union contract for their older, previously organized mines without insisting that they also sign up their newly organized mines.[14]

Nor did Lewis ever convincingly refute the charge that the national organization had failed to give adequate support to the Somerset County strikers in 1922. Financially, it did contribute a total of $107,000 to the seventeen-month-long effort, but District 2, which went it alone for most of the strike, had spent $1,186,915. It is also undeniable that the International's organizers who came into the county did so only after district leaders had challenged Lewis's leadership. When they came, their purpose was to denounce Brophy and the February referendum to continue support for the strike. They did nothing constructive to aid the strikers.[15]

Obviously, Windber-area strikers were the victims, in part, of intra-union rivalries, red-baiting, and politics, especially after the *UMWJ* began a systematic attack on Brophy, radicalism, and District 2 during the spring of 1923. The union's newspaper did not report the Hirshfield Committee's findings at the time, or mention the February referendum, the spring campaign, or the calling-off of the strike in August 1923. Perhaps the most damning criticism of the International, however, was that it never at any time made an unequivocal public declaration of its support of the strike, a fact that the nonunion operators duly noted and used.

Windber miners felt betrayed by the union's top leadership in 1922, not by unionism per se. On September 14, 1922, Powers Hapgood had written his mother that he and other organizers were consciously tempering their own critical views of Lewis and the settlement in talking to the miners of Somerset: "The coal operators are spreading the report among the miners that Lewis has gone back on them and that they might as well go back to work. While we know that Lewis has gone back on us, we have to defend him before the men to encourage them because if they will only stick solid they can win without help from the international union."[16] Not long afterward, however, Hapgood, Joe Foster, and others involved in the Windber strike became vocal critics of Lewis. One result of the strike was that, throughout the 1920s, such men continued to work toward the goal of finding a viable alternative to Lewis and his conservative policies.

The charge that Lewis was corrupt and had "sold out" Windber miners in 1922 was commonplace at the time and in retrospect cannot simply be dismissed out of hand. Because the negotiations at Cleveland were secret and conducted almost solely by Lewis, it is impossible to determine exactly what, if any, deals were made by him to exclude certain mines from the settlement. Yet, as far as Windber is concerned, the possibility that a corrupt deal existed cannot be ruled out.

In their biography of Lewis, Melvyn Dubofsky and Warren Van Tine cited contemporary rumors that their subject's rise to power within the UMWA was linked to unscrupulous activities. From at least 1917 on, Lewis had maintained a close personal and business association with the notorious Alfred Reed Hamilton, who had offered John Mitchell and John Brophy secret deals in 1906 and 1918 respectively. In exchange for various favors, Hamilton had asked the men to "protect" Berwind-White mines from unionization. As late as the 1920s, John Walker, a former District 12 president from Illinois, and other prominent UMWA officials, continued to charge that Hamilton, who died in 1927, had proposed secret deals and that he had interfered in many union affairs with Lewis's approval and with disastrous consequences for the union and rank-and-file. The argument that Lewis (or other national leaders) might have made such a deal in Windber's case is circumstantially supported by the fact that Lewis had previously rejected so many requests to cooperate with the district to inaugurate an organizing campaign in Somerset County.[17]

In any event, Lewis's Cleveland settlement turned out to be a Pyrrhic victory at best. The International UMWA, which had maintained higher wages in 1922 for the old, unionized mines and then negotiated the Jacksonville agreement in February 1924 to sustain them, became greatly weakened in the interval between 1922 and 1927, when it was on the verge of extinction. By 1925, union operators had begun to openly repudiate the Jacksonville contract, more mines operated on an open-shop basis, and the nonunion share of the nation's coal production increased from 28 percent in 1919 to 60 percent.[18]

Moreover, the UMWA was bitterly divided. Disgruntled rank-and-file miners had joined with the communist Trade Union Educational League (TUEL) leaders to form the Progressive International Committee of the UMWA in February 1923, after the calling off of the Connellsville strike. At a June conference, the coalition pressed for implementation of the UMWA's radical 1919 platform, supported recognition of Soviet Russia, and endorsed a Farmer-Labor party. Lewis responded by blasting the Progressive Committee as a "dual" union, but the intraunion fighting continued.[19]

In 1926, District 2 President John Brophy responded to requests from various quarters to lead a "Save the Union" movement, which challenged Lewis's conservative policies and his candidacy for president of the International. Brophy lost not only the election, which he and many others claimed had been stolen from him, but also his position as District 2 president. An era in which district miners, including those from Windber, might have gone to various centers to attend union-sponsored educational programs and labor chautauquas based on "The Miners' Program" was coming to an end.[20]

Meanwhile, the strike of 1922 marked the end of the golden era of Berwind-White's operations in Windber. The epoch of relative steady work and plentiful jobs that had prevailed locally in the first decades of the century gradually came to an end in the years immediately following the strike. The crises within the bituminous industry—the overproduction of coal, the superabundance of miners, decreased markets, the increased use of alternative fuels such as oil, the introduction of new technology, the shifting of production away from Pennsylvania to West Virginia and nonunion fields—were long-term serious problems that affected both nonunion and union mines in District 2.[21]

In April 1924 the *Era* noted that Windber mines were working only two or three days a week. State records indicate that in subsequent years, with a couple of exceptions, local mines worked no more than that until World War II. At the same time, the number of employees varied significantly. The high point of employment in the 1920s was reached before the strike in 1921, when the company employed 3,498 people. The strike itself had disrupted entire families, some of whom never returned to Windber, and afterward, miners never again reached their prestrike numbers. The decade's low point was attained in 1929, when the work force totaled 2,353. The result was that unemployment, underemployment, and economic imperatives pushed young people and miners, who had always been migrating people, to move in increased numbers to New York, Detroit, Cleveland, or other large cities in search of work, opportunity, and better jobs.[22]

Meanwhile, the demise of union locals and the disintegrating position of the UMWA in the declining bituminous industry in the 1920s was of more than passing interest to resident Windber miners. In its traditional manner, Berwind-White sought to retain its work force and remain competitive by basing its wage scale on the rates negotiated by the region's unionized miners. Windber's nonunion miners therefore benefited indirectly through higher wages whenever the union was strong, but suffered when the reverse was true.

Given this context, the *Era*'s antiunion articles and calls for wage cuts and other concessions by union miners in the mid-1920s had dark, if indirect, implications for Windber miners. At the same time, competition, even between nonunion companies, was fierce, and wages fluctuated accordingly. Thus, in April 1925, when the district's nonunion operators reduced their wage rates to near 1917 levels, Berwind-White followed suit. Although the company raised rates slightly in November 1926, the wages of Windber miners continued to fall below the Jacksonville scale, and work was slack.[23]

Merchants as well as miners suffered from the economic slump in the coal trade. During the summer of 1925, the depressed local economy had once again led businessmen to raise the issue of forming a Chamber of Commerce or other business organization. Immediately the old divisions within the business community over the new organization's purpose and its relationship to Berwind-White's monopolistic control of industry surfaced.

When a group met to found the Business Men's Association of Windber on July 8, 1925, the *Era* commented: "The sole purpose of the organization is to create a better feeling among the business men, get them acquainted with each other, and to convince the buying public of Windber and vicinity that they can get better values for their dollars in Windber than elsewhere." However, others at the founding meeting disagreed about the new organization's purpose. The *Era* noted: "Some of the men who came to the meeting were under the impression that the Association has formed to try and get other industries to Windber, but this thought was soon dispelled after listening to some of the speakers, who pointed out that industries are not needed in Windber."[24]

Whenever the subject of bringing in new industries arose, company loyalists and the *Era* always opposed it. Thus, the Business Men's Association was moribund from the start, and another business organization, formed one year later for the express purpose of obtaining new industries for the community, quickly failed.[25] Every such movement throughout the nonunion period met Berwind-White's resistance because independence and diversification of industry threatened the company's autocratic control of the town.

Moreover, both the miners' wage increase of November 1926 and the optimism it engendered were short-lived. In February 1927, the company announced a cut in the wage scale, back to the old, pre-November 1926, rates, effective February 1.[26] This wage cut took place in the context of an anticipated lengthy regional coal strike and wholesale assault on the union.

The Strike of 1927

The Jacksonville contract, which had been widely violated and ignored in 1925 and 1926, was set to expire on April 1, 1927. Meanwhile, the Pittsburgh Coal Company, a major producer that had traditionally signed union contracts to guarantee coal for its steel mills, had decided to close down its mines and reopen them nonunion, thereby signaling a general drive by coal operators in central and southwestern Pennsylvania to destroy the union. The expected strikes, lockouts, and overt class warfare soon materialized.[27]

It was in this context—economic depression, the termination of the Jacksonville contract, class conflict, recent wage cuts, declining union power, and intraunion dissension—that the issue of unionization arose again in Windber in April 1927. As in 1922, grass-roots initiatives intersected with a larger regional and national movement. It was not accidental that the discontented miners who met to form the United Front Committee of Windber in early April invited Powers Hapgood—1922 organizer, intimate of John Brophy, and activist in the "Save the Union" committee of the UMWA—rather than the current District 2 officers who were supporters of John L. Lewis, to come to Windber.

On April 10, 1927, Hapgood wrote his parents: "Some of the miners at Windber, growing discontented under the worsening conditions in the Berwind-White Coal Company's mines and towns, asked help from the progressive bloc in the union and I went down to see them."[28] He was surprised at what he found: "Today there is to be another mass meeting, bringing in more of the miners of the company to consider the question. The company may concede all or part of the demands or there may be a strike. We don't know yet, but in a couple of days we will."[29]

In their efforts to get miners to come out to mass meetings, Windber-area activists distributed flyers that emphasized the wage issue and that called for a return to older, higher wage rates, pay for deadwork, and honest weights. The flyers also announced that Hapgood, Tony Minerich (a Slavic miner who had long been active in the progressive miners movement), and speakers of various languages would be present.[30]

Later Hapgood informed his parents: "When Tony and I called our first meeting, we advised against a strike even though the men wanted it. The next meeting, however, it was no use. Most of the 500 men attending were young and enthusiastic and would not listen to anything except strike." He added, "Throughout our speeches we were interrupted with shouts of 'Call us out and we'll come,' 'The hell with lifting three and a half feet of bottom for nothing,' 'No work tomorrow.' " He considered his role in these meetings: "My mistake, if any, was not in precipitating a strike but in not opposing it. While I was doubtful of the outcome, I was still much influenced by the men's unshakable conviction that if they walked out the other 2,000 would follow."[31]

Hapgood, who subsequently described the strike as "premature and thus unfortunate," explained what happened next: "So out they came, but the others, with the memory of 18 months of a lost strike in 1922 and 1923 still fresh in their minds, kept on working."[32] Thus, the local walkout of 1927—another expression of the miners' desire for unionization—proved to be a faint echo of the 1922 strike.

That many of the 500 youthful militants were known to have supported alternatives to Lewis's policies at the time may have deterred a few older miners who feared dual unionism from joining the strike. More important, however, for many others the conditions for winning union recognition seemed much less favorable than in either 1906 or 1922. Joe Zahurak reported that when Hapgood and others had asked the 1922 strike leaders for their opinion about a strike in 1927 he and other leaders had answered:

> We told them it was no use. The miners were still frustrated at what happened that they had to go back without the union. They'd lost and in fact they wasn't about to go out on another strike. It was no use trying it. . . . You can't do it. You're not gonna get a shutdown. You just better let things alone, the way they are, because the people can't take it here.
>
> The union was still not financially able to take care of the strikers. So finally they tried it anyhow a couple of times, but it didn't work. The people kept on working and going back. . . . They went right

> through the picket lines. It didn't mean nothing because they was frustrated with what happened before.[33]

For the duration of the partial three-day strike, however, those who were on strike had to confront state troopers or "Cossacks," Coal and Iron Police, other gunmen, and the Somerset County sheriff and deputies, who tried to intimidate them and prevent picketing. But Hapgood, who had assumed charge of the strike, knew his legal rights and insisted on the right to picket, threatening a lawsuit against opposing officials if necessary. Nonetheless, despite the militance of the young, it quickly became clear to the organizers that the strike was not crippling production. They then sought to make the most of the situation and end it on the best terms possible.[34]

Hapgood, who periodically met and argued with Superintendent Newbaker on the picket line, asked for a conference. Thus, eight Windber miners met with the company official on April 13. Hapgood wrote his parents about the result: "They said after a long argument Newbaker had agreed to establish the status quo, giving every one back his job regardless of strike activities, if we would call off the strike and stop picketting. So this was agreed to and the strike was called off."[35]

But Newbaker, like E. J. Berwind in 1922, subsequently betrayed the promise, claiming he had never made any such agreement. Hapgood explained: "The next day Newbaker went back on his word. About a dozen men, including all the committee, were not allowed to go back. When they applied at the office for reemployment, they were told 'never again.' " In one last move, the miners' committee printed up and distributed a circular, "Company's Dishonesty Exposed," which presented their case.[36]

Ultimately, many of the twelve leaders left Windber. Hapgood wrote his mother: "Most of the young fellows who got fired didn't want their jobs back in the mines anyway. When I got back from New York, some of them had already left for Detroit, Dayton, and other factory centers."[37] For the minority who remained, however, reinstatement became possible only in 1933, when UMWA locals took up the cause after they had successfully unionized.

Despite the failure of the 1927 strike, the union cause in Windber was not hopeless. Hapgood reported: "Not one of those who stayed at work was hostile to us. We'd meet them on the street, on the picket lines going to and from work, and always they would say they knew we were right but that they couldn't make the sacrifice just now." He noted that it had taken some miners a year and a half or more to find work after the strike of 1922.[38]

Hapgood later concluded that the Windber strike had done a lot of good. "It has shown," he said, "in the first place, that with a little pep and energy a great deal more can be done in the non-union fields than the officers are doing. Secondly it has spread a lot of union ideas among the non-union miners themselves." He also believed, erroneously, that "the company will hardly cut wages again very quickly, as operators in other sections are doing."[39]

Yet Hapgood was right when he wrote his parents that the Windber miners were "union men at heart."[40] In 1927, though, the majority of miners had concluded that this was not yet the "favorable opportunity" to unionize they were awaiting. Until such time, they pursued their everyday lives and struggles, which included enduring the lean years of the 1920s and the depression as best they could.

Meanwhile, neither the UMWA nor the "Save the Union" faction within it could stem the deterioration of the union or the demoralization of wages and mining life. Lengthy strikes to save the union throughout Central and Western Pennsylvania were smashed by operator intransigence, federal antilabor injunctions, and the Coal and Iron Police. By 1928, the UMWA had virtually ceased to exist, not only in Pennsylvania but in most other states outside Illinois.[41]

On April 1, 1928, the "Save the Union" movement met in conference in Pittsburgh. Rank-and-file miners voted to continue the ongoing strikes and adopted traditional radical resolutions, such as nationalization of the mines. In May, Lewis responded by expelling movement members from the UMWA on the grounds of "dual" unionism. John Brophy and Powers Hapgood were among those ousted. For many, it now seemed impossible to work for progressive change from within the UMWA.[42]

On September 9, 1928, a group of communists and other dissidents from "Save the Union" met to found a new rival union, the National Miners Union (NMU), to replace the UMWA. Immediately, several hundred of Lewis's loyal followers physically attacked those assembled at the conference in Pittsburgh, but that did not deter the new organization or rebuild the UMWA, which was, in effect, defunct. From 1928 to 1933, it was the NMU that represented the cause of mass unionism in Pennsylvania's soft-coal fields.[43]

The Limits of Politics and Protest

Working people and other citizens who hoped to use the political system to register protests against existing conditions continued to confront many obstacles in Windber in the mid-1920s. Given Berwind-White's autocratic controls over the workplace and the community, genuine political democracy and an independent franchise did not exist, as the Civic Association and other critics had charged earlier. Moreover, the town government–company collusion in subverting civil liberties, including the freedoms of assembly and speech, made an ongoing mockery of the U.S. Constitution and the political process. Those who had vested interest in the company town had a stake in maintaining an undemocratic America.

One example from this period highlights the issue of constitutional liberties. It reveals the company's fears that the exercise of the rights of assembly and free speech would inevitably lead to unionization, and hence loss of the corporation's autocratic control, and it shows that divisions existed within the immigrant communities. The incident described took place in the context of new immigrants mobilizing throughout the nation in an effort to counter ongoing nativist attacks and to defeat proposals for additional federal nativist legislation. In the process, radical groups and ethnic fraternal societies became more active politically, and mining areas where large numbers of the foreign-born lived were likely arenas in which to enlist support.

In July 1927, several months after the failed strike, Jeanette D. Pearl, a field organizer for the National Council for the Protection of Foreign-Born Workers, a predecessor organization of the American Committee for the Protection of the Foreign Born, began to solicit the new post-Brophy leadership of District 2. She wanted help to set up meetings of the foreign-born to organize efforts against a number of proposed anti-alien legislative measures. On July 11, she happily wrote District 2 President James Mark: "I have arranged a mass protest meeting against alien registration for July 17, 1927, evening at Slovak Hall, Windber, Pa. The hall was given on condition that I absolutely refrain from mentioning unionism in my talk."[44]

Her letter suggested that Windber miners had considerable interest in the subject but that there was conflict within ethnic societies over the wisdom of organizing. She wrote: "Yesterday, I addressed four different language organizations and everywhere the men were most enthusiastic.

I also addressed the executive committee of the Slovak Hall and for almost two hours the President fought his committee and struggled to keep them from giving me the hall." In the end, she reported, "The vote was eight against two."[45]

Despite an announcement in the *Era,* the proposed meeting never took place. In ways that remain unclear, Berwind-White enforced its old strictures against meetings of independent, potentially prolabor forces, even when activities were confined to narrow political ends. On July 14 a disappointed Pearl wrote Mark: "None of the halls are open to us in Windber. They will not even allow the use of the street. If we could secure a lot near Windber that would be wonderful. We could then throw into Windber several thousand leaflets. That would stimulate some thought."[46] Mark, a veteran of the union's struggle in Windber in 1922, replied: "This is just what I expected. The Company has always had a complete monopoly of Windber streets and everything else, and I was very much surprised that you had been able to arrange a meeting there."[47]

New immigrants and the American-born had few means to organize in ways that the company did not approve, and overt protests were rare. Nevertheless, in retrospect, it seems that working people were beginning to make some inroads in the political process in the mid-to-late 1920s.

The presidential election of 1928 was a harbinger of the New Deal realignment elections of the 1930s. In 1928, as usual, Berwind-White and the *Era* enthusiastically supported the Republican nominee, Herbert Hoover, for President, and Hoover did carry Windber. In the process, however, Al Smith, the Democratic candidate, garnered a far greater percentage, nearly 42 percent, of the vote than any previous Democratic presidential candidate had ever done.[48]

Given Windber's political history, it seems that the new high mark of Democratic presidential strength in 1928 was significant. In 1924 the Democratic candidate had won only 11.7 percent of the vote. Meanwhile, Robert LaFollette, running on the Farm-Labor ticket, actually finished second in the race, with 14.5 percent of the vote, while Calvin Coolidge won with a commanding 71.6 percent of the total vote. But even in 1916, when Woodrow Wilson was seeking a second term and the Democratic party enjoyed relative popularity, the incumbent Democrat polled only 29.5 percent of the town's vote.[49]

The New York Democrat's religion—Roman Catholic—had become a major issue during a nasty campaign in which the Ku Klux Klan and other organizations portrayed Smith as the representative of un-American, alien, and "wet" sinister forces. Naturalized immigrant Catholics and their voting-age American-born children turned out at the nation's polls to support his candidacy. Windber was no exception to the general trend. Two of four Windber precincts, where many foreign-born resided, voted for Smith in 1928 in what seems to have been a protest against the nativist, anti-Catholic attacks on the New Yorker.[50]

In the Windber area, working people who voted for Smith and the Democrats had to summon up courage to ignore the company's routine intimidations. But the election's results showed that many must have done so. One who did was Pearl Leonardis, a 23-year-old daughter of Italian immigrants and a miner's wife. Leonardis subsequently became one of the first ethnic women—indeed, of any women—in Windber to take an active part in politics. She has been an ongoing activist for labor and the Democratic party since the 1930s. In an interview, Leonardis described her experience of voting at her first presidential election:

> When we lived out in the country [on a company farm], I voted [for the first time]. . . . The boss came, and he says to me, "Do you vote?"
>
> I says, "I am going to go to vote."
>
> "Did you register?"
>
> I did register. I wanted to go to vote that day.
>
> And he says, "I'm going to pick you up [to drive her to the polls]."
>
> I says, "Okay. I don't care who picks me up."
>
> Well, he took me to 42 [a mine location and polling site] because we had to vote up at 42 at that time. . . . We lived out in the country, and I couldn't come to Windber. I had to go to Cambria County. See, I was in that county.
>
> I went to vote. He wanted to come in and show me how to vote.
>
> I says, "Wait a minute! I didn't go to school too far, but I know something." I says, "Nobody's going to be with me when I vote. Not you, nor somebody else, nor even my husband."
>
> And he says, "Thanks [nastily, sarcastically]" to me.
>
> I said, "Listen [firmly]." I went in and I voted. . . .
>
> When I come out, he says, "Who did you vote for?"
>
> I says, "What do you care who I voted for. I voted the way—for who[m] I wanted."
>
> He says, "Now, you have to walk home. I'm not taking you home."

Pearl Camille Leonardis, age 90, at her home in Windber, 1995.

> I says, "I'll have you arrested." I went to him. "You'd better take me home or you're going to find out from my husband." I says, "And then you'll find out."
>
> He got scared.[51]

The man who had picked up Leonardis was a Berwind boss, manager of Eureka's Mine 42 company store. He was the same man who had once visited her to investigate why she had bought so many canning jars and to threaten her that she and her family should not sell farm produce to compete with the company store.[52]

Meanwhile, other changes were in the making. The "tribal twenties," nativist attacks, the Sacco-Vanzetti case, the brutal murder of John Barcoski by Coal and Iron Police in 1929, cast a pall over foreign-born communities. The depression in coal and the destruction of the labor movement had added to the woes of ethnic miners and their families, but even middle-class ethnic people experienced discrimination. In Windber, the ethnic leaders who traditionally supported Republican candidates had been unable to win in local primaries against American-born candidates. They also held little power themselves within the larger party. Ethnic voters of all classes were increasingly restless and eager to gain political influence.

Signs that changes were to come first appeared in the mid-1920s. In 1926 a committee of prominent Italian-Americans from western Pennsylvania met to endorse a gubernatorial candidate, John S. Fisher, whom Andrew Mellon, Pittsburgh industrialist and financier, supported. Albert Torquato of Windber was one of the committee's founding members. The movement itself signaled a new effort by ethnic leaders, especially Italians, to seek greater political influence, influence that also went beyond the local setting.[53]

A side effect of Fisher's victory in November 1926 was that Albert's son, John R. Torquato, received an appointment as a chainman for the Highway Department. Thus, the political career of the younger Torquato, who later became a central figure in the region's Democratic party, was launched during the Republican party's reign, before the realignment of the 1930s.[54]

By the time of the stock market crash in October 1929, after Hoover's victory, it was already clear that the prosperity the *Era* had promised during the electoral campaign was not materializing. When the nation entered the Great Depression, economic conditions, which were already bad in Windber, only worsened.

From 1929 to 1933, local mines continued to work only two or three days a week. When the Industrial Workers of the World sent reporters to the coal fields in August 1932, they visited Windber and found emaciated children and widespread hunger. People interviewed have painted the same general picture of the times. Meanwhile, in the region, the National Miners Union tried to mobilize despairing miners in a series of massive strikes for economic justice and for abolition of the Coal and Iron Police in what has often been called an era of "coal wars."[55]

For many Americans, Franklin D. Roosevelt's candidacy in the 1932 presidential election provided an alternative to the laissez-faire policies of Hoover and offered hope that government could help alleviate misery and bring about national recovery. One of the great ironies of the national realignment election was that a majority of the electorate in Pennsylvania, alone among the major industrialized states, voted for Hoover.[56]

The percentage of the Windber electorate voting for the Democratic nominee in 1932 changed but slightly from that of 1928. In November 1932, Roosevelt won 43 percent of the town's vote, compared with Smith's near 42 percent in 1928.[57] Despite widespread dissatisfaction and misery, the big changes in politics, unionization, and democratization in Windber had to await the national victory of Roosevelt and the practical impact of the New Deal.

The New Deal

On May 15, 1933, during the earliest days of the New Deal, several thousand men, women, and children marched under the auspices of the United Unemployed Councils of Somerset County to the courthouse in Somerset to present nineteen demands to the Emergency County Relief Board. They came from Windber, Scalp Level, Holsopple, and many other towns. Of the 4,800 families representing 24,000 individuals who were on the county's relief rolls at the time, 3,101 were families of unemployed miners. The protesters' demands included removal and reorganization of the committee in charge of relief, a range of payments depending on the age and marital status of those in need, cash payments instead of merchant orders in order to be fair to all businesses, free medical aid, assistance without a work requirement for those over 60 years of age, free milk for children with compensation to the dairy producers, no poll tax for the unemployed, no evictions of those unable to pay rent, no discrimination against those taking part in the hunger march. From 1931 to 1933, similar marches occurred in many places in the United States.[58]

Poverty became a political issue during the Great Depression. Charity, voluntarism, and local and state public relief programs could not meet the needs, which were enormous. In general, those in power provided minimal aid only to those who they determined were the deserving poor. Thus, Windber residents on borough relief who had chosen to take part in the Somerset hunger march were denied relief work and food assistance afterward. According to a Somerset newspaper's account, Burgess

Barefoot and Chief of Police Ray Holsopple told them they had "no relief orders for men who preferred the hunger march to their jobs."[59]

Such was the context in which Franklin Roosevelt and the New Dealers in Washington set to work to try to bring recovery to the nation's economy. During the administration's first "hundred days," Congress passed an impressive array of bills in an effort to cope with the crisis. Meanwhile, John L. Lewis and rank-and-file miners were not idly standing by.

The militant John L. Lewis, who initiated a UMWA organizing drive among the unorganized during the spring of 1933, who lobbied the federal government to have organized labor represented in key decision-making bodies, and who fought to have the interests of labor represented in the National Industrial Recovery Act (NIRA) was certainly not behaving as the cautious business unionist of the 1920s had. Beyond sheer expediency or newfound idealism, many miners, scholars, and others, then and since, have found it difficult to explain the contradiction, the enigmatic change, in his behavior, but the change itself is undeniable.

Melvyn Dubofsky and Warren Van Tine are among those who have stressed the existence of certain continuities between Lewis's policies in the 1920s and 1930s. In their view, the opportunistic labor leader favored federal government intervention in the economy in both decades. The argument is that, in the mid-1920s, he had expected the government to enforce the Jacksonville agreement, stabilize the coal industry, and protect the UMWA. In 1931 and 1932 he had also lobbied, unsuccessfully, for the passage of the Davis-Kelly bill to stabilize the industry, and his speeches during this period presaged New Deal policies and his own calling for the federal government's active intervention in the economy and in labor affairs.[60]

By contrast, Curtis Seltzer has recently described an opportunistic Lewis as a man who consistently favored free-market, capitalistic-oriented business unionism throughout his lifetime. Thus, it was only when these bankrupt policies had resulted in the near demise of the UMWA during the 1920s that he temporarily turned to the federal government to intervene to stabilize the industry in order to save coal capitalism. This view discounts the image of Lewis as a heroic labor leader of the 1930s and 1940s and considers his collaboration with the Bituminous Coal Operators' Association in the 1950s as a continuation of his earlier commitment to laissez-faire capitalism.[61]

Whatever motivated Lewis or whatever he really believed, Windber-area miners could certainly attest to the view that the UMWA leader was a paradox. The very man who had betrayed Somerset County strikers by excluding them from the strike settlement of 1922 ironically became a hero to many in the 1930s. History did not end in 1922 or 1923, and Lewis subsequently redeemed himself in their eyes. Decades after what one author has characterized as the "turbulent thirties," most elderly Windber miners preferred to remember the militant Lewis of the 1930s and 1940s, the bold UMWA president and founder of the Congress of Industrial Organizations, not the UMWA leader of the 1920s and author of the 1922 strike agreement.[62]

In any event, John L. Lewis sent UMWA Vice-President Philip Murray into the bituminous coal fields in April 1933 to launch a new UMWA organizing offensive. The large crowds that turned out to hear him speak were promising. Meanwhile, rank-and-file activists who viewed Roosevelt's election in November as a hopeful sign were already at work trying to rebuild the union. Joe Jasway, one of the first presidents of Windber's UMWA Local 6186, was one of these men and one of the many miner activists who provided a direct link between the local efforts of 1922 and 1933.[63]

The democratic hopes and aspirations of ordinary miners and their families for greater control over their own lives were given an enormous boost by passage of the National Industrial Recovery Act (NIRA) on June 16, 1933. These pragmatic people who knew the realities of class struggle from personal experience immediately grasped the significance of section 7(a) of the NIRA, which guaranteed workers the right to organize and to engage in collective bargaining. The correlation of forces between capital and labor had now changed significantly. The UMWA's slogan, "President Roosevelt wants you to join a union," met such a responsive chord among Windber miners and other American workers precisely because they appreciated the changed role of the state and of the law. In the process Roosevelt became the popular embodiment of collective hope.[64]

The UMWA came to the Windber area without the drama of the strikes of 1906 and 1922, or even 1927, and with a rapidity that should not have been surprising. The UMWA organizing campaign had met with rapid success in Pennsylvania's noncaptive mines. By July 4, 1933, Lewis was claiming that all of Central Pennsylvania, nearly 100 percent of the Pittsburgh district, and formerly nonunion Westmoreland County were "fully organized." Windber's Local 6186 received its charter on July 22, 1933; Scalp Level's Local 5229 did so on December 9, 1933. In a last-minute effort to forestall unionization, Berwind-White raised wages, but it never tried to set up a company union because officials presumably understood that the effort would be useless.[65]

Nevertheless, miners had to mobilize to take advantage of the new law, and their first organizational meetings had to be held in the woods. One Slovak union leader, who stressed the significance of the law in breaking down fear and apathy, commented: "At first we couldn't get a meeting place. No club or no hall would give us a meeting place. I'm a Slovak. We had a big hall. . . . I knew the officers of the Hall Association, and they were good union men, strong." Joe Novak then explained: "See, they were all afraid of Berwind, to be honest about it, and I said, 'Aw, the company ain't going to do nothing no more about this. *This is the law!*' " Besides, he added, " 'Why don't you let the union meet in your hall here? You'll make a few bucks. Not only that, they'll come down to the club and spend money.' That was the first organization in Windber that gave the miners a place to meet."[66]

Another organizer and subsequently a Local 6186 president, Joe Zahurak, also stressed the key importance of the legal changes brought about during the New Deal. He characterized the nonunion era in Windber as one in which industrial and political freedom had been lacking, and the period from 1927 to 1933 as especially bleak because of the depression, the many desperate industrial mining conflicts, and the poor status of the UMWA. Here is his description of the union's coming to Windber:

> Later on . . . finally after all these strikes from 1927 to 1933, the dark clouds disappeared. Oh, the miners in Windber, we got our sympathetic president who got in there, Franklin Delano Roosevelt, and they got that law passed where you could organize. I'm telling you, that law was passed, and the union came in.
>
> The United Mine Workers was short on funds and money, so they borrowed $500,000 from the American Federation of Labor to put men out in the field for expenses, to go out and talk to these miners. They brought cards for us to sign the miners up, who all wanted the union, and I'm telling you them cards went fast. Everybody was signing up. That's 1933 when they knew that they was finally safe. It could be done.
>
> After all that good and strong signing up, why, we held a meeting then. And Mr. Booker [Newbaker], General Manager at that time,

> called us in, and they [company officials] said that they will do anything that the President—Roosevelt—wants to do.[67]

One of the first acts of the new Windber UMWA local was to elect checkweighmen, who were installed on company tipples by August 1933. Another was to get the workers who had been discharged in the 1927 strike reinstated. On the everyday work level, the miners jealously guarded honest weights, fought for the company's obligation to buy bits for new jackhammer machines, and took up multiple grievances about dirty coal and other matters. To have a say about wages and working conditions was a new and welcome experience.[68]

On September 21, 1933, thanks in large part to the militancy of rank-and-file miners, Lewis won President Roosevelt's approval of the first NIRA code—a bituminous coal code that gave bituminous miners a contract that covered all soft-coal districts, stabilized wages, guaranteed the eight-hour day, abolished scrip, ended compulsory trading at company stores, outlawed child labor, and secured the checkoff and the miners' right to elect checkweighmen. In organized mines, contracts accompanied the code, which became effective on October 2, 1933. After these initial victories, the UMWA president reconciled with John Brophy and various other radicals, who continued to play important roles of their own in the struggles of the 1930s.[69]

Winning union recognition and a union contract in 1933 was a heady experience for Windber miners. Democratic participation in regular UMWA activities was a particularly exciting phenomenon for those who had been denied such possibilities before. The minutes of Local 6186 indicate that union meetings in the 1930s were anything but quiescent bureaucratic affairs or confined to narrow trade union issues. Conflicts and policies were vigorously debated, and frequent elections for officers were held. At times, UMWA officials enforced a discipline of their own on unruly members who violated contract terms or constitutional procedures. A union officer described the early meetings of the 1930s in the following way:

> See, after we got the union then, everybody was so thrilled and everything. We'd start our union meetings at 7 o'clock. They met Friday night. Every night they'd start at 7 o'clock, and we was there sometimes until midnight because everybody was so happy and thrilled that all our problems was coming out.

Four Windber miners, happy to have a union—finally.

> They didn't want to do anything that they're not gonna get paid by the union. They felt then, we got the union now. They can get the baseball club and hammer the foreman or boss or anyone over the head. We're on top now.
>
> But we had a hard time like you see in the minute books, and some of them meetings were pretty strident and lasted a long time. . . . You still had a constitution and contract to follow. But the guys gradually got used to it and everything.[70]

Politics had become a new central concern of workers with the arrival of unionization. By 1934, Windber-area miners had become staunch supporters of Roosevelt and the New Deal, and the UMWA locals mobilized to support sympathetic candidates in the 1934 election. Unlike other industrialized states, it was the 1934—not 1932—elections that turned out to be the critical contests that brought Pennsylvania into the New Deal camp.

In November 1934, George Earle was elected the first Democratic governor of the state since 1890, while the UMWA International's secretary-treasurer, Thomas Kennedy, was elected lieutenant governor. New Deal partisans also won many state and local offices. During the year, various prominent ethnic leaders and politicians, including Charles Margiotti, a Punxsutawney attorney and subsequently Pennsylvania's attorney general in the Earle administration, followed their working-class constituents and switched party allegiances from Republican to Democrat. Among them was John R. Torquato, who served as secretary of the Workmen's Compensation Board during Earle's governorship from 1935 to 1939, a period popularly known as "Pennsylvania's Little New Deal."[71]

The UMWA's mobilization for political goals took many forms. In 1934 many Windber miners and their wives were not yet citizens and could not vote. The union offered informal help. On February 6, 1934, District 2's organizer, Abe Martin, who oversaw Local 6186, called for "the attention of those members that haven't yet got their citizenship papers to turn their names in and we would give them a helping hand in any way possible." Meanwhile, the local requested that the contract and union constitution be translated into foreign languages for its membership and sent resolutions to congressmen to support the new Wagner bill.[72]

While planning "the biggest picnic that was ever held by labor in this field" for Labor Day festivities in the fall of 1934, the miners held pertinent political discussions at their weekly union meetings. On September 11, 1934, Local 6186 appointed a committee of three to study and report on the records of congressional and other candidates. On September 22, 1934, members joined with the St. Michael local to celebrate Code Day and the New Deal. On September 25, 1934, L. G. Lichliter, chairman of the county Democratic Committee, and numerous candidates for office, attended the local's meeting in Slovak Hall. The common theme was the necessity for labor to support the New Deal, which had many enemies.[73]

UMWA locals then planned and sponsored a big rally on October 31 at Delaney Field with Thomas Kennedy, UMWA official and fellow miner, the featured speaker. Another rally was held on November 3. To ensure that Berwind-White officials did not intimidate the men, the union declared and took a miners' holiday on Election Day on

November 6, 1934. To enforce it, Local 6186 unanimously voted to fine working members $5.00 each and later called violators to a meeting to explain their noncompliance.[74]

The UMWA's political mobilization paid off at the state and local levels. In Windber and Scalp Level, Democrats Joseph Guffey (a candidate for U.S. Senator) and Earle carried the boroughs with more than 66 percent of the total vote, a massive departure from previous Republican victories. On November 20, 1934, UMWA Local 6186 President Joe Jasway proposed a massive celebration as "a climax to our New Deal victory in Windber." The result was a gala New Year's Eve party—a victory ball. The miners also paraded on FDR's birthday on January 30, and in his honor the UMWA locals organized the first annual community-wide President's Day Ball to aid the victims of infantile paralysis in the local community and in Warm Springs, Georgia.[75]

From the outset, the union did much more than fulfill the traditional narrow functions of a trade union. As part of a broad-based social movement, Locals 6186 and 5229 became involved in many political and social issues of importance to the community and the nation. Their members organized union-relief committees to help the neediest; distributed Christmas baskets; helped truckers, clerks, and steel workers to organize; lobbied for the Wagner Act, the Townsend plan, Social Security, and other progressive legislation; fought nativist, antilabor, and anti–New Deal legislative proposals at the state and federal levels. By the mid-1930s, local UMWA leaders were leading community-wide efforts to bring new industries to town and lobbying to secure a fair share of unemployment and relief projects for the local population. Even as they negotiated new contracts, they took time to help members and their families to become citizens or to apply for worker compensation. The range of the union's activities was impressive.[76]

But New Deal victories did not come easily, given staunch corporate opposition. In May 1935 the Supreme Court struck down the National Industrial Recovery Act; in October, Berwind-White joined with four other major national coal corporations to fight the Guffey bill; and from the outset of the Earle administration, the Republican-dominated Pennsylvania State Senate blocked key New Deal reforms that had the support of the State House and the governor. At the state level, organized labor was able to achieve only two of its many goals: the definitive abolition of the much-hated Coal and Iron Police and more restrictive child-labor laws.[77]

It was in this context that Windber miners continued to mobilize politically to defend workers' rights, their union, and the New Deal. Despite an impressive victory in the 1934 election, the electoral experience had brought home the fact that many miners were not yet citizens and could not vote. On November 20, 1934, after Local 6186 discussed the problem, Joe Novak and Mike Syrian volunteered to offer classes of instruction to members interested in becoming citizens. On December 18, 1934, B. A. Sciotti, a Johnstown attorney and counsel for the UMWA in Indiana, Pennsylvania, promised to help by sending copies of questions and answers used in naturalization hearings.[78]

The first UMWA-sponsored citizenship class began on Saturday morning, March 2, 1935, with thirty miners in attendance. Novak taught the class, and the union paid for materials and provided space. By April, the classes were so popular that more space was needed. The local then formed a committee to see if the classes could be moved to the public schools, but after consulting with the school superintendent decided on the status quo for the time being.[79]

The UMWA classes filled a real void in the community. No one else was performing this service. Moreover, Novak had discovered that many immigrants could not afford to become citizens. Despite various laws that mandated nominal fees, the custom had been for applicants for citizenship to pay $50 or $100 to fraternal society officers or elected officials who had helped in the process of application. By contrast, the UMWA taught immigrants basic English and the fundamentals of American history and government free of charge. Novak himself wrote away for booklets in various languages, taught many of the initial classes, and took many people to Somerset for the citizenship hearings. Hundreds of foreign-born men and women became citizens in the late 1930s and early 1940s in this manner. Eventually public school teachers took over the citizenship classes, which had been originally sponsored by the UMWA.[80]

Joe Novak was one of the candidates selected by Local 6186 to run for borough office in 1935. The union and other progressive forces were interested in challenging the company's monopoly of political power on the local level. Berwind-White had always been associated with the Republican party and with the less progressive elements within it. Novak, a popular union activist and Boy Scout leader, seemed to be a good choice to run as a Democratic party candidate for the Windber council, but he and another union-sponsored candidate lost. One friend later confessed to Novak that some miners, including himself, had been paid

Cover of a naturalization booklet, 1911.

to vote for his rival. But the more serious problem of so many working people being noncitizens and unable to vote gave the activist added incentive to continue the citizenship classes.[81]

The national election of 1936 was popularly viewed as a referendum on Roosevelt and the New Deal. In March 1936, UMWA District 2's Board Member John Ghizzoni, organizer of Labor's Non-Partisan League, visited Windber and Scalp Level to organize "Miners Roosevelt Clubs," which functioned as organizations separate from the union locals but with close ties to them.[82]

Miners Roosevelt Club
WINDBER, PA.

Letterhead for "Miners Roosevelt Club" stationery, Windber, 1935.

Local 6186's president, Joe Jasway, became president of the Windber group and Novak secretary. In subsequent months, the clubs coordinated local broad-based efforts to secure Roosevelt's reelection. Still, it is significant that this widespread support for Roosevelt and the New Deal did not preclude discussion in union meetings about the possible necessity of forming a Labor Party. The climax of the Miners Roosevelt Club's work in Windber was a big rally on October 29 in which Attorney General Margiotti and John Torquato spoke. Meanwhile, the company and the *Era* blasted the New Deal and actively supported the Republican opposition.[83]

The Democratic party/labor/ethnic coalition bore fruit in Windber. In 1936, for the first time in the history of the town, the Democratic party's presidential candidate carried the borough in what was the largest voter turnout in its history. Roosevelt beat Al Landon 1,867 to 1,146. Also, for the first time since 1871, Democrats won control of the State Senate and were thus in command of both houses of the Pennsylvania legislature. Governor Earle had already pledged to fulfill the promises of his thwarted 1934 platform.[84]

Another event that marked the passing of an era had occurred during the course of the electoral campaign. On August 18, 1936, E. J. Berwind, who had retired from active business in 1930, died in New York City at the age of 88.[85] It was his successors, not he, who had to deal with the company's long-dreaded unionization of his mines and other social changes.

In January 1937 in a speech before a business group, Governor Earle declared: "I am determined that economic serfdom and its chief symbol, the company town, must go." At that time, the Commonwealth of Pennsylvania counted within its borders 1,200 company towns with more than 300,000 people living in them. Many of these Windbers, ruled by a monopolistic coal or steel company, lacked even elementary civil liberties.[86]

New Deal legislation at the state and federal levels in the late 1930s struck down many of the measures that enabled corporations to control one-industry towns and workplaces autocratically. In Pennsylvania, from 1936 to 1938, a coalition of organized labor, civic groups, Democrats, and others secured passage of a landmark series of reforms that corresponded to those at the federal level. Working people and the larger society thus gained by the passage of a Labor Relations Act, similar to the Wagner Act, which guaranteed workers the right to organize and to bargain collectively; abolished privately paid deputy sheriffs; increased worker compensation; put limitations on the use of antilabor injunctions, whereas there had been no limitations before; reduced the maximum work week for women and children from 54 to 44 hours; regulated industrial home-work and abolished sweatshop conditions; and prohibited company unions, industrial spying, blacklisting, certain strikebreaking practices, and racial discrimination.[87]

Genuine political democracy and the constitutional rights of free speech and assembly finally came to Windber in the 1930s through unionization and political mobilization. Ordinary working people had organized and acted to secure passage of major labor reforms, and it was their actions and their organizations that made the observance of constitutional rights realities instead of dead letters. Other classes benefited as a result. The reforms sought and gained through the labor movement finally enabled Windber's small-business owners to form their first lasting, successful organization—the Windber Business Men's Association—in 1938.[88]

Foreign-born and American-born miners and their families had initially responded to the UMWA and to Roosevelt and the New Deal because they seemed to offer a viable option for changing—not merely accepting—American industrial capitalism. The New Deal did not ultimately end the depression, abolish poverty, or bring about the justice that many working people wanted, but it was the best option available to them at the time. As organized labor and the New Deal came under increasing attack after 1937, working people continued to be mainstays of the New Deal coalition.

From at least World War I on, new immigrants had been faced with the question of "What does it mean to be an American?" Unlike many corporate employers who had sought to ensure an open shop by an Americanization "from above," Berwind-White had never promoted "Americanization" per se. Instead, its antiunion policy included the promotion of segregated ethnic enclaves and other measures that divided and maintained a docile multiethnic work force. Until the New Deal, to become an "American" in the context of Windber, an autocratic company town, had meant becoming, or remaining, a second-class citizen without any meaningful political, economic, or civil rights.

Windber immigrants confronted another version of what it meant to be an American in the 1920s. The Ku Klux Klan had followed on the heels of other nativist groups who considered "true" Americans to be a homogeneous people (white Anglo-Saxons), who spoke one language (English), who shared a unified religion (Protestantism), who were loyal to the government over a long period of time (multiple generations), and who shared a uniform group mentality based on common racial traditions. A "good" American was racially superior, antilabor, anti-Catholic, and anti-immigrant and did not tolerate cultural diversity or ideas that challenged the "group mind." Immigration exclusion was the logical culmination of this mind-set, of a "100 percent Americanism" that divorced people from their own cultures and histories. There was simply no room for Windber miners or their families in this or similar nativist definitions of what it meant to be an American.

In Windber, the version of "Americanization" that foreign-born working people ultimately accepted in the 1930s was not a justification of the autocratic status quo or an acknowledgment of racial, national, or cultural inferiority. They chose instead the one viable alternative that allowed them to retain self-respect, obtain basic human rights, maintain cultural diversity, and reform American society.

The UMWA had provided them with a definition of "Americanism" that was both self-affirming and empowering. To be an American did not mean subjection to endless exploitation or hunger, and it did not mean adherence to rigid standards of cultural, religious, or mindless uniformity. All Americans, workers included, had rights and deserved to live lives of dignity. The UMWA's citizenship schools are perhaps the most concrete and direct link between the labor movement and this broad-based, inclusive "Americanization" that occurred "from below."

Miners and their wives attended the citizenship schools because the union, their fellow workers, ran them. They wanted to vote for Roosevelt and ensure the political, economic, and civil gains made during the New Deal. The foreign-born saw no contradiction in attending UMWA meetings or Labor Day celebrations, which unified all nationalities, and in frequenting Verhovay, Jednota, Polish Falcons, or Sons of Italy events and festivities.

Cultural pluralism coexisted with class integration in an American setting on an everyday level. These working people continued their struggles to gain greater control over their lives, to reform industrial capitalism, and to improve American society beyond the 1930s, but it was the American labor movement—not the company, churches, schools, government, or the passage of time—that ultimately offered them hope of a more egalitarian and democratic America.

12

The Achievements and Limits of Worker Protest

Throughout the nonunion era, Windber's new immigrant miners were neither docile sufferers who would endure any measure of capitalistic exploitation, nor lawless rioters, incipient anarchists, or ready-made unionists. To be sure, these pragmatic, thoughtful human beings had brought their diverse cultural heritages and their own conceptions of justice with them from Europe. Neither the cultures nor the rural peasant societies from which they came were static. New class formations, the ideology of rights, and the uneven development of capitalism had already made an impact on their lives. In certain circumstances, these working people were responding to dynamic changes in Europe with passivity, while in others they were rebelling and actively using their cultures as resources for change.[1]

Without discounting the importance of their premigration backgrounds, then, it is nevertheless critical to stress that it was the immigrants' experiences at an autocratic workplace in a company town controlled by one of the world's largest mining corporations which ultimately forged Windber's diverse miners into a working class in twentieth-century industrial America. Those who came to Windber to find work immediately confronted absolute company rule, unequal power relations, and the need to consider what options they might have for gaining greater control over their lives. It was not without reason that union activists in the 1890s were already remarking on Windber's resemblance to slavelike conditions.

The scant evidence we have about the town's earliest days shows that, in 1899, new immigrants reacted to their new situation by joining with English-speaking miners in a movement to try to secure the existence of Windber's first ephemeral UMWA local. A much greater body of evidence from 1906 suggests that it was new immigrants—not the American-born—who led the movement to create a second ephemeral union. Their leadership and massive participation in the strike of 1906—as big and as important a strike, in many ways, as the strike of 1922—was duly noted by the Dillingham Commission, the nativist public, and Berwind-White, which rejoiced at the repressive measures taken by the Pennsylvania State Police after its hired gunmen had massacred three foreign-born people and a child.

The behavior of southern and eastern European immigrants in unionizing and striking in 1906 shows not only that new immigrants could be organized under certain circumstances, but also that their original motivation—to work, save money, and return to Europe—was not the single, overriding obstacle to unionization that some labor historians have noted. Their challenges to autocratic rule were not dependent on the arrival of World War I or their decisions to take up permanent residency. Differing situations in industrial America could lead to a variety of immigrant responses.

Moreover, it can reasonably be argued that these early immigrants had more options than many settled, American working-class families. Such factors as the assumed transitory nature of their work stints, their extensive kinship networks, the migratory nature of mining and immigrant life, the predominance in many industrial centers of single males or married men temporarily unburdened by the presence of wives and dependent children, and the lack of home ownership or any material stake in one-industry settings made it possible for those who wanted to protest conditions to do so by voting with their feet, migrating to other towns, or returning to Europe. During the strike of 1906, many

Windber miners chose to move elsewhere rather than stay and become strikebreakers.

At the same time, Windber's miners were deeply influenced by the environment around them and by major national and regional events. Their struggles for unionization were necessarily uneven, but it was no accident that in every instance their major efforts—in 1899, 1903–1904, 1906, 1919, 1922, 1927, and 1933—took place in conjunction with the active mobilization of District 2 and the International UMWA. That they chose these particular times to act, and not others, shows how closely attuned they were to the changing fortunes of the labor movement, the varying degrees of union activity, and the significance of major national and regional events. Despite ethnic divisions, they joined together in conjunction with larger social movements to end isolation, gain higher wages, and achieve greater justice.

Along with many other working people, Windber miners and their families had experienced a general rise in expectations during World War I. From 1918 until the strike of 1922, class-conscious unionists thus repeatedly beseiged the UMWA to come in and "get Windber Free for Democracy." In 1922 the masses of local miners joined the nation's union miners in the largest coal strike in American history. Their heroic near seventeen-month strike for a union, despite having been excluded from the national settlement by Lewis, was a testimonial to the depth of their commitment and their desire for freedom.

After an abortive strike in 1927, they successfully unionized in 1933 during the New Deal. That this last effort in the nonunion era was the logical extension of previous efforts, a continuum in a long-term, uneven struggle that dated from the town's origins, should not be surprising. After all, the same grievances with regard to work and community that miners and their families had voiced in the 1890s remained issues in 1933. On a daily level, throughout the intervening period, male and female immigrants had sought to expand their spheres of freedom wherever possible: by bypassing the company store and buying produce from a farmer; by questioning the tonnage of the company's weighmaster who they believed had cheated them; by discussing comparative conditions in union and nonunion mines with trusted friends; and by voting against the company's chosen candidates when they became citizens.

Struggles and strikes were family and community affairs. Without the active support of women and children, the men could not have waged their ongoing battles for unionization. During strikes, women not only fulfilled their traditional caretaking roles, but also picketed, shamed strikebreakers, swore out affidavits against the Berwind police, and then campaigned for Franklin Roosevelt and the New Deal. Throughout the nonunion period, women had also made a critical if often overlooked monetary contribution to the miners' family economy through the boardinghouse system, which was built on female labor in the same way that mining was based on male labor. Boardinghouses were essential, not tangential, to Berwind-White's development of the mines, because a domestic female presence was necessary to recruit and retain a male work force.

Women also bore the primary onus of the care and maintenance of that work force, reproduced the next generation of mine workers, and otherwise ensured family survival in a town where jobs other than mining and housework were scarce or nonexistent. They built Windber's many ethnic institutions, took care of livestock and gardens to ensure against hard times, and tried to feed their families while living through the horrors of evictions and tent-colony life during strikes. The company's monopoly of jobs, housing, stores, and credit had effectively brought the entire family within its tentacles and given wives and daughters reasons of their own for supporting their husbands, sons, and brothers in struggles that challenged the Berwind system.

For these working people, the one viable option for ending economic misery, abolishing autocratic controls, and democratizing their lives lay in the imperfect United Mine Workers. The much-vaunted land of freedom was not so free in company-town America, where political democracy and civil liberties were truly foreign. Restrictions on free speech and the right of free assembly were imposed on others besides miners. As the *Journal* editor discovered in 1903, Slavic fraternal organizations in 1914, and small-business owners in 1919, no genuinely independent organization could thrive in an environment in which the company's monopoly of power remained intact and the miners were unorganized. Political democracy and basic civil liberties became realities for all citizens and all classes in Windber because of the labor movement.

Nonetheless, the wealthy Berwinds, absentee owners living in New York and Philadelphia, could not have ruled the town or the workplace without a lot of help from privileged sectors of the American middle classes and other local residents, American-born and foreign-born. Through monopoly ownership, deed restrictions, interlocking directorates, and many other structural devices, the company had taken great pains at the outset to establish the open shop in its mines and a pervasive domination of the town, but it was the local managers, foremen, ethnic bosses, interested business people, borough councilmen, company spies, and others who perpetuated the system from which they benefited at the expense of the working people. This middle-class society, which had a vested interest in maintaining the autocratic status quo, also helped Berwind-White foster the myth, still heard in town today, that Windber was a "model" town in which workers were uniquely happy and obedient, strikes never happened, and tent colonies never existed.

Throughout the nonunion era, working miners of all nationalities confronted a company-town establishment that was predominantly, if not exclusively, American-born. Class structure thus intersected with ethnicity and religion in a hierarchical class system that placed white Anglo-Saxon Protestants and northern Europeans at the top and the masses of new immigrants at the bottom. At the same time, ethnic communities were never monolithic, and miners had to deal with the important ethnic bosses who upheld the company's arbitrary rule, thereby betraying the class interests of the mine workers. Class struggle took place within each of the immigrant communities too.

Ethnic diversity posed a host of practical problems for union organizers, but ultimately it was much less of an obstacle to unionization than undemocratic restrictions on civil liberties and the massive economic, legal, and political resources that capital could mobilize against organized labor. Unlike American-born workers, new immigrant miners had to confront nativist legislation, the POSA and the KKK, and various forms of informal as well as formal discrimination. Yet both American-born and foreign-born miners had to confront the company's wealth and a hostile state.

Although members of the Dillingham Immigration Commission of 1911, the U.S. Industrial Relations Commission of 1915, a U.S. Department of Labor mediator in 1919, and New York City's Hirshfield Committee of 1922 were among those who offered trenchant critiques of the Berwind system, the company made no substantial reforms. Only successful unionization and organized labor's political mobilization in the 1930s secured New Deal legislation that corrected some of the most glaring abuses of corporate power. Until then, Berwind-White and other open-shop corporations had easy access to labor injunctions, the Coal and Iron Police, and a strikebreaking state police force. The United Mine

Workers thus faced many practical difficulties, other than linguistic or cultural differences, in their efforts to organize closed company towns.

Despite bitter internal union conflicts, questionable UMWA policies of the 1920s, and a long depression, Windber's diverse foreign-born and American-born miners again turned to the imperfect UMWA in the 1930s in the hope of redressing their many grievances. Under John L. Lewis's new aggressive leadership, the union seemed to offer the best—perhaps the only—hope for achieving the ending of exploitation. Through the union, and through Roosevelt's New Deal, these working people sought to reform American capitalism and affirm their own dignity both as workers and as ethnics.

By focusing on labor's rights, the UMWA had provided foreign-born and American-born miners with a self-affirming and empowering definition of what it meant to be an American. Throughout its history, the UMWA had, in theory if not always in practice, been an integrating force that united the members of all nationalities and races on a class basis. In general, it had not tried to impose cultural uniformity or to destroy the immigrant communities in which so many of its members lived. Despite its imperfections, the UMWA was perhaps unique in simultaneously providing working-class immigrants with a vision of a more egalitarian America and a strong incentive for them to become American citizens. During the New Deal, Windber miners and their families thus embraced the union's citizenship schools and organized politically, thereby becoming an integral part of the New Deal coalition.

At the same time, it is important to note that Berwind-White was one of a number of major U.S. corporations that did not engage in an "Americanization" effort "from above," in the manner of Henry Ford and others, even after the strike of 1922. Although it participated in Liberty Bond and other patriotic drives during World War I, engaged in rhetoric that termed labor agitators "un-American," and exploited nativism for its own purposes, the company's basic antiunionization policies were designed to keep what it viewed as docile foreign labor both docile and foreign, with little or no exposure to ideas and organizations that linked Americans and the rights of labor together. Through segregated, controlled ethnic enclaves, it promoted cultural pluralism in order to maintain its divide-and-conquer rule, while other corporations used a self-serving version of "Americanization" to achieve the same goal.

In many ways, the history of Windber and of its working class is a microcosm of the history of industrial America itself. As such, this study suggests the limitations as well as the merits of the New Deal. Certainly union recognition, collective bargaining rights, and government-supported social programs were progressive measures that benefited the miners and their families, who had fought hard and paid dearly for these gains. But these limited reforms always had adamant opposition and did not alter the basic inequalities inherent in capitalism or resolve its internal contradictions.

Despite unionization and the New Deal reforms, therefore, Berwind-White continued to retain ownership of all mineral rights in the area "in perpetuity," as well as ownership of vast tracts of surface lands. It also jealously guarded management's monopoly on decision-making and would therefore dictate the timing and terms of future mine closings, company investments, and diversification, within a system in which profit, not human considerations, remained the motivating force. Although the distance between the extremely unequal power relationships of the two classes in the company town did narrow somewhat, temporarily, in the 1930s, "The Miners' Program" or other proposals that had offered genuine alternatives to coal capitalism never materialized.

The political mobilization of labor, more effective in the 1930s and 1940s than previously, was also limited. Helping immigrants to become citizens was a lengthy process. As would become clear, it was often easier in Windber for the union to mobilize support for Roosevelt and national candidates than it was to oust certain mainstays at the local level. Blaine Barefoot, the company's chosen Republican candidate, thus served as burgess of the company town from 1918 to 1949. Moreover, in retrospect, we now know that the zenith of the New Deal coalition's influence at the national level was reached during the 1930s and 1940s.

None of these limitations or realities, however, detract from the many achievements of the generations of immigrant and American mining families, who from the 1890s to the 1930s fought many battles against the odds for greater control over their own lives and workplaces. These modest working people helped make New Deal reforms and a more humanitarian restructuring of capitalism possible. In the process, they not only built industrial America but also made Windber and the United States more democratic and egalitarian than either had been before.

The history of such extraordinary people deserves to be remembered and honored, not forgotten, demeaned, ignored, or romanticized. There are no permanent or absolute victories, and progress does not neatly unfold in a straight, linear line. In a day when the New Deal consensus and organized labor have come under sharp attack, it is important to understand how and why miners and other citizens mobilized in the 1930s, and what they did accomplish as well as what they did not. Since that time, historical conditions on all levels—local, national, and international—have changed greatly, and the ongoing struggle of working people to democratize their lives and secure social justice has necessarily entered new phases. But that is a history beyond the scope of this study, and one that Windber workers and others are still writing.

Epilogue

At a community meeting in Windber in 1992, a citizen commented that he wanted me to carry my history of Windber up to the present, because he did not think things were very different for working people now than they were in the nonunion era. He seemed to suggest that local power relationships were as unequal and as arbitrary today as they were then. I pondered this sobering thought at the time and have been reflecting on it ever since.

Without trying to cover the second fifty years of Windber's history—the subject for another book—I want to suggest that even a cursory survey of this later period calls for serious scholarly research into deindustrialization and the contested power relations that occurred during this process in Windber and in other industrial communities. The issues involved go far beyond one community, but it is useful to remember that area miners and the town of Windber faced a crisis of capitalist deindustrialization long before many other places and long before widespread talk of a restructured global economy became commonplace. Those who are familiar with mine and plant closings elsewhere know that, while industrial decline may be inevitable, what people, corporations, and governments do about it—even in other capitalist countries—is not.

Change may be more or less humane. Corporations may act in a socially responsible manner or they may ignore the human suffering that results from decisions designed to maximize the profits of shareholders. Government programs may guarantee the dislocated new jobs, give them educational opportunities or job training but without any guarantee of a job, or do nothing at all. Workers and unions may be passive in the face of change or organize in new ways. During these periods of forced unemployment and occupational transition, the general public, through taxation and other measures, may or may not provide the uprooted with a decent level of income, health benefits, and basic security. Comparative studies of deindustrialization are potentially as valuable as those that have focused attention on industrialization.[1]

To be sure, deindustrialization is often a long-term process. The soft coal industry has been declining since at least the 1920s, when Windber's population was at its height. Seven of the thirteen Berwind-White mines in the area had closed before World War II, but the permanent closings of the six biggest mines occurred after the war, which had created a huge temporary demand for coal. This short-lived local prosperity ended when the war did.[2]

Meanwhile, Windber miners took part in wartime and postwar wildcat strikes. In 1946 they joined the nation's miners in a strike that led to the federal government temporary takeover of the mines. During this period, the government reached an agreement with the United Mine Workers to establish a precedent-setting miners' welfare and pension fund. The fund was to be financed by mine owners paying a royalty on every ton of coal the miners mined. The strike also led to a federal survey of medical conditions in the bituminous fields, which brought investigators from the Department of Interior to the Windber area in 1946 to document existing conditions and needs.[3]

In the meantime, however, the national political climate that had tolerated a militant labor movement began to change, especially after the Republicans won victories in the congressional elections of 1946. The U.S. Congress, which had recently been debating proposals for full employment and national health care, began to defeat Fair Deal measures that continued the New Deal. Over President Harry Truman's veto, it also passed the antilabor Taft-Hartley Act, which modified the Wagner Act and which John L. Lewis denounced as a "slave labor" statute. The power of organized labor was further weakened in the late 1940s by antialien legislation and by the anticommunist McCarthy era crusades that purged radicals from the labor movement and punished dissent.

Windber's own working class reached the culmination of its local political victories around this time, in 1949 and 1950, when Dr. Matthew J. Klena finally replaced Burgess Blaine Barefoot as mayor of Windber. The electoral triumph of the miners in 1949 was undercut, however, by the company's decision to curtail production abruptly, with a view to closing its mines entirely. Thus, the miners and their families found they had achieved their greatest degree of local political power at the same time that they were inheriting all the problems associated with capitalist deindustrialization.

From 1949 to 1962, Berwind-White management systematically disengaged from active mining and shut down its underground operations in Pennsylvania and West Virginia. It gradually sold off company houses and

The coal cleaning plant at Mine No. 37, 1946.

donated certain other properties, which could undoubtedly be claimed as sizable tax write-offs. It sold the last Eureka Store in Windber to new owners in 1969, and in a step that angered many residents, merged the Windber Hospital with Conemaugh Valley Hospital in Johnstown in 1970. In the process, the company averted decline and became more profitable than ever. From 1960 to 1962, Guaranty Bank, which had been appointed a co-trustee of the company under the terms of Edward Berwind's 1936 will, intervened to increase company profits by ousting the Berwind management, which had been in power from 1930 to 1960, and by installing a new aggressive management that diversified, acquired other companies and leased out coal holdings.[4]

The terms on which this "limited deindustrialization" took place in Windber, and the options available to mining families, were shaped by

Children at play at Mine No. 37, 1946.

the unequal power relationships that continued to prevail. For a brief period during the 1930s and 1940s, the gap between the power held by the coal company and that held by the miners through their union had narrowed somewhat. At their height in the 1940s, each of the two UMWA locals, Local 6186 in Windber and Local 5229 in Scalp Level, had about 2,000 members. Labor Day celebrations were big community events, and no one doubted the efficacy of the union's "No contract, no work" slogan. Even then, though, the miners' influence was not great enough to prevent nonunion inroads into strip-mining or to keep the old company-town guard from dominating Windber's 50th anniversary publication of the town's history in 1947, when only scant mention was made of the union, ethnic miners, or the struggles to make the autocratic town freer and better.[5]

Meanwhile, the UMWA locals continued their long-term quest to try to bring new industries to Windber. As early as March 1903, months before his open break with Berwind-White, Will Hendrickson, editor of the *Windber Journal*, had warned the public that Windber needed new industries. In an editorial, he argued for an independent business association that would launch an aggressive campaign to bring in new industries, because "that community which has the greatest diversity of industries is the most independent and also the most prosperous." Realizing that even large coal resources were finite, and that the industry is subject to great fluctuations, he prophesied: "Were the mining industry of this vicinity, for any reason, to close down for any length of time, this entire business community would be paralyzed."[6] Despite the efforts of the union, a Community Development Association that had been founded in 1945, and other subsequent organizations, few alternative industries, with the exception of Bestform Foundations of Pennsylvania, have since located in the town.[7]

What happens to a company town whose raison d'être was coal when coal is no longer king and no other industries have replaced it? Certainly the mine closings in the 1950s and 1960s forced Windber-area miners and their families to confront a formidable array of new problems as well as perennial ones. The specter of unemployment, which with the exception of a few prosperous war years had haunted the community from at least the 1920s on, now threatened the very existence of the town and of its working class.

Some of the younger displaced miners found jobs in Johnstown's steel mills or other regional industries. A few mines continued to operate in the surrounding area, and for a period Berwind-White leased out Mine 40 to another coal company, which employed some miners. Many others who could find no such jobs migrated out of the region in the same manner as their European parents or grandparents had once done. Using their kinship ties and networks, they sought work in other places. Young people had few alternatives but to join the military or otherwise leave.

In the 1950s and 1960s, UMWA Locals 6186 and 5229 and certain politicians tried to exert political influence to ameliorate the human suffering caused by the mine closings. They worked to extend unemployment compensation and did make surplus food available for the local mining population. The International UMWA, however, was no longer displaying the militance it had in the 1930s and 1940s. In another apparent change of behavior in the early 1950s, John L. Lewis began to make cooperative agreements with big coal producers that did not always take into account the interests of the rank-and-file. The union's top leadership thus returned to the business unionism of an earlier era and approved technological changes that led to additional job losses, greater health risks, and more mine closings. In addition to having to deal with all their usual enemies, miners who were fortunate enough to get UMWA pensions soon found themselves facing the problems of union corruption, including the brutal murder of a union presidential candidate and a generation gap that separated them from younger miners.

To a large extent, Berwind-White was able to dictate the terms of the limited deindustrialization that occurred in Windber. As the owner of so much private property, the company had resources and influence that ordinary miners or their union lacked. Consequently, when the company sold off its company houses, it not only retained the mineral rights but also put into the deeds of sale clauses that released the company and any successor companies or lessees from any liability for any future environmental problems, cave-ins or other property damage, and health or safety hazards related to coal-mining operations past or future. Churches such as St. John Cantius, which in the 1970s or 1980s sought the removal of all sorts of other autocratic deed restrictions dating from the early century, continue to find their institutions—not the company—liable under similar conditions. Homeowners, business people, and the borough council find that they often have to get the company's approval in property or right-of-way matters.[8]

One undeniable result of the mine closings is that, by 1990, Windber's urban population had dropped to half of what it had been at its height in 1920. The broader mining area has lost 45 percent of the population it had at its height in 1940. Decline has been gradual but ongoing. In 1990, Windber's population of 4,756 fell below the threshold number of 5,000, with the result that the Greyhound bus no longer stops at the drug store.[9]

The trend to outmigration would have been even larger were it not for another phenomenon. In recent decades, a number of workers who had left the area have returned to their former community to spend their retirement days. The remembrance of community, as working-class people define it, and the relatively cheap cost of living have contributed to this return migration. Consequently, the current population is not only smaller than in the past but also older. According to the census of 1990, the median age of urban Windber's population is 41 years.[10]

Despite the company's mine closings and sale of houses, it would be highly inaccurate to say that Berwind-White ever *left* Windber. The owners of textile mills and auto plants may decide to uproot factories and move elsewhere, but minerals and other natural resources are in fixed locations. Consequently, even though the renamed Berwind Corporation's main base is in Philadelphia, Windber still remains the locus of the company's major field operations. It has held onto ownership of vast tracts of surface lands and owns virtually all mineral rights in the region "in perpetuity." According to its own publicists: "Although the Natural Resources division of Berwind today concentrates on leasing its mines and coal reserves to other companies, it is still one of the largest owners of coal reserves in Appalachia."[11]

Thus, the citizen who compared today's unequal power relationships in Windber to those in an earlier nonunion era had a point. The combination of outmigration, the declining number of miners, organized labor's decreased influence, and the town's failure to bring in new industries and diversify have given the old company management and allied business interests, who have a vested material and psychological stake in perpetuating a modified version of the old company town, a greater power today than they have perhaps had for some time. It is thus fair to ask: How independent and how free is Windber today?

The legacies of an undemocratic autocratic town live on in other ways as well. As I sadly learned during the course of my oral history interviews, the name of Berwind can still strike fear in the hearts of some elderly miners and their families. That poor, elderly, vulnerable people in the late 1980s vividly remembered the not-so-good old days and firmly believed that the coal company would have their black-lung benefits taken away if it were known that they told their life stories or criticized it in any way is an ongoing indictment of the present-day power of the Berwind system.

Meanwhile, tourism has become the latest hope for a revival of Windber and the region's economy. In 1988, Congressman John Murtha got Congress to appropriate millions of dollars to promote industrial heritage and tourism projects in a depressed nine-county region in central and southwestern Pennsylvania, which includes the Windber area. Plans for a mining museum in Windber and a tour of Mine 40 are currently being developed. Whether these efforts will be more successful than previous ones in creating economic growth remains to be seen.

At the same time, what version of history will emerge for the public to see in the industrial exhibits, and what it is exactly that residents will be celebrating during the town's centennial in 1997 (coincidentally the 75th anniversary of the landmark 1922 strike), remain open questions. There are clear choices. In 1994 certain heritage and business groups honored Edward J. Berwind by placing his name in the Johnstown Chamber of Commerce Hall of Fame as a "captain of industry." These same forces will undoubtedly try to honor the company in local heritage exhibits and glorify the autocratic company town as its "model" benevolent creation. During the centennial, citizens can even dress up and parade in the costumes of the Coal and Iron Police or others who beat up on miners a century ago. They can ignore the labor movement and all substantive historical issues. But there is another choice. An alternative version of history could celebrate the Joe Zahuraks, the Joe Novaks, and the Pearl Leonardises, the ethnic and American working people who built industrial America and whose struggles for social justice led them to unionize and, in the process, make Windber "Free for Democracy."

Whose history will be commemorated? Whose version of history will prevail? The questions themselves remind us that there are no permanent or absolute victories in this world, while the answers to these questions tell us a great deal about ourselves and our current society.

Appendixes

Appendix A
Native-Born and Foreign-Born Populations, 1910–1950

	1910	1920	1930	1940	1950
Windber					
Total population	8,013	9,462	9,205	9,057	8,010
Native-born whites	57%	71%	78%	82%	85%
of native parents	31%	29%	32%	—[a]	—[a]
of foreign or mixed parentage	26%	41%	45%	—[a]	—[a]
Foreign-born whites	43%	29%	22%	18%	15%
Number of blacks	13	6	17	6	4
Number of other races	3	3	3	0	2
Scalp Level					
Total population	1,424	1,690	1,875	1,950	1,756
Native-born whites	51%	—[a]	77%	82%	87%
of native parents	27%	—[a]	28%	—[a]	—[a]
of foreign or mixed parentage	24%	—[a]	49%	—[a]	—[a]
Foreign-born whites	49%	—[a]	23%	18%	13%
Number of blacks	0	0	0	0	0
Number of other races	0	0	0	0	0
Paint Borough					
Total population	999	1,283	1,336	1,700	1,547
Native-born whites	68%	—[a]	86%	87%	90%
of native parents	53%	—[a]	55%	—[a]	—[a]
of foreign or mixed parentage	15%	—[a]	31%	—[a]	—[a]
Foreign-born whites	32%	—[a]	14%	13%	10%
Number of blacks	0	0	0	0	0
Number of other races	0	0	0	0	0
Combined population total	10,436	12,435	12,416	12,707	11,313

SOURCES: Department of Commerce, Bureau of the Census, *Thirteenth Census,* vol. 3: *Population* (Washington, D.C., 1913), 601; Bureau of Census Manuscript Schedules, 13th Census 1910: Pennsylvania (1910; Washington, D.C.: National Archives, 1978), microfilm rolls 1324, 1420, 1421; Department of Commerce, Bureau of the Census, *Fourteenth Census,* vol. 3: *Population* (Washington, D.C., 1923), 882; Department of Commerce, Bureau of the Census, *Fifteenth Census: 1930 Population,* vol. 3, part 2: *Reports by States* (Washington, D.C., 1932), 699, 722, 741; Department of Commerce, Bureau of the Census, *Sixteenth Census: 1940 Population,* vol. 2, part 6: *Characteristics of the Population* (Washington, D.C., 1943), 92, 126, 156; Department of Commerce, Bureau of the Census, *Seventeenth Decennial Census: 1950,* vol. 2, part 38: *Characteristics of the Population* (Washington, D.C., 1952), 160, 182, 183.

[a] Complete figures are not provided by the Census.

Appendix B

Leading Nationalities, Windber Area, 1910 (percents)

	Windber	Scalp	Paint	#37	#38	Area
Native	57.1	50.6	67.9	30.1	49.9	55.0
Foreign	42.9	49.4	32.1	69.9	50.1	45.0
Slovaks	12.7	11.9	0.7[a]	11.2	20.3	11.9
Magyars	7.8	17.4	19.5[b]	26.8	15.2	11.1
Poles	6.0	11.2	8.1	19.4	7.5	7.7
Italians	10.1	0.2	0.1	0.0	2.0	6.9
Others	6.3	8.7	3.7	12.7	5.1	7.4

SOURCE: Bureau of Census Manuscript Schedules, 13th Census 1910: Pennsylvania (1910; Washington, D.C.: National Archives, 1978), microfilm rolls 1324, 1420, 1421.

[a] This percentage includes 6 people, or 0.6%, whose nationalities were not identified in the census. All had been born in Hungary, and all had names that were more typically Slovak than Hungarian.

[b] This percentage includes 59 individuals, or 5.9%, whose nationalities were not formally identified in the census. All 59 had been born in Hungary, had typical Magyar names, lived in houses with other Magyars, or were located in Hungarian church records.

Appendix C
Less Numerous Nationalities, Windber Area, 1910

	U.S. Census Figure	Dillingham Figure
English	196	58
Germans	152	201
Lettish	74	0
Romanians	68	300
Swedes	62	100
Syrians	31	114
Welsh	30	60
Greeks	27	17
Lithuanians	28	174
Russians	24	354
Yiddish (Hebrew)	22	95
Norwegians	18	37
Scots	13	61
Croatians	13	46
French	10	31
Bohemians and Moravians	10	10
Irish	7	56
Danes	3	0
Chinese	3	0
Ruthenes	3	15
Servians	3	0
Europeans (unidentified)	1	0
Armenians	1	0
Born at sea	1	0
Unknown nationality	2	0
Slovenian	0	6
Tyrolean	0	15
Total	802	1,750

SOURCES: Bureau of Census Manuscript Schedules, 13th Census 1910: Pennsylvania (1910; Washington, D.C.: National Archives, 1978), microfilm rolls 1324, 1420, 1421; Immigration Commission, *Reports*, vol. 6: *Immigrants in Industries*, part 1, *Bituminous Coal Industry* (1911; reprint, New York: Arno and New York Times, 1970), 6:479.

Appendix D
Windber-Area Employees Inside and Outside Mines, 1897–1918

Year	Employees	Inside	Outside
1897	320	73%	27%
1898	1,438	93	7
1899	2,158	93	7
1900	3,013	92	8
1901	3,086	92	8
1902	3,034	90	10
1903	4,159	93	7
1904	3,993	93	7
1905	4,154	90	10
1906	3,789	88	12
1907	4,782	88	12
1908	5,026	89	11
1909	4,896	89	11
1910	4,976	94	6
1911	4,548	94	6
1912	4,159	96	4
1913	3,711	94	6
1914	4,281	93	7
1915	3,640	86	14
1916	3,200	85	15
1917	3,435	84	16
1918	2,543	78	22
Average	3,561	89%	11%

SOURCES: Pennsylvania Department of Internal Affairs, Bureau of Mines, *Report* (Harrisburg, Pa., 1898–1903), annual reports, 1897–1902; Pennsylvania Department of Mines, *Report* (Harrisburg, Pa., 1904–1920), annual reports, 1903–1918.

Appendix E

Days Worked, Berwind-White Mines, Windber, 1900–1921

Year	Berwind-White Mines, Windber Area	Bituminous Mines, Pennsylvania
1900	301	219
1901	221	216
1902	282	221
1903	278	216
1904	256	190
1905	248	216
1906	276	239
1907	285	238
1908	295	184
1909	261	210
1910	291	217
1911	257	220
1912	278	241
1913	294	251
1914	224	202
1915	233	207
1916	293	240
1917	296	239
1918	292	259
1919	279	194
1920	291	246
1921	212	157
Average	270	219

SOURCES: Pennsylvania Department of Internal Affairs, Bureau of Mines, *Report* (Harrisburg, Pa., 1901–1903), annual reports, 1900–1902; Pennsylvania Department of Mines, *Report* (Harrisburg, Pa., 1904–1925), annual reports, 1903–1921.

Appendix F
Fatalities, Bituminous Mines, Pennsylvania, 1898–1940

Years	Number of Fatalities	Average per Year	Employees per Fatality	Tonnage per Fatality
1898–1900	723	241.0	401	301,253
Total or average	723	241.0	401	301,253
1901–1905	2,174	434.8	340	235,579
1906–1910	2,900	580.0	327	240,649
1911–1915	2,427	485.4	395	325,835
1916–1920	2,249	449.8	406	370,688
Total or average	9,750	487.5	367	293,188
1921–1925	1,766	353.2	540	377,413
1926–1930	1,958	391.6	384	356,084
1931–1935	816	163.2	749	532,465
1936–1940	804	160.8	799	629,539
Total or average	5,344	267.2	618	473,875

SOURCES: Pennsylvania Department of Internal Affairs, Bureau of Mines, *Report* (Harrisburg, Pa., 1899–1903), annual reports, 1898–1902; Pennsylvania Department of Mines, *Report* (Harrisburg, Pa., 1904–1959), annual reports, 1904–1940.

Appendix G

The Composition of Berwind-White's Labor Force, Windber Area, 1922 and 1923

	January 1, 1922	January 1, 1923	% Losses or Gains
Americans	411	507	+23
Slovaks	753	275	−63
Poles	495	264	−47
Hungarians	545	358	−34
Italians	415	179	−56
Spaniards	23	90	+290
Mexicans	6	15	+150
All others	246	189	−23
Total employees	2,894	1,877	−35

SOURCE: Heber Blankenhorn, *The Strike for Union* (1924; reprint, New York: Arno and New York Times, 1969), 261.

Notes

Introduction

1. *New York Times*, 19 August 1936, 1; Arthur D. Howden Smith, *Men Who Run America: A Study of the Capitalistic System and Its Trends Based on Thirty Case Histories* (Indianapolis: Bobbs-Merrill, 1935), 301–309; *Dictionary of American Biography*, 1958 ed., s.v. "Berwind, Edward Julius," 11:37–38.

2. *Encyclopedia Americana*, 1984 ed., s.v. "Coal: 6 Coal-Mining Industry," by C. L. Christenson; *Encyclopædia Britannica*, 1973 ed., s.v. "Coal and Coal Mining," by W. A. McCurdy et al.

3. J. Cutler Andrews, "The Gilded Age in Pennsylvania," *Pennsylvania History* 34 (1967): 11; George Miller, *A Pennsylvania Album: Picture Postcards 1900–1930* (University Park: The Pennsylvania State University Press, 1979), 10, 64.

4. U.S. Immigration Commission, *Reports of the Immigration Commission*, vol. 6, *Immigrants in Industries*, part 1, *Bituminous Coal Industry* (1911; reprint, New York: Arno and New York Times, 1970), 473. (Hereafter: IC *Reports*, 6.)

5. Richard O. Boyer and Herbert M. Morais, *Labor's Untold Story* (New York: United Electrical, Radio and Machine Workers of America, 1955), 221.

6. Richard C. Keller, *Pennsylvania's Little New Deal* (New York: Garland, 1982), 6–7.

7. Herminie, Pennsylvania, for example, was named after the wife of one of the Berwind brothers.

8. Leroy R. Hafen, "Colorado Cities: Their Founding and the Origin of Their Names," *Colorado Magazine* 9 (September 1932): 172. After Windber's founding, another "Berwind" came into existence early in the century when Berwind-White entered the West Virginia coal fields and established another company town. E. J. Berwind's entry into this new field is briefly described in W. P. Tams, Jr., *The Smokeless Coal Fields of West Virginia* (Morgantown: West Virginia University Library, 1963), 77–79.

9. Ewa Morawska, *For Bread with Butter: The Life-Worlds of East Central Europeans in Johnstown, Pennsylvania, 1890–1940* (Cambridge: Cambridge University Press, 1985), 19, 20.

10. This argument is most vividly expressed in David Brody, *Steelworkers in America: The Nonunion Era* (New York: Harper and Row, 1960), 96–111, 135–136.

11. This argument is contained in a recent ethnocentric interpretation offered by Thomas Göbel, "Becoming American," *Labor History* 29 (Spring 1988): 173–198.

12. David Montgomery, "The 'New Unionism' and the Transformation of Workers' Consciousness in America, 1909–22," *Journal of Social History* 7 (Summer 1974): 516–517.

13. There is a vast interdisciplinary literature on the definition and significance of class and class-consciousness. For a few classic interpretations, see Georg Lukács, *History and Class Consciousness: Studies in Marxist Dialectics*, trans. R. Livingstone (Cambridge, Mass.: MIT Press, 1971); E. P. Thompson, *The Making of the English Working Class* (New York: Vintage, 1963), esp. 9–13; Eric Hobsbawm, "The Making of the Working-Class, 1870–1914," in Hobsbawm, *Worlds of Labour: Further Studies in the History of Labour* (London: Weidenfeld and Nicolson, 1984). For recent treatments of these subjects and the "crisis" in labor history, see Ira Katznelson and Aristide R. Zolberg, eds., *Working-Class Formation: Nineteenth-Century Patterns in Western Europe*

and the United States (Princeton: Princeton University Press, 1986); Lenard R. Berlanstein, ed., *Rethinking Labor History: Essays on Discourse and Class Analysis* (Urbana: University of Illinois Press, 1993); Leon Fink, *In Search of the Working Class: Essays in American Labor History and Political Culture* (Urbana: University of Illinois Press, 1994). One of the most trenchant critiques of the "new" labor history is Lawrence T. McDonnell's "'You Are Too Sentimental': Problems and Suggestions for a New Labor History," *Journal of Social History* 17 (Summer 1984): 629–654. The most readable survey of a century of American labor historiography is John Schacht's, "Labor History in the Academy: A Layman's Guide to a Century of Scholarship," *Labor's Heritage* 5 (Winter 1994): 4–21.

14. John Higham, *Strangers in the Land: Patterns of American Nativism 1860–1925* (New Brunswick, N.J.: Rutgers University Press, 1955), 35–67; Henry Carter Adams, "Relation of the State to Industrial Action," *Publications of the American Economic Association* 1 (January 1887): 465–549.

15. John Bodnar, *The Transplanted: A History of Immigrants in Urban America* (Bloomington: Indiana University Press, 1985), xv–xxi, 206–216.

Chapter 1: From Berwind to Windber

1. James W. Siehl and Lois J. Siehl, "They Came to Scalp Level to Sketch Nature's Beauties," *The Era, Johnstown Tribune Democrat* supplement, 28 February 1984, 3; E. Howard Blackburn and William H. Welfley, *History of Bedford and Somerset Counties, Pennsylvania*, ed. W. H. Koontz (New York: Lewis Publishing Co., 1906), 2:x; "Somerset County in History," *Somerset County Democrat*, 10 July 1895.

2. Sylvester K. Stevens, *Pennsylvania: Titan of Industry* (New York: Lewis Historical Publishing, 1948), 1:204; Howard N. Eavenson, *The First Century and a Quarter of American Coal Industry* (Pittsburgh: privately published, 1942), 447; Blackburn and Welfley, *History*, 2:660; Fred C. Doyle, ed., *50th Anniversary Windber, Pa.* (Windber, 1947), 25.

3. The term "native-born" is used in this work as the Census Bureau used it—that is, to mean born in the United States. U.S. Department of the Interior, Census Office, *Compendium of the Eleventh Census: 1890*, 3 parts (Washington, D.C., 1892–1897), 3:46; 1:503–504.

4. Ibid., 3:46–47; 2:664–665.

5. Ibid.

6. Blackburn and Welfley, *History*, 2:478, 486–487, 494–495.

7. *Somerset County Democrat*, 11 November 1896, 30 June 1897, 25 April 1900, 1 August 1900. For a history of the Republican party's long-term dominance and subsequent decline in the 1920s in Pennsylvania, see Samuel John Astorino, "The Decline of the Republican Dynasty in Pennsylvania, 1929–1934" (Ph.D. diss., University of Pittsburgh, 1962).

8. *Somerset County Democrat*, 10 July 1895.

9. Stevens, *Pennsylvania*, 1:297–298.

10. Ibid., 1:296–301, 320–323; Eavenson, *First Century*, 419; *Encyclopedia Americana*, 1984 ed., s.v. "Coal: 6 Coal-Mining Industry," by C. L. Christenson; Donald G. Puglio, "Production, Distribution, and Reserves of Bituminous Coal in Pennsylvania," in *Pennsylvania Coal: Resources, Technology, and Utilization*, ed. S. K. Majumdar and E. W. Miller (Easton, Pa.: Pennsylvania Academy of Science, 1983), 33–35; Charles L. Best, "History of Coal Transportation in Pennsylvania," in *Pennsylvania Coal*, 281–292; Greater Pennsylvania Council, *The Decline of the Bituminous Coal Industry in Pennsylvania* (Harrisburg, Pa., 1932), 3, 6–8.

11. Greater Pennsylvania Council, *Decline*, 4; Wilbert G. Fritz and Theodore A. Veenstra, *Regional Shifts in the Bituminous Coal Industry with Special Reference to Pennsylvania* (Pittsburgh:

University of Pittsburgh Bureau of Business Research, 1935), 21–23; John F. Reese and James D. Sisler, *Bituminous Coal Fields of Pennsylvania*, part 3, *Coal Resources* (Harrisburg, Pa.: Pennsylvania Geological Survey, Fourth Series, 1928), 7.

12. *Somerset County Democrat*, 18 March 1903.

13. Pennsylvania Department of Mines, *Annual Reports*, 1910–1930 (Harrisburg, Pa., 1911–1931); Frank Paul Alcamo, *The Windber Story: A 20th Century Model Pennsylvania Coal Town* (n.p.: n.p., 1983), 75; Fritz and Veenstra, *Regional Shifts*, 20; *Windber Era*, 19 February 1925, 25 February 1926, 3 March 1927.

14. A. Tappan Sargent, "Enduring Coal Enterprise," *Black Diamond* 96 (28 March 1936): 3–4.

15. Doyle, *50th Anniversary*, 28; Alcamo, *Windber Story*, 127; Richard D. Adams, "Pennsylvania and North Western," *The Keystone* 10 (Winter 1977): 2; Sargent, "Enduring Coal," 23–24; and John H. Ingham, *Biographical Dictionary of American Business Leaders, A–G* (Westport, Conn.: Greenwood, 1983), s.v. "Berwind, Edward J."

16. See *New York Times*, 13 June 1906, 1; David Hirshfield, Committee on Labor Conditions at the Berwind-White Company's Coal Mines in Somerset and Other Counties, *Statement of Facts and Summary of Committee Appointed by Honorable John F. Hylan, Mayor of the City of New York, to Investigate the Labor Conditions of the Berwind-White Company's Coal Mines in Somerset and Other Counties, Pennsylvania* (New York: M. B. Brown, December 1922).

17. Tams, *Smokeless Coal Fields*, 77. E. J. Berwind was intensely disliked by many West Virginia operators, one of whom frequently described Berwind as two of the "three greatest liars in New York." For such assessments and a partial account of Berwind-White's activities in West Virginia, see Joseph T. Lambie, *From Mine to Market: The History of Coal Transportation on the Norfolk and Western Railway* (New York: New York University Press, 1954), esp. 74, 184–185.

18. *New York Times*, 19 August 1936, 21; 22 August 1936, 13.

19. Sargent, "Enduring Coals," 4–9; Elizabeth B. Berwind, *Gateways to the Past* (Philadelphia: n.p., 1978), 310. Even today, long after the permanent closing of Windber mines, the town remains a prime focus of Berwind's activities because Windber is the locus of many of the Berwind Corporation's profit-making diversified subsidiaries and because many of its other company towns no longer—or barely—exist. For a recent advertisement and official, uncritical, in-house glorification of the company's history, see *The History of Berwind* (Philadelphia: Berwind Group, 1993).

20. Sargent, "Enduring Coals," 5, 9.

21. *Philadelphia Times*, 4 June 1899.

22. Sargent, "Enduring Coals," 5.

23. Ibid.

24. There was also a confusing political movement that failed in its aim to create a new county from existing counties. See Alcamo, *Windber Story*, 40, 46, 48; *Somerset County Democrat*, 21 April 1897, 4 February 1903, 18 February 1903, 11 March 1903, 18 March 1903, 25 March 1903; *Windber Journal*, 24 February 1903; *Windber Era*, 12 March 1903.

25. Doyle, *50th Anniversary*, 38–39.

26. Ibid., 29.

27. Somerset County Deed Book, Somerset County Courthouse, Somerset, Pa., 82:273–275; J. L. Fehr, comp., *Windber–Scalp Level and Vicinity: Eureka Mines, Berwind-White Coal Mining Co. Operators* (Windber, December 1900), 19, 32, 35.

28. Somerset County Deed Book, 97:95–100.

29. Dr. Chauncey Hoffman, interview by author, tape recording, Geistown, Pa., 11 July 1989.

30. Somerset County Deed Book, 80:463–470.

31. Fehr, *Windber–Scalp Level*, 10.

32. *Somerset County Democrat,* 26 July 1893, 9 August 1893.

33. *Somerset County Democrat,* 6 December 1905, 13 June 1906, 27 June 1906, 22 August 1906, 21 November 1906, 28 November 1906, 16 January 1907. See also *Windber Journal,* 27 August 1903; "Berkey v. Berwind-White Coal Mining Company," in Pennsylvania Courts, *The District Reports of Cases Decided in All the Judicial Districts of the State of Pennsylvania During the Year 1907* (Philadelphia, 1907), 16:429–439; "Berkey v. Berwind-White Coal Mining Company, Appellant," in Pennsylvania Supreme Court, *Pennsylvania State Reports 1908* (New York: Banks Law Publishing Co., 1910), 220:65–81; "Berkey v. Berwind-White Coal Mining Company, Appellant," in Pennsylvania Supreme Court, *Pennsylvania State Reports 1910* (New York: Banks Law Publishing Co., 1911), 229:417–429; "Weaver v. Berwind-White Coal Company, Appellant," in Pennsylvania Supreme Court, *Pennsylvania State Reports 1906 and 1907* (New York: Banks Law Publishing Co., 1909), 216:195–205; "Hoffman v. Berwind-White Coal Mining Co., Appellant," in Pennsylvania Supreme Court, *Pennsylvania State Reports 1919 and 1920* (Philadelphia: George T. Bisel Co., 1920), 265:476–486.

34. *Somerset County Democrat,* 24 January 1894, 25 July 1894, 30 January 1895, 6 May 1896, 20 January 1904; *Windber Era,* 30 October 1902; *New York Times,* 18 May 1906, 2; 19 May 1906, 1; 24 May 1906, p.1; 25 May 1906, 1; 25 May 1906, 10; 26 May 1906, 1; 29 May 1906, 4; 6 June 1906, 5; 7 June 1906, 3; 9 June 1906, 1; 11 June 1906, 1; 13 June 1906, 1; 14 June 1906, 5; 15 June 1906, 5; 20 June 1906, 6.

35. Benjamin F. G. Kline, Jr., *"Stemwinders" in the Laurel Highlands: The Logging Railroads of Southwestern Pennsylvania* (n.p.: n.p., 1973), 1302; *Somerset County Democrat,* 10 March 1897, 17 March 1897, 11 August 1897; Blackburn and Welfley, *History,* 2:525; and Doyle, *50th Anniversary,* 50.

36. *Somerset County Democrat,* 20 October 1897. That Berwind-White and Babcock might have made similar agreements in the New River field in West Virginia too is worthy of future investigation.

37. Fehr, *Windber–Scalp Level,* 21; *Somerset County Democrat,* 25 June 1902; Alcamo, *Windber Story,* 24–27.

38. *Somerset County Democrat,* 20 November 1901.

39. Fehr, *Windber–Scalp Level,* 2.

40. Ibid., 10–12; Alcamo, *Windber Story,* 75.

41. *Somerset County Democrat,* 9 March 1898; 19 October 1898; Fehr, *Windber–Scalp Level,* 10; IC *Reports,* 6:473.

42. Fehr, *Windber–Scalp Level,* 10–16, 38; "The Largest Collieries in the United States," *Coal Trade Journal* 39 (21 March 1900): 185–186.

43. *Somerset County Democrat,* 25 August 1897.

44. *Somerset County Democrat,* 1 September 1897, 6 October 1897, 12 January 1898, 22 March 1899.

45. John Brophy, *A Miner's Life,* ed. J. O. P. Hall (Madison: University of Wisconsin Press, 1964), 61–62. For information on the corporation's operations, see Industrial Relations Commission, *Final Report and Testimony Submitted to Congress by the U.S. Commission on Industrial Relations Created by the Act of August 23, 1912* (Washington, D.C., 1916), 8:7580–7582.

46. Somerset County Deed Books, 129:633, 138:342–343, 150:102–103, 154:261–262, 160: 305–306, 627–629; *Somerset County Democrat,* 17 January 1906.

47. Fehr, *Windber–Scalp Level,* 1, 3, 25, 34, 35.

48. *Somerset County Democrat,* 10 January 1900, 28 February 1900, 4 July 1900.

49. "Windber Borough Charter," Somerset County Deed Book, 109:420–425.

50. It was possible to locate 272 individual petitioners in the census. Of these, 152 clearly owned their own homes, with or without a mortgage; 108 rented, including 25 boarders; and the status of 12 was unknown.

Of the 272, some 170 had been born in the United States and 102 in foreign countries. Of the foreign-born, 76 came from northern Europe, 25 from southern or eastern Europe, and 1 from South America. Naturalized citizens totaled 58, those with first papers 10, aliens 14, and those for whom there was no information on citizenship 20. Of the naturalized, 47 were from northern Europe and 11 from southern or eastern Europe.

The census identified 101 of the 272 known signers simply as miners; another 15 were engineers or skilled mine workers (car builders, electricians, etc.), and 12 were mine supervisors or railroad bosses. In addition, there were 38 contractors, carpenters, or building tradesmen, 35 merchants, 27 laborers, 18 professionals (doctors, druggists, publishers, ministers, etc.), and 16 clerks who signed. The remainder held various jobs not already mentioned or were without profession. These figures were tabulated from the manuscript schedules of the 1900 U.S. Census for Paint Township. See U.S. Bureau of Census Manuscript Schedules, 12th Census 1900: Pennsylvania (1900; Washington, D.C.: National Archives, 1978), microfilm roll 1486.

51. Alcamo, *Windber Story,* 107; and Fehr, *Windber–Scalp Level,* 2–3.

52. Official state election results from 1896 to 1920 are contained in the yearbooks. See *Smull's Legislative Hand Book and Manual of the State of Pennsylvania* (Harrisburg, Pa., 1897–1921).

53. *Somerset County Democrat,* 25 April 1900.

54. Windber Borough Council, Windber Borough Council Minute Books, Meetings, 24 July 1900, 3 August 1900, and 10 August 1900, Windber Borough Building.

55. Ibid., 28 August 1900, 25 September 1900, 7 November 1900, 20 November 1900, 5 December 1900, 8 January 1901.

56. Ibid., 23 January 1901, 4 June 1901, 2 July 1901, 29 July 1901.

57. *Somerset County Democrat,* 12 December 1900.

58. Fehr, *Windber–Scalp Level.*

59. Of the 106 entries, 31 were in their 20s, 41 in their 30s, 15 in their 40s, 6 in their 50s, 1 in his 70s; 12 were unknown. While 86 had been born in the United States, 76 of these had been born in Pennsylvania. Somerset County births numbered 15, Cambria County 14, Clearfield County 12, and Bedford County 6. The rest were scattered throughout the state. Only the religions of 32 were identified—29 were Protestant, 2 were Roman Catholic, and 1 was Jewish.

Chapter 2: From Europe to Windber

1. IC *Reports,* 6:509.

2. Industrial Relations Commission, *Final Report,* 8:7597.

3. Ibid., 7596, 7749.

4. IC *Reports,* 6:473, 476–477, 481.

5. Dan Lennon, "Secretary Lennon Sends Some Interesting News from the District," *United Mine Workers Journal* (hereafter *UMWJ*), 16 April 1891; Fehr, *Windber–Scalp Level,* 19, 32, 35.

6. Alcamo, *Windber Story,* 93–94, 261; IC *Reports,* 6:481.

7. IC *Reports,* 6:481, 502.

8. For a critique of the commission's assumptions and methodology, see Oscar Handlin, introduction to *Abstracts of Reports of the Immigration Commission* (1911; reprint, New York: Arno and New York Times, 1970), xxii–xxiv. See also Philip S. Foner, *History of the Labor Movement*

in the United States, vol. 3: *The Policies and Practices of the American Federation of Labor 1900–1909* (New York: International Publishers, 1964), 264–265; Isaac A. Hourwich, *Immigration and Labor: The Economic Aspects of European Immigration to the United States* (New York: G. P. Putnam's Sons, 1912), 18–21, 414–443.

9. IC *Reports*, 6:479; U.S. Department of Commerce, Bureau of the Census, *Thirteenth Census . . . 1910*, vol. 3: *Population Reports by States* (Washington, D.C., 1913), 537, 550.

10. Department of Commerce, Census Office, *Twelfth Census . . . 1900*, vol. 1: *Population* (Washington, D.C., 1901), 334, 346.

11. Department of Commerce, Bureau of the Census, *Fourteenth Census . . . 1920*, vol. 1: *Population: Number and Distribution* (Washington, D.C., 1921), 599; Department of Commerce, Bureau of the Census, *Sixteenth Census . . . 1940 Population*, vol. 2, part 6: *Characteristics of the Population* (Washington, D.C., 1943), 92, 126.

12. IC *Reports*, 6:473, 479, 529.

13. *Thirteenth Census* (see note 9 above), 3:601.

14. Ibid., 563, 582.

15. These calculations are based on the Bureau of Census Manuscript Schedules, 13th Census, 1910: Pennsylvania (1910; Washington, D.C.: National Archives, 1978), microfilm rolls 1324, 1420, 1421. See also IC *Reports*, 6:481.

16. *The Catholic Encyclopedia*, 1909 ed., s.v. "Greek Catholics in America," by A. J. Shipman.

17. IC *Reports*, 6:527.

18. Leonard Winograd, "The Horse Died at Windber, A History of Johnstown Jewry" (D.H.L. diss., Hebrew Union College–Jewish Institute of Religion, 1966), 203–204.

19. *Windber Era*, 17 December 1903, 16 November 1905; *Windber Daily Era*, 19 December 1903; Winograd, "The Horse Died," 133–134, 251, 301–302.

20. Paul Robert Magocsi, "Carpatho-Rusyns" and "Russians," *Harvard Encyclopedia of American Ethnic Groups*, 1980 ed., 200–202, 885; Paul Robert Magocsi, *Our People: Carpatho-Rusyns and Their Descendants in North America* (Toronto: Multicultural History Society of Ontario, 1984), 5–8.

21. [Edward Surkosky], *Anniversary Book of the Diamond Jubilee of St. Mary's Byzantine Church, Windber, Pennsylvania, 1900–1975, October 12, 1975* (Windber, 1975); *Windber Era*, 17 May 1951; Byzantine Slavonic Rite Catholic Diocese of Pittsburgh, *Silver Jubilee 1924–1949* (Pittsburgh, 1949).

22. Emily Greene Balch, *Our Slavic Fellow Citizens* (1910; reprint, New York: Arno Press and New York Times, 1969), 53–54; and Bodnar, *The Transplanted*, 3–13, 34, 54–56.

23. M. Mark Stolarik, "Slovaks," in *Harvard Encyclopedia of American Ethnic Groups*, 1980 ed., 928; Humbert S. Nelli, "Italians," in ibid., 545; Balch, *Our Slavic Fellow Citizens*, 34.

24. Magocsi, *Our People: Carpatho-Rusyns and Their Descendants in North America*, 4–5.

25. Arthur Chervin, *L'Autriche et la Hongrie de demain: Les différentes nationalités d'après les langues parlées* (Paris: Berger-Levrault, 1915), 28, 43–45; Balch, *Our Slavic Fellow Citizens*, 122.

26. Balch, *Our Slavic Fellow Citizens*, 99–100, 236–237.

27. Julianna Puskás, *From Hungary to the United States 1880–1914* (Budapest, 1982), 33–36, 60–62.

28. Balch, *Our Slavic Fellow Citizens*, 112.

29. Victor R. Greene, *The Slavic Community on Strike: Immigrant Labor in Pennsylvania Anthracite* (Notre Dame: University of Notre Dame, 1968), xv–xvi, 35.

30. Puskás, *From Hungary*, 29, 56, 62.

31. Ibid., 48; Marian Mark Stolarik, "Immigration and Urbanization: The Slovak Experience, 1870–1918," (Ph.D. diss., University of Minnesota, 1974), 1–4.

32. Balch, *Our Slavic Fellow Citizens*, 137–138.

33. Victor Greene, "Poles," in *Harvard Encyclopedia of American Ethnic Groups*, 1980 ed., 790–791; Bodnar, *The Transplanted*, 10; Puskás, *From Hungary*, 47.

34. Stolarik, "Immigration and Urbanization: The Slovak Experience, 1870–1918," 4–6; Puskás, *From Hungary*, 49.

35. Balch, *Our Slavic Fellow Citizens*, 125, 138–139; Greene, "Poles," 790–791; John J. Bukowczyk, *And My Children Did Not Know Me: A History of the Polish-Americans* (Bloomington: Indiana University Press, 1987), 10.

36. Puskás, *From Hungary*, 55; Morawska, *For Bread with Butter*, 61, 66–67; Stolarik, "Immigration and Urbanization," 7–8, 23.

37. Puskás, *From Hungary*, 21, 27, 55–62.

38. Morawska, *For Bread with Butter*, 72–74; Puskás, *From Hungary*, 33–36.

39. Stolarik, "Immigration and Urbanization," 35–36, 110–112; IC *Reports*, 6:34, 44.

40. *Thirteenth Census*, 3:601; *Fourteenth Census*, vol. 3: *Population: Composition and Characteristics* (Washington, D.C., 1923), 882; *Thirteenth Census*, vol. 1: *Population 1910* (Washington, D.C., 1913), 1018; Magocsi, *Our People: Carpatho-Rusyns and Their Descendants in North America*, 11; Puskás, *From Hungary*, 38, 42.

41. Puskás, *From Hungary*, 38, 41; IC *Reports*, 6:507.

42. Puskás, *From Hungary*, 42.

43. Of 166 native-born miners, 56.6% were married and 42.8% were single. IC *Reports*, 6:504–505. The census figure on the marital status of boarding men is an estimate based on the manuscript schedules of the 1910 census, which contained many errors. For example, data from the East Ward of Windber, with a population of 4,050, had to be discounted because the enumerator had routinely listed all boarders as single. Thus, an incredible 1,030 of 1,046 boarders were marked as unmarried. In sharp contrast, an apparently more careful enumerator in the West Ward, where there were 544 boarders enumerated out of a population of 3,963, found that 34% of the boarding population were married.

44. According to the manuscript schedules of the 1910 census, the local groups with the highest percentage coming to the United States after 1900 were Romanians (96%), Magyars (87%), Poles (85%), Italians (79%), Lettish (78%), Slovaks (70%), Syrians (68%), Lithuanians (64%), Russians (59%), Germans (54%), and Yiddish (50%). By contrast, only 17% of the area's Welsh, 13% of the Swedes, 12% of the English, 6% of the Norwegians, and none of the Irish or Scots had emigrated to the U.S. during the decade. Most of these groups had come earlier.

45. *Catholic Encyclopedia*, 1909 ed., s.v. "Greek Catholics in America," by Shipman, 6:745, 750.

46. Immigration Commission, *Reports of the Immigration Commission*, vol. 1, *Abstracts of Reports of the Immigration Commission* (1911; reprint, New York: Arno and New York Times, 1970), 185 (hereafter: IC *Reports*, 1); William I. Thomas and Florian Znaniecki, *The Polish Peasant in America* (New York: Alfred A. Knopf, 1927), 2:1489–1503; Puskás, *From Hungary*, 27–28; Morawska, *From Bread with Butter*, 72; Bodnar, *The Transplanted*, 44–45, 52–53.

47. IC *Reports*, 1:24.

48. *Windber Era*, 4 December 1902, 3 December 1903; *Windber Journal*, 31 March 1903.

49. *Windber Era*, 26 November 1903; *Windber Journal*, 9 July 1903.

50. IC *Reports*, 6:505–506.

51. Ibid.

52. *Windber Era*, 4 August 1904. Most national Slovak fraternal records are located at the Immigration History Research Center of the University of Minnesota in Minneapolis–St. Paul.

53. Immigration Commission, *Reports of the Immigration Commission*, vol. 2, *Abstracts of Reports of the Immigration Commission, with Conclusions and Recommendations of the Minority* (1911; reprint, New York: Arno and New York Times, 1970), 426. (Hereafter: IC *Reports*, 2.)

54. IC *Reports*, 6:516–517.

55. IC *Reports*, 2:426–427.

56. Andrew Molnar [pseud.] and Elizabeth Molnar [pseud.], interview by author, tape recording, Windber, 23 February 1984.

Chapter 3: The Work of Mining

1. The cross-tabulations were based on information from the manuscript schedules of the 1910 census. See also IC *Reports*, 6:509.

2. These figures resulted from cross-tabulations of occupations and nationalities as enumerated in the manuscript schedules of the 1910 census. See ibid., 485.

3. These cross-tabulations of nationalities and occupations were also derived from the manuscript schedules of the 1910 census.

4. *Johnstown Democrat*, 7 April 1899; John H. Rohalley, interview by author, tape recording, Windber, 2 March 1984. The Windber Museum contains a notebook of ads placed by Berwind-White in various ethnic newspapers in 1920.

5. I am indebted to Sewell Oldham, the late Berwind-White engineer, for his recollections about the traveling labor agent.

6. John Toth [pseud.], interview by author, tape recording, Windber, 30 September 1986.

7. IC *Reports*, 6:476.

8. W. R. Jones to E. V. McCandless, Esq., 25 February 1875, quoted in James Howard Bridge, *The Inside History of the Carnegie Steel Company: A Romance of Millions* (New York: Aldine, 1903), 81–82.

9. Rohalley, interview, 2 March 1984; Toth, interview, 30 September 1986; IC *Reports*, 6:491.

10. IC *Reports*, 6:477, 529.

11. Bridge, *Inside History*, 81–82.

12. IC *Reports*, 6:499, 546.

13. Pennsylvania Department of Internal Affairs, Bureau of Mines, *Report* (Harrisburg, Pa., 1898–1903), annual reports, 1897–1902; and Pennsylvania Department of Mines, *Report of the Department of Mines of Pennsylvania* (Harrisburg, Pa., 1904–1920), annual reports, 1903–1918.

14. Carter Goodrich, *The Miner's Freedom: A Study of the Working Life in a Changing Industry* (Boston: Marshall Jones Co., 1925), 15–55; Walton H. Hamilton and Helen R. Wright, *The Case of Bituminous Coal* (New York: Macmillan, 1925), 304.

15. Goodrich, *Miner's Freedom*, 53–55; IC *Reports*, 6:476. Merle Travis put folklore and music together when he composed the song "Sixteen Tons," which contained the phrase "a mind that's weak and a back that's strong." For a history of the song and its various recordings, see Archie Green, *Only a Miner: Studies in Recorded Coal-Mining Songs* (Urbana: University of Illinois Press, 1972), 294–316.

16. Goodrich, *Miner's Freedom*, 30–38, 47.

17. "Talks in Many Tongues," *UMWJ*, 12 April 1906, 8.

18. Goodrich, *Miner's Freedom*, 13–14, 19–20, 47–50, 55, 100, 181.

19. Ibid., 44, 55. The figures for Windber mines were taken from Pennsylvania, Bureau of Mines, and Department of Mines annual reports (see note 13 above).

20. Pennsylvania General Assembly, *Laws of the General Assembly of the Commonwealth of Pennsylvania . . . Passed at the Session of 1911 . . .* (Harrisburg, Pa., 1911), 762. Calculations of the percentages employed as foremen and assistant foremen in Windber-area mines were based on employment statistics taken from Pennsylvania Bureau of Mines and Department of Mines annual reports, 1897–1902 and 1903–1918.

21. Bureau of Census Manuscript Schedules, 13th Census 1910: Pennsylvania (1910; Washington, D.C.: National Archives, 1978), microfilm rolls 1324, 1420, 1421.

22. Ibid.

23. Ibid.; Pennsylvania Department of Mines annual reports, 1914–1918.

24. Bureau of Census Manuscript Schedules, 1910; Pennsylvania Department of Mines annual reports, 1914–1918.

25. These calculations are based on employment statistics in Pennsylvania Department of Mines, *Report . . . 1914,* part 2, *Bituminous* (Harrisburg, Pa., 1915), 26; and Pennsylvania Department of Mines, *Report . . . 1918,* part 2, *Bituminous* (Harrisburg, Pa., 1920), 11.

26. IC *Reports,* 6:44; William Graebner, *Coal-Mining Safety in the Progressive Period: The Political Economy of Reform* (Lexington: University Press of Kentucky, 1976), 118.

27. Joseph Zahurak, interview by author, tape recording, Windber, 7 October 1986.

28. Ibid.

29. IC *Reports,* 6:476–477, 529.

30. Pennsylvania Department of Mines, *Report . . . 1906,* part 2, *Bituminous* (Harrisburg, Pa., 1907), 37.

31. Calculations are based on information on the number of days worked in Pennsylvania Bureau of Mines and Department of Mines annual reports, 1900–1921 and 1928–1940.

32. Chris Evans, *History of United Mine Workers of America,* 2 vols. (Indianapolis: n.p., 1918–1921), 2:17, 114–117, 125–126; Andrew Roy, *A History of the Coal Miners of the United States from the Development of the Mines to the Close of the Anthracite Strike of 1902 . . .* (1905; reprint, Westport, Conn.: Greewnwood, 1970), 360–364; W. B. Wilson, "8-Hour Day and 11 Per Cent Advance," *UMWJ,* 26 March 1903, 5; IC *Reports,* 6:500.

33. "General Labor Review," *Coal Age* 11 (13 January 1917): 103; "Coal and Coke News," *Coal Age* 11 (9 January 1917): 1015.

34. Susie Shuster and Katherine Rodish, interview by author, tape recording, Windber, 12 March 1984; Mike Jankosky [pseud.], interview by author, tape recording, Windber, 24 February 1984.

35. Jankosky, interview, 24 February 1984.

36. Shuster, interview, 12 March 1984.

37. UMWA Local 6186, Windber, Pennsylvania, Minute Books, Meetings, 12 May 1936, 19 May 1936, 26 May 1936, 2 June 1936.

38. Joseph J. Novak, interview by author, tape recording, Windber, 21 May 1985.

39. Stephen R. Washko, interview by author, tape recording, Windber, 29 September 1984.

40. Pennsylvania, General Assembly, "An Act to protect miners in the bituminous coal region of the Commonwealth," in Pennsylvania Department of Mines, *Report . . . 1904,* part 2, *Bituminous* (Harrisburg, Pa., 1905), 732. UMWA Windber Local 6186 received its charter from the national on July 22, 1933, and it hangs in Slovak Hall in Windber. Checkweighmen were on the tipples in August 1933. Berwind-White Coal Mining Company, Time Records, Checkweighmen All Mines, August–December 1933, Berwind Corporation, Windber.

41. Department of Commerce, *Historical Statistics of the United States Colonial Times to 1970* (Washington, D.C., 1975), 607.

42. Calculations are based on information on accidents in Windber mines in Pennsylvania Bureau of Mines and Department of Mines annual reports, 1897–1902 and 1903–1921.

43. See note 42.

44. Washko, interview, 29 September 1984; *Windber Journal,* 24 September 1903; *Johnstown Democrat,* 3 August 1900; *Windber Era,* 10 December 1903.

45. *Windber Era,* 6 November 1902, 25 December 1902, 22 January 1903, 26 March 1903;

Windber Borough Council Meeting, 4 September 1900; Pennsylvania General Assembly, "An Act relating to bituminous coal mines and providing for the lives, health, safety and welfare of persons employed therein," in Pennsylvania Department of Mines, *Report . . . 1904,* part 2, 773.

46. Pennsylvania Department of Mines, *Report . . . 1905,* part 2, *Bituminous* (Harrisburg, Pa., 1906), xx–xxi.

47. Pennsylvania Department of Internal Affairs, Bureau of Mines, *Report . . . 1897* (Harrisburg, Pa., 1898), xiii.

48. Graebner, *Coal-Mining Safety,* 121–122.

49. Pennsylvania Bureau of Mines and Department of Mines annual reports, 1897–1902 and 1903–1921.

50. Graebner, *Coal-Mining Safety,* 1–10, 112–115, 142–145, 155–175.

51. Ibid., 146–147; IC *Reports,* 6:499; Zahurak, interview, 7 October 1986.

52. Jankosky, interview, 24 February 1984.

53. Pennsylvania Department of Mines, *Report . . . 1906,* part 2, 6–14, 29–30; *Windber Era,* 9 May 1912, 27 June 1912; Graebner, *Coal-Mining Safety,* 148–152; Pennsylvania Department of Mines, *Report . . . 1911,* part 2, *Bituminous* (Harrisburg, Pa., 1912), v–xvii; *Windber Era,* 18 January 1917.

54. Novak, interview, 21 May 1985.

55. Ibid.; Toth, interview, 30 September 1986; and Pennsylvania General Assembly, *Laws Passed at the Session of 1911,* 807.

56. Novak, interview, 19 May 1985.

57. Pennsylvania Department of Mines, *Report . . . 1906,* part 2, 27.

58. Joseph Keleschenyi, William Kosturko, and Mary Kosturko, interview by author, tape recording, Windber, 13 March 1984.

59. Shuster, interview, 12 March 1984; *Sixteenth Census,* vol. 2, part 6, 156.

60. Fehr, *Windber–Scalp Level,* 10–16; Hamilton and Wright, *Case,* 135; and Novak, interview, 21 May 1985. Conveyors represented a major reorganization of work in that teams of three to four men replaced the buddy system, but conveyors were slow to come into use in Windber mines. In 1928 the first conveyor was installed in a local Berwind-White mine; in 1930 only thirty employees were engaged in related work. Conveyors became widespread in the 1940s. Pennsylvania Department of Mines, *Report . . . 1927–1928,* part 2, *Bituminous* (Harrisburg, Pa., 1929), 339; and Pennsylvania Department of Mines, *Report . . . 1929–1930,* part 2, *Bituminous* (Harrisburg, Pa., 1932), 291, 613.

61. Pennsylvania General Assembly, "An Act relating to bituminous coal mines," 746; Pennsylvania General Assembly, *Laws Passed at the Session of 1911,* 766.

62. Pennsylvania Department of Internal Affairs, Bureau of Mines, *Report . . . 1899* (Harrisburg, Pa., 1900), 739; Pennsylvania Department of Mines, *Report . . . 1918,* part 2, *Bituminous* (Harrisburg, Pa., 1920), 1258.

63. IC *Reports,* 6:509; Pennsylvania Department of Mines, *Report . . . 1904,* part 2, x–xi; "The Enduring Conflict," *Coal Age* 7 (26 June 1915): 1108–1109.

64. Calculations are based on Pennsylvania Bureau of Mines and Department of Mines annual reports, 1897–1902 and 1903–1918.

65. "Mining School Started at Windber, Pa.," *Coal Trade Bulletin* 48 (16 March 1923): 320; *Windber Era,* 19 June 1924.

66. Industrial Relations Commission, *Final Report,* 8:7584.

67. Ibid., 7588, 7590–7592.

Chapter 4: Women's Work

1. Shuster, interview, 12 March 1984.

2. *Thirteenth Census*, 3:601; Ann G. Gary, interview by author, tape recording, Windber, Pennsylvania, 29 February 1984.

3. Novak, interview, 19 May 1985.

4. Shuster, interview, 12 March 1984.

5. Puskás, *From Hungary*, 38, 41, 42. Calculations are based on data in *Fourteenth Census*, 3:882; *Fifteenth Census . . . 1930*, vol. 3, part 2, *Reports by States* (Washington, D.C., 1932), 699; and *Sixteenth Census*, vol. 2, part 6, 126.

6. Shuster, interview, 12 March 1984.

7. U.S. Department of Labor, Women's Bureau, *Home Environment and Employment Opportunities of Women in Coal-Mine Workers' Families* (Washington, D.C., 1925), 13. (Hereafter: *Home Environment.*)

8. Ibid., 11–13.

9. These calculations are based on information in the Bureau of Census Manuscript Schedules, 1910: Pennsylvania.

10. Ibid.; Jankosky, interview, 24 February 1984.

11. These calculations are based on information in Bureau of the Census Manuscript Schedules, 1910: Pennsylvania.

12. The 54 native-born servants worked exclusively for American or northern European families or businesses. Of 30 southern and eastern European servants, only 2 Slovaks and 1 Magyar worked for an American or northern European family. Ibid.

13. *Home Environment*, 7; Katherine Jankosky [pseud.], interview by author, tape recording, Windber, 8 March 1984; Agnes Mucker, interview by author, tape recording, Windber, 14 March 1984.

14. Katherine Jankosky, interviews, 24 February 1984, 8 March 1984.

15. *Home Environment*, 5.

16. Bodnar, *The Transplanted*, 84.

17. Donna Gabaccia, "*The Transplanted*: Women and Family in Immigrant America," *Social Science History* 12 (Fall 1988): 249.

18. Calculations are based on information in Bureau of the Census Manuscript Schedules, 1910: Pennsylvania.

19. Ibid. Slovaks probably had fewer boarders than the other groups because a greater number of them had been in America longer. According to the 1910 census, 30% of Windber's Slovaks had entered the United States before 1900, but only 21% of the town's Italians, 15% of its Poles, and 13% of its Magyars had. Numerous Slovaks had emigrated to Windber via the anthracite region, where they had already had experience with the boarding system. Boarding was often a temporary phenomenon linked to recency of arrival, marital and family status, or stage of life, and some Slovaks interviewed suggested that their fathers—men who had had previous experience as boarders—hoped to avoid having boarders in their own homes when they themselves had families. See Novak, interview, 19 May 1985.

20. Margaret Byington, *Homestead: The Households of a Mill Town* (1910; reprint with new intro. by Samuel P. Hays, Pittsburgh: University of Pittsburgh, 1974), 142–143; Dorothy Schwieder, *Black Diamonds: Life and Work in Iowa's Coal Mining Communities 1895–1925* (Ames: Iowa State University Press, 1983), 99, 109.

21. *Home Environment*, 5.

22. Pearl Leonardis, interview by author, tape recording, Windber, 5 March 1984.

23. Zahurak, interview, 7 October 1986.

24. *Windber Journal*, 11 February 1904.

25. *Home Environment*, 16–17.

26. For a discussion of Berwind-White's policies toward ethnic communities, see Mildred A. Beik, "The Competition for Ethnic Community Leadership in a Pennsylvania Bituminous Coal Town, 1890s–1930s," in Second International Mining History Congress, *Towards a Social History of Mining* (Munich: Beck, 1992).

27. Mike Jankosky, interview, 24 February 1984.

28. *Home Environment*, 17–20.

29. IC *Reports*, 6:493–496; *Home Environment*, 6; Andrew Molnar [pseud.] and Elizabeth Molnar [pseud.], interview by author, tape recording, Windber, 23 February 1984.

30. Molnar and Molnar, interview, 23 February 1984.

31. Leonardis, interview, 5 March 1984.

32. Molnar and Molnar, interview, 23 February 1984.

33. Ibid.

34. John H. Rohalley, interview, 2 March 1984.

35. Morawska, *For Bread with Butter*, 135.

36. Zahurak, interview, 7 October 1986.

37. Rohalley, interview, 2 March 1984.

38. Mike Jankosky, interview, 24 February 1984.

39. Leonardis, interview, 5 March 1984.

40. Washko, interview, 29 September 1984.

41. Shuster, interview, 12 March 1984.

42. Ibid.; *Sixteenth Census*, vol. 2, part 6, 156.

43. *Windber Journal*, 29 October 1903.

Chapter 5: Ethnic Communities and Class

1. IC *Reports*, 6:499, 511.

2. Ibid.

3. A few works that cite the central importance of the immigrant church include Marian Mark Stolarik, "Immigration and Urbanization: The Slovak Experience, 1870–1918" (Ph.D. diss., University of Minnesota, 1974), 69; James S. Olson, *Catholic Immigrants in America* (Chicago: Nelson-Hall, 1987), 220–221; and Raymond A. Mohl and Neil Betten, "The Immigrant Church in Gary, Indiana: Religious Adjustment and Cultural Defense," *Ethnicity* 8 (March 1981): 1–2.

For conflicting views on national parishes, see Olson, *Catholic Immigrants*, 103, 183, 185–186, 203, 218; William J. Galush, "Faith and Fatherland: Dimensions of Polish-American Ethnoreligion, 1875–1975," in *Immigrants and Religion in Urban America*, ed. Randall M. Miller and Thomas D. Marzik (Philadelphia: Temple University Press, 1977), 87; Victor R. Greene, *For God and Country: The Rise of Polish and Lithuanian Ethnic Consciousness in America* (Madison: State Historical Society of Wisconsin, 1975), 1–13; Rudolph J. Vecoli, "Prelates and Peasants: Italian Immigrants and the Catholic Church," *Journal of Social History* 2 (Spring 1969): 222, 228–229, 239, 268; Timothy L. Smith, "Lay Initiative in the Religious Life of American Immigrants, 1880–1950," in *Anonymous Americans: Explorations in Nineteenth-Century Social History*, ed. Tamara K. Hareven (Englewood Cliffs, N.J.: Prentice-Hall, 1971), 216, 241–243; Harold J. Abramson, *Ethnic Diversity in Catholic America* (New York: Wiley, 1973), 3, 173; and Michael Novak, *The Rise of the Unmeltable Ethnics: Politics and Culture in the Seventies* (New York: Macmillan, 1972).

4. Puskás, *From Hungary*, 212–213. The number of national parishes grew rapidly from 1900 until 1940, when their number reached a zenith. In 1900 there were 737; in 1920, 1,759; and in 1940, 2,006. See Olson, *Catholic Immigrants*, 122.

5. *Windber Era*, 21 May 1903; *Windber Journal*, 28 May 1903, 18 February 1904.

6. *Windber Journal*, 3 September 1903, 15 September 1904; *Windber Era*, 16 September 1904; *Johnstown Democrat*, 13 January 1905; Urban J. Peters, comp., *Memorial Volume of the Dedication of the Cathedral of the Blessed Sacrament Altoona, Pa.* (Altoona, Pa., 1930), 55, 99.

7. *Windber Journal*, 8 September 1904; *Johnstown Democrat*, 9 September 1904.

8. *Windber Era*, 8 June 1905, 18 November 1909, 25 November 1909; SS. Cyril and Methodius Parish, *Golden Jubilee SS. Cyril and Methodius Parish, 1906–1956, Windber, Pennsylvania* (Windber, 1956).

9. SS. Cyril and Methodius Parish, *Golden Jubilee*; *Windber Era*, 20 July 1905.

10. Vecoli, "Prelates and Peasants," 222–231, 235–238, 266.

11. Peters, *Memorial Volume*, 59–60; St. Anthony's Church, *40th Anniversary St. Anthony's Church, Windber, Pennsylvania, 1908–1948* (Windber, 1948); St. Anthony of Padua Parish, *Golden Jubilee 1908–1958 St. Anthony of Padua Parish, Windber, Pennsylvania* (Windber, 1958).

12. Peters, *Memorial Volume*, 59–60; St. Anthony's Church, *40th Anniversary*; St. Anthony of Padua Parish, *Golden Jubilee*; Leonardis, interview, 8 March 1984.

13. Paula Benkart, "Hungarians," *Harvard Encyclopedia of American Ethnic Groups*, 1980 ed., 462, 467; St. Mary's Hungarian Roman Catholic Church, *60th Anniversary 1914–1974, St. Mary's Hungarian Roman Catholic Church, Windber, Penna.* (Windber, 1974); *Windber Era*, 22 June 1905.

14. *Johnstown Democrat*, 24 June 1910; *Windber Era*, 30 May 1912, 17 October 1912.

15. *Windber Era*, 26 December 1912, 5 June 1913.

16. *Windber Era*, 9 January 1913, 23 January 1913, 30 January 1913, 15 May 1913; St. Mary's Hungarian Roman Catholic Church, *60th Anniversary*; *Szabadság*, 21 August 1920.

17. Peters, *Memorial Volume*, 68.

18. Somerset County Deed Book, 159:334.

19. Ibid., 381:433, 754:540, 920:282–284.

20. Ibid., 129:179, 121:71, 139:258, 111:367, 111:338.

21. *Somerset County Democrat*, 6 September 1911; *Windber Era*, 7 November 1912, 23 October 1924; and Elizabeth Houghton, "Miners Declare 'Model Town' Is Ruled by Tyrants," *The World*, 22 October 1922, in Powers Hapgood Papers, "Scrapbook 1921–1927," Manuscripts Department, Lilly Library, Indiana University, Bloomington, Indiana.

22. *Dongó*, 15 July 1920.

23. For discussions of this problem, see John Higham, "Integration vs. Pluralism: Another American Dilemma," *Center Magazine* 7 (July–August 1974): 67–73; and Philip Gleason, "American Identity and Americanization," *Harvard Encyclopedia of American Ethnic Groups*, 1980 ed., 31–58.

24. Leonardis, interview, 5 March 1984.

25. Ibid.

26. SS. Peter and Paul Orthodox Greek Catholic Church, *Twenty-Fifth Anniversary of SS. Peter and Paul Orthodox Greek Catholic Church, Windber, Pennsylvania, 1936–1951* (Windber, 1951), 34–35.

27. Zahurak, interview, 7 October 1986.

28. IC *Reports*, 6:499, 511; Fehr, *Windber–Scalp Level*, 5–6; *Windber Era*, 21 February 1924, 26 February 1925, 27 August 1925, 3 June 1926.

29. Zahurak, interview, 7 October 1986; Victoria Dembrinsky, Mary Chmaj, Mary Zaczek, interview by author and the Rev. Father Stanley J. Zabrucki, tape recording, Windber, 5 October 1986.

30. I am indebted to the Most Reverend James J. Hogan, bishop of the Altoona-Johnstown Diocese, retired, for pointing out the Nanty-Glo example.

31. Josef Barton has definitively shown that fraternal societies were not unique to the United States, as once thought, and that many ethnic associations originated in Europe. See Josef J. Barton, *Peasants and Strangers: Italians, Rumanians, and Slovaks in an American City 1890–1950* (Cambridge, Mass.: Harvard University Press, 1975); and Josef J. Barton, "Eastern and Southern Europeans," in *Ethnic Leadership in America*, ed. J. Higham (Baltimore: Johns Hopkins University Press, 1978), 150–175. See also John H. Pankuch, "Fifty Years of Progress," *Pamätnica N.S.S. 1890–1940, Národný Kalendár* 47 (1940): 86, 88.

32. Toth, interview, 30 September 1986.

33. Benkart, "Hungarians," 467. Local Slovak records show that the St. John the Baptist Society, Branch 292 of the First Catholic Slovak Union, was founded on August 21, 1898; St. Thomas Slavonic Society, Branch 304 of the National Slovak Society, on December 18, 1898; and the St. John the Baptist Beneficial Society, Branch 48 of the Pennsylvania Slovak Roman and Greek Catholic Union, on September 4, 1899. See Spolok Sväti Jána Krstitela, číslo 292, Prvá Katolícka Slovenská Jednota, Windber, "Zápisnica," Meeting, 21 August 1898; Sväti Tomas, Odbor 304, Národný Slovenský Spolok, Windber, "Zápisnica," Meeting, 18 December 1898; and Spolok Sväti Jána Krstitela, číslo 48, Pennsylvánska Slovenská Katolícka Jednota, Windber, "Zápisnica," Meeting, 21 August 1898. Courthouse records include "St. Michael Beneficial Society of Windber, Pa.," charter, 25 September 1901, Somerset County Deed Book, 111:591–593; "Italian Workman's Beneficial Association of Windber," charter, 22 February 1904, Somerset County Deed Book, 130:445–446; "St. John's Cantius Beneficial Society of Windber," charter, 8 September 1902, Somerset County Deed Book, 125:402–404.

34. "St. John's Cantius Beneficial Society of Windber," charter, 402.

35. SS. Cyril and Methodius Parish, *Golden Jubilee.*

36. A parish history of St. Mary's listed eight such organizations, including Hungarian and Carpatho-Russian ones, still listed as linked to the church in 1951. See St. Mary's Greek Catholic Church, *Golden Jubilee of St. Mary's Greek Catholic Church, Windber, Penna.* (Windber, 1951). Other societies transferred allegiance to SS. Peter and Paul after the church split.

37. Marian Mark Stolarik has described Slovak religious divisions in various places, including *The Role of American Slovaks in the Creation of Czecho-Slovakia, 1914–1918* (Cleveland, Ohio: Slovak Institute, 1968), 9–21. On the national level, American Slovaks were approximately 75% Roman Catholic and 15% Lutheran; the rest were Reformed and Greek Catholic. See Stolarik, "Immigrants, Education, and the Social Mobility of Slovaks, 1870–1930," 113.

38. "Hungarian Reformed Beneficial Society of Windber, Pa., and Vicinity," charter, 21 March 1907, Somerset County Deed Book, 149:439–441; "St. Anthony Catholic Beneficial Society of Windber," charter, 5 October 1908, Somerset County Deed Book, 154:627–630; "Windber and Surrounding District St. Mary's Roman and Greek Catholic Hungarian Church and Sick Benefit Society," charter, 25 March 1918, Somerset County Deed Book, 208:305–338.

39. "Italian Literary, Musical and Beneficial Association of Windber Borough," charter, 21 March 1907, Somerset County Deed Book, 149:436–438; *Windber Era*, 4 December 1902, 9 October 1913.

40. SS. Cyril and Methodius Parish, *Golden Jubilee*; St. Mary's Greek Catholic Church, *Golden Jubilee*; National Slavonic Society, *Constitution and By-Laws . . . Adopted at the VII. Convention . . .*

1899 (Pittsburgh, 1899), 51–52; National Slovak Society, *Constitution and By-Laws . . . Adopted at the XI. Regular Convention . . .* (1914), 66–70.

41. *Windber Era,* 8 April 1909, 15 April 1909, 6 May 1909, 13 May 1909, 14 October 1909, 8 May 1913, 10 October 1912, 17 October 1912.

42. The *Windber Era* contained several important articles on the serious personal impact of the war on Windber-area immigrants but noted a general lack of ill-feeling between the nationalities residing there. See *Windber Era,* 6 August 1914, 10 September 1914, 15 October 1914.

43. Stolarik, *Role of American Slovaks,* 19; Stolarik, "Immigration and Urbanization: The Slovak Experience, 1870–1918," 201–202; Constantine Čulen, "Beginnings of the Slovak League of America," in *60 Years of the Slovak League of America,* ed. Joseph Paučo (Middletown, Pa., 1967), 26–36; *Windber Era,* 13 September 1917.

44. *Windber Era,* 27 September 1917, 1 November 1917.

45. National Slavonic Society, *Constitution,* 47–48; "Slavish Educational Club," charter, 19 May 1902, Somerset County Deed Book, 125:8–10; Slovenský Národniský Politický Klub [Slovak Political Club], "Zápisnica," Meeting, 13 June 1901; *Johnstown Democrat,* 18 October 1901; Slovenský Neodvislý Politický Klub [Slovak Independent Political Club], "Zápisnica," Meeting, 22 September 1907; *Windber Era,* 10 October 1907. The Dillingham Commission estimated that 100 of 150 Slovaks in the two Slovak political clubs were naturalized citizens by 1909. Slovaks also led in voting among Windber's foreign-born. See IC *Reports,* 6:525–527.

46. "American Polish Educational Association," Petition for Incorporation, 12 June 1911, Somerset County Deed Book, 176:193–196; *Windber Era,* 16 May 1912; Benkart, "Hungarians," 469–470. Only three of eight petitioners for the incorporation of the Hungarian Reformed Beneficial Society were citizens, and some but not all of the fifteen petitioners for St. Mary's Beneficial Association were naturalized. See "Hungarian Reformed Beneficial Society of Windber, Pa., and Vicinity," Petition for Incorporation, 21 March 1907, Somerset County Deed Book, 149:439–441; "Windber and Surrounding District St. Mary's Roman and Greek Catholic Hungarian Church and Sick Beneficial Society," charter, 25 March 1918, Somerset County Deed Book, 208:335–338.

47. *Fourteenth Census,* 3:860; *Thirteenth Census,* 3:601. In 1912, Albert Torquato, president of the Italian-American Citizens' Beneficial Association, claimed there were only 45 Italian citizens in Windber before his organization was founded a year or so before, but that his fraternal society had increased the number of Italian citizens to 140, while another 58 had received their first papers. See *Windber Era,* 16 May 1912.

48. "Abruzzi Workman's Beneficial Society of Windber," charter, 24 February 1913, Somerset County Deed Book, 182:496–500; "Sicilian Italian Mutual Beneficial Society," charter, 2 September 1918, Somerset County Deed Book, 210:394–395; and Alcamo, *Windber Story,* 254–255.

49. "Slovak Gymnastic Union Sokol, Assembly 60, Windber, Pennsylvania," charter, 25 June 1913, Slovak Hall, Windber; "Abruzzi Cooperative Association," Certificate of Organization," 16 February 1920, Somerset County Deed Book, 233:38–39.

50. *Windber Journal,* 28 July 1904; *Windber Era,* 30 May 1907.

51. *Windber Era,* 22 April 1915, 6 May 1915, 13 September 1917, 31 October 1912.

52. *Johnstown Democrat,* 6 July 1900; *Somerset County Democrat,* 11 July 1900, 3 October 1900.

53. Leonardis, interview, 5 March 1984. An unnamed prominent Windber official, probably attorney George H. Somerville, wrote a letter to the *Somerset County Democrat* to defend the town's reputation a year after the strike of 1906, when nativists lumped foreigners, labor unrest, drunkenness, crime, and Windber together. In the letter, he defended the foreign-born, who,

he said, had "no more than the average of 'undesirable' citizens." See *Somerset County Democrat*, 6 November 1907.

54. IC *Reports*, 6:499, 511, 528–529.

55. Hungarians chartered their first fraternal under the English name of St. Michael's Beneficial Society of Windber, but it was known in Hungarian circles as Szt. Mihály Magyar Bányász Betegsegélyező Egylet (St. Michael's Hungarian Miners' Sick Relief Society). See its listing in Julianna Puskás, *Kivándorló Magyarok az Egyesült Államokban, 1880–1940* (Budapest, 1982), 513; "Italian Workman's Beneficial Association of Windber," 130:445; "Slovak Workingman's Benevolent Association," charter, 14 November 1914, Somerset County Deed Book, 192:192–194.

56. *Windber Journal*, 29 October 1903; H. L. Kerwin, Washington, D.C., to John Brophy, Clearfield, Pa., 19 February 1919, UMWA, District 2, UMWA Records, formerly at District 2, Ebensburg, now at University Archives, Indiana University of Pennsylvania, Indiana, Pa.; Industrial Relations Commission, *Final Report*, 8:7584, 7592. For more on Hendrickson and the 1919 investigation, see Chapters 7 and 9.

57. Zahurak, interview, 7 October 1986.

58. Ibid.

59. *Windber Era*, 23 June 1927.

60. Slovenská Národná Budova, "Zápisnica," Meetings, 5 April 1914, 26 April 1914, 3 May 1914, 17 May 1914, 31 May 1914, 8 June 1914, Windber, Slovak Hall, Windber.

61. Ibid., 17 May 1914, 15 June 1914, 21 June 1914, Somerset County Deed Book, 198:297–299.

62. Slovenská Národná Budova, Meeting, 25 June 1914.

63. Ibid., Meetings, 28 June 1914, 27 September 1914.

64. "The Slovak National Hall Association of Windber, charter," 14 September 1914, Somerset County Deed Book, 192:14–16; *Windber Era*, 24 September 1914, 30 May 1935, and Slovenská Národná Budova, Meeting, 14 January 1922.

65. Novak, interview, 19 May 1985.

66. Victor Greene, "'Becoming American': The Role of Ethnic Leaders—Swedes, Poles, Italians and Jews," in *The Ethnic Frontier: Essays in the History of Group Survival in Chicago and the Midwest*, ed. M. G. Holli and P. d'A. Jones (Grand Rapids, Mich.: Eerdmans, 1977), 145; Barton, *Peasants and Strangers*; Barton, "Eastern and Southern Europeans."

67. Ewa Morawska stressed the importance of seniority in residence as a symbol of status within the immigrant community. See her "Internal Status Hierarchy in the East European Immigrant Communities of Johnstown, Pa., 1890–1930's," *Journal of Social History* 16 (Fall 1982): 86, 90.

68. Kurt Lewin, *Resolving Social Conflicts*, ed. G. W. Lewin (New York: Harper and Brothers, 1948), 194–197; John Higham, "The Forms of Ethnic Leadership," in *Ethnic Leadership in America*, ed. J. Higham (Baltimore: Johns Hopkins University Press, 1978), 1–16.

69. *Windber Era*, 28 May 1903, 13 June 1912. For a general overview of the role of these bankers, see IC *Reports*, 2:413–438.

70. George Dolak, *A History of the Slovak Evangelical Lutheran Church in the United States of America, 1902–1907* (St. Louis: Concordia, 1955), 51; "Naši pioneri," *Národný Kalendár* 24 (1927): 143–146.

71. United States Naturalization Records, Somerset County, August 1900 to December 1903, Somerset County Courthouse, Somerset, Pa.

72. *Somerset County Democrat*, 26 April 1911, 3 May 1911, 10 July 1912; *Windber Era*, 26 September 1912.

73. Alcamo, *Windber Story*, 151; *Windber Era*, 11 April 1912.

74. *Windber Era*, 16 May 1912, 8 October 1914, 26 December 1912.

75. Morawska, "Internal Status Hierarchy," 87; *Windber Era*, 15 January 1903, 17 September 1925, 25 November 1925. Torquato later became active in state politics, and his son, John Torquato, became an important Democratic party leader in the New Deal years. Alcamo, *Windber Story*, 117.

76. *Windber Era*, 17 January 1935, 8 July 1915; Peters, *Memorial Volume*, 60; *Johnstown Democrat*, 25 December 1903; *Windber Journal*, 31 December 1903.

77. *Johnstown Democrat*, 21 August 1908; Toth, interview, 30 September 1986; Zahurak, interview, 7 October 1986.

78. Steve Conjelko, interview by author, tape recording, 13 March 1984, Windber.

79. *Windber Journal*, 3 November 1904; *Windber Era*, 17 January 1935, 14 October 1909; *Windber Era*, 5 July 1928.

80. Leonardis, interview, 5 March 1984; Zahurak, interview, 7 October 1986.

81. "The Enduring Conflict," *Coal Age* 7 (26 June 1915): 1108–1109.

82. A convenient "Scrapbook" of these newspaper ads is located at the Windber Museum, Windber. See also *Dongó*, 15 July 1920; *Szabadság*, 5 August 1920; *Magyar Bányászlap*, 15 July 1920; *Amerikai Magyar Népszava*, 13 July 1920; *Ameryka Echo*, 1 August 1920, 8 August 1920; *Keleivis*, 4 August 1920; *Slovák v Amerike*, 12 August 1920; *Jednota*, 25 August 1920; *Amerikánsko-Slovenské Noviny*, 8 September 1920; *Italo-Americano*, 11 August 1920. The Immigration History Research Center of the University of Minnesota has major collections and runs of foreign-language newspapers.

Chapter 6: First Stirrings

1. McAlister Coleman, *Men and Coal* (New York: Farrar and Rinehart, 1943), xix.

2. Andrew Roy, *A History of the Coal Miners of the United States from the Development of the Mines to the Close of the Anthracite Strike of 1902, Including a Brief Sketch of Early British Miners* (1905; reprint, Westport, Conn.: Greenwood, 1970), 278–279.

3. This account of the founding and early history of the United Mine Workers is from ibid.; Coleman, *Men and Coal*; Chris Evans, *History of United Mine Workers of America*, 2 vols. (Indianapolis, Ind., 1918–1921); and Philip S. Foner, *History of the Labor Movement in the United States*, vol. 2: *From the Founding of the AFL to the Emergence of American Imperialism* (New York: International Publishers, 1955).

4. Foner, *From Founding of AFL*, 58; Coleman, *Men and Coal*, 55. For a more critical evaluation of the Knights' policy toward these nationalities, see Victor Greene, *Slavic Community on Strike*, 86. William B. Wilson, the first U.S. Secretary of Labor, and Thomas Haggerty, an important UMWA organizer, were two of the most famous of a number of important regional union leaders who began their apprenticeship under the Knights in the 1880s.

5. Victor R. Greene, "A Study in Slavs, Strikes, and Unions: The Anthracite Strike of 1897," *Pennsylvania History* 31 (April 1964): 199–215.

6. Dan Lennon, "An Important Mass Meeting at Houtzdale, Pennsylvania," *UMWJ*, 15 October 1891, 1.

7. Evans, *History of UMWA*, 2:257.

8. Elsie Glück, *John Mitchell* (1929; reprint, New York: AMS Press, 1971), 72.

9. Roy, *History of Coal Miners*, 314–315; Greene, *Slavic Community on Strike*, 150, 212; Thomas W. Davis, "Return of Davis," *UMWJ*, 21 April 1892, 1; Thomas W. Davis, "The Campaign," *UMWJ*, 28 April 1892, 5.

10. Roy, *History of Coal Miners*, 324–330; Frank Julian Warne, *The Slav Invasion and the Mine Workers: A Study in Immigration* (Philadelphia: Lippincott, 1904); Evans, *History of UMWA*, 2:317–333, 335–357, 360; Coleman, *Men and Coal*, 58.

11. "News from Illiterate, and a Few Pertinent Questions Asked," *UMWJ*, 28 December 1893, 3; Thomas A. Bradley, "Grand Result of the Delegate Convention in Central Pa.," *UMWJ*, 4 January 1894, 1; "Official Report of Fifth Annual Convention of District 2," *UMWJ*, 1 March 1894, 1; Thomas R. Davis, "Courage," *UMWJ*, 10 May 1894, 1; "Editorial," *UMWJ*, 28 June 1894, 4; Thomas A. Bradley, "The Situation in Central Pennsylvania Fully Explained in a Circular Issued by Officers," *UMWJ*, 26 July 1894, 4; "Acknowledgments," *UMWJ*, 30 August 1894, 5.

12. William Lockyer, "Mining Notes," *UMWJ*, 29 November 1894, 8.

13. "Conditions in the Clearfield Region as Told by the Miners to the Legislators," *UMWJ*, 20 May 1897, 1.

14. Harry Donald Fox, Jr., "Thomas T. Haggerty and the Formative Years of the United Mine Workers of America" (Ph.D. diss., West Virginia University, 1975), 116–117; Paul W. Pritchard, "William B. Wilson, Master Workman," *Pennsylvania History* 12 (April 1945): 108.

15. Pat McBryde, "McBryde's Address," *UMWJ*, 16 April 1896, 8.

16. Katherine Harvey, *The Best-Dressed Miners: Life and Labor in the Maryland Coal Region 1835–1910* (Ithaca, N.Y.: Cornell University Press, 1969), 276.

17. Pritchard, "William B. Wilson, Master Workman," 103–105.

18. Dan Lennon, "Secretary Lennon Sends Some Interesting News from the District," *UMWJ*, 16 April 1891, 5.

19. Pritchard, "William B. Wilson, Master Workman," 105; Fox, "Thomas T. Haggerty and the Formative Years of the United Mine Workers of America," 39–47; Dan Lennon, "Spice from Lennon," *UMWJ*, 29 October 1891, 5; Dan Lennon, "Lennon Reviews," *UMWJ*, 7 January 1892, 2.

20. "A Worthy Cause," *UMWJ*, 9 August 1894, 8; "Acknowledgments," *UMWJ*, 30 August 1894, 5; William Lockyer, "Mining Notes," *UMWJ*, 29 November 1894, 8; Scratch, "'Scratch' Writes Up Things as He Views Them Around Horatio, Pennsylvania," *UMWJ*, 6 December 1894, 1; Remember, "'Remember'—A Sad Outlook in Central Pennsylvania," *UMWJ*, 20 December 1894, 1.

21. John Brophy, "The Reminiscences of John Brophy" (transcript, Oral History Research Office, Columbia University, New York, N.Y., 1957; New York, N.Y.: Columbia University, 1972, microfilm), 68, 134–139, 156–164; Brophy, *A Miner's Life*, 60–66.

22. Roy, *History of Coal Miners*, 349–364.

23. Chris Evans, "Chris Evans Reviews the Situation in the Central Pennsylvania Coal Field," *UMWJ*, 2 December 1897, 1; "Proceedings," *UMWJ*, 2 December 1897, 8; and "Proceedings," *UMWJ*, 24 February 1898, 5.

24. "Official Circular," *UMWJ*, 17 March 1898, 8; "Proceedings," *UMWJ*, 7 April 1898, 2; "Proceedings of Convention," *UMWJ*, 14 April 1898, 1; James W. Killduff, "Secty. Killduff," *UMWJ*, 19 May 1898, 5; James W. Killduff, "Detectives," *UMWJ*, 30 June 1898, 1; James W. Killduff, "Killduff," *UMWJ*, 7 July 1898, 1; James Killduff, "A Strike," *UMWJ*, 21 July 1898, 3; "The Strike in District 2," *UMWJ*, 21 July 1898, 4; *Johnstown Democrat*, 29 July 1898; "Conditions in District Two," *UMWJ*, 11 August 1898, 4; W. C. Pearce, "Secretary's Report," *UMWJ*, 18 January 1900, 3; Evans, *History of UMWA*, 2:706–707.

25. John Mitchell to E. R. Smith, Joliet, Ill., 13 July 1899 in John Mitchell Papers, Department of Archives and Manuscripts, Catholic University of America, Washington, D.C. (Glen Rock, N.J.: Microfilming Corporation of America, 1974), Series I, reel 1.

26. Chris Evans, "Chris Evans," *UMWJ*, 2 March 1899, 8.

27. "Official Call," *UMWJ*, 9 March 1899, 4; "Proceedings," *UMWJ*, 6 April 1899, 7.
28. "Official Call," *UMWJ*, 2 March 1899, 1; Chris Evans, "Chris Evans," *UMWJ*, 6 April 1899, 4.
29. Chris Evans, "Organizer Evans," *UMWJ*, 15 June 1899, 1.
30. Chris Evans, "Chris Evans," *UMWJ*, 6 April 1899, 4; *Somerset County Democrat*, 28 February 1899; *Johnstown Democrat*, 14 April 1899.
31. *Somerset County Democrat*, 22 March 1899, 12 April 1899, 26 April 1899; *Johnstown Democrat*, 5 May 1899.
32. *Johnstown Democrat*, 14 April 1899; *Somerset County Democrat*, 15 October 1902.
33. *Johnstown Democrat*, 7 April 1899.
34. Chris Evans, "Chris Evans," *UMWJ*, 20 April 1899, 8; Chris Evans, "The Missing Link," *UMWJ*, 27 April 1899, 1.
35. For more on these businessmen, see Chapter 9.
36. *Johnstown Democrat*, 17 May 1899; "Proceedings," *UMWJ*, 11 May 1899, 7; Chris Evans, "Not for Years," *UMWJ*, 4 May 1899, 1; Chris Evans, "Central Pennsylvania," *UMWJ*, 18 May 1899, 4; "Organizer Warner," *UMWJ*, 1 June 1899, 2; Evans, *History of UMWA*, 2:732, 821.
37. Chris Evans, "Philanthropy," *UMWJ*, 29 June 1899, 1; Chris Evans, "Tramping Ties," *UMWJ*, 6 July 1899, 1.
38. William Warner to T. W. Davis, 29 June 1899, in John Mitchell Papers, Series I, reel 1.
39. Elizabeth Catherine Morris, Indianapolis, Ind., to John Mitchell, Fresno, Calif., 1 July 1899, in John Mitchell Papers, Series 1, reel 1.
40. Evans, "Philanthropy," 1.
41. Chris Evans, "Chris Evans," *UMWJ*, 27 July 1899, 1; Evans, *History of UMWA*, 2:731–732.
42. Evans, "Tramping Ties," 1.
43. "Coal Notes," *UMWJ*, 11 January 1900, 3; "Official Proceedings," *UMWJ*, 28 December 1899, 7; Evans, *History of UMWA*, 2:760–762.
44. "Another 'Model' Town," *UMWJ*, 22 August 1901, 3; *Windber Journal*, 7 April 1903.
45. John Mitchell to R. M. Easley, New York, 23 November 1901, in John Mitchell Papers, Series I, reel 3.
46. Ed McKay, Versailles, Pa., to John Mitchell, Indianapolis, Ind., 27 December 1901, in John Mitchell Papers, Series I, reel 3; Ed McKay, Lilly, Pa., to John Mitchell, 5 March 1902, in John Mitchell Papers, Series I, reel 4.
47. John Mitchell to Ed McKay, Versailles, Pa., 23 December 1902, in John Mitchell Papers, Series 1, reel 5.
48. Novak, interview, 19 May 1985.
49. "Official Proceedings of District No. 2," *UMWJ*, 21 March 1901, 2; "Matters in District 2," *UMWJ*, 5 December 1901, 4.
50. Warne, *Slav Invasion and the Mine Workers*, 134–135, 153, 193–194.
51. Rambler, "From Coalport," *UMWJ*, 10 July 1902, 7; *Johnstown Democrat*, 27 June 1902; "From District 2," *UMWJ*, 17 July 1902, 3.
52. *Johnstown Democrat*, 18 July 1902; "Proceedings of Convention," *UMWJ*, 11 September 1902, 3.
53. Richard Gilbert, "Financial Statement of Sub-Districts," *UMWJ*, 7 May 1903, 7; W. B. Wilson, "8-Hour Day and 11 Per Cent Advance," *UMWJ*, 26 March 1903, 5; *Johnstown Democrat*, 27 March 1903; Patrick Gilday, [Letter], *UMWJ*, 2 July 1903, 5; "The Mining Trade," *UMWJ*, 22 October 1903, 3; "Editorial," *UMWJ*, 15 October 1903, 4.
54. Patrick Gilday, "Important Circular," *UMWJ*, 22 October 1903, 7.
55. *Johnstown Democrat*, 27 February 1903; *Windber Journal*, 10 March 1903, 24 March 1903, 7 April 1903; *Windber Era*, 2 April 1903.

Chapter 7: Friends and Enemies

1. Herbert G. Gutman, *Work, Culture, and Society in Industrializing America: Essays in American Working-Class and Social History* (New York: Vintage, 1966), 254–255.

2. Higham, *Strangers in the Land,* 57.

3. Patriotic Order Sons of America, State Camp of Pennsylvania, *Pennsylvania State Camp and Subordinate Camp Constitutions . . . National Camp Constitution and General Laws . . .*, rev. ed. (Philadelphia, 1894), 3, 5.

4. Ibid.; *Windber Era,* 5 March 1903, 6 August 1903; *Windber Journal,* 10 November 1904, 25 February 1904; U.S. Immigration Commission, *Reports of the Immigration Commission,* vol. 41: *Statements and Recommendations Submitted by Societies and Organizations Interested in the Subject of Immigration* (1911; reprint, New York: Arno and New York Times, 1970), 323–334; Higham, *Strangers in the Land,* 187–188; Industrial Relations Commission, *Final Report,* 8:7749.

5. *Windber Journal,* 4 June 1903.

6. Philip S. Klein and Ari Hoogenboom, *A History of Pennsylvania,* 2nd ed. (University Park: The Pennsylvania State University Press, 1973), 522–523; Balch, *Our Slavic Fellow Citizens,* 367–368.

7. *Windber Journal,* 3 March 1904, 17 March 1904; Windber Borough Council, Meeting, 1 March 1904.

8. *Windber Journal,* 1 October 1903, 8 October 1903, 19 November 1903; *Windber Era,* 24 October 1907, 2 December 1909.

9. *Windber Era,* 4 March 1909.

10. Higham, *Strangers in the Land,* 72–73.

11. *Windber Journal,* 19 November 1903.

12. Ibid.; *Windber Journal,* 26 November 1903; "Farmers Ask State to Probe Coal Grabs," *UMWJ* 2 August 1906, 6.

13. *Windber Era,* 7 January 1909.

14. Windber Borough Council, Meetings, 13 August 1902, 7 November 1905, 6 February 1906.

15. IC *Reports,* 6:511.

16. *Windber Journal,* 9 July 1903; *Windber Era,* 7 July 1904.

17. Higham, *Strangers in the Land,* 72; Windber Borough Council, Meeting, 4 April 1905; *Windber Era,* 6 April 1905.

18. *Windber Era,* 4 December 1902, 28 February 1903, 19 March 1903, 26 March 1903; *Windber Journal,* 17 February 1903.

19. *Somerset County Democrat,* 26 February 1908, 11 March 1908.

20. *Windber Era,* 15 June 1905, 22 June 1905; Windber Borough Council, Meeting, 6 June 1905.

21. *Windber Era,* 8 April 1909, 22 April 1909, 29 April 1909; Windber Borough Council, Meeting, 6 April 1909.

22. *Windber Era,* 8 December 1904.

23. *Windber Era,* 10 October 1907, 10 December 1903.

24. *Windber Journal,* 10 December 1903.

25. *Windber Era,* 21 September 1905, 19 October 1905, 30 November 1905.

26. *Somerset County Democrat,* 6 May 1908; *Windber Era,* 17 January 1909, 9 September 1909; [H. M. J. Klein], *The Pennsylvania Young Men's Christian Association: A History, 1854–1950* (Kennett Square, Pa.: Kennett News and Advertiser Press, 1950), 68, 69, 77.

27. IC *Reports,* 6:499.

28. *The World*, 11 October 1922.

29. *Windber Journal*, 29 October 1903.

30. Ibid.

31. *Windber Journal*, 5 November 1903.

32. Ibid.

33. *Windber Journal*, 12 November 1903, 10 December 1903, 28 January 1904, 24 December 1903, 3 December 1903.

34. "Editorial," *UMWJ*, 5 November 1903, 4; "Editorial," *UMWJ*, 19 November 1903, 4; "Editorial," *UMWJ*, 26 November 1903, 4.

35. Robert Salmond, "Letter to the Editor," *UMWJ*, 19 November 1903, 5; *Windber Journal*, 12 November 1903, 26 November 1903, 17 December 1903; "Pennsylvania Notes," *Coal Trade Journal*, 16 December 1903, 901.

36. *Windber Journal*, 3 December 1903, 10 December 1903.

37. *Windber Era*, 8 January 1904.

38. *Windber Journal*, 17 December 1903, 24 December 1903, 21 January 1904, 4 February 1904, 28 January 1904, 3 March 1904, 10 March 1904, 27 October 1904.

39. *Windber Era*, 9 March 1905, 23 March 1905.

40. *Windber Journal*, 12 November 1903, 14 April 1904.

41. *Windber Journal*, 28 January 1904.

42. *Windber Journal*, 18 February 1904; "Official Notice," *UMWJ*, 18 February 1904, 1; "Senator Hanna Is Dead," *UMWJ*, 18 February 1904, 1.

43. *Windber Journal*, 5 May 1904; *Somerset County Democrat*, 27 August 1904; Leonardis, interview, 8 March 1984; Conjelko, interview, 13 March 1984; Washko, interview, 16 March 1984.

44. *Windber Journal*, 17 March 1903.

45. *Windber Journal*, 18 June 1903, 10 September 1903, 4 February 1904, 30 June 1904.

46. Windber Borough Council, Meetings, 7 April 1903, 23 April 1903, 10 November 1904.

47. Ibid., Meetings, 22 June 1904, 23 June 1904, 25 June 1904, 21 May 1912, 11 June 1912, 2 July 1912.

48. *Windber Era*, 11 August 1904; *Windber Journal*, 23 June 1904, 11 August 1904.

49. Windber Borough Council, Meeting, 24 July 1900.

50. Ole S. Johnson, *The Industrial Store: Its History, Operations, and Economic Significance* (Atlanta: School of Business Administration, University of Georgia, 1952), 6, 11, 23–33, 40–48, 72; Industrial Relations Commission, *Final Report*, 8:7586–7587.

51. Johnson, *Industrial Store*, 5–6, 52, 70, 91; "Company and Other Stores," *Coal Age* 8 (October 30, 1915): 716.

52. IC *Reports*, 6:496–499.

53. Ibid.

54. Ibid.

55. Rohalley, interview by author, 2 March 1984; Novak, interview, 19 May 1985; Leonardis, interview, 5 March 1984.

56. *Windber Era*, 4 December 1902, 8 January 1903; *Windber Journal*, 8 September 1904.

57. Zahurak, interview, 7 October 1986; Molnar and Molnar, interview, 23 February 1984; Conjelko, interview, 13 March 1984.

58. Molnar and Molnar, interview, 23 February 1984; Conjelko, interview, 13 March 1984.

59. Novak, interview, 19 May 1985; Zahurak, interview, 7 October 1986; Washko, interview, 16 March 1984; Leonardis, interview, 5 March 1984; Molnar and Molnar, interview, 23 February

1984; Windber Borough Council, Meetings, 15 September 1902, February 2, 1903; *Windber Era,* 26 September 1907.

60. *Windber Journal,* 11 February 1904.

61. Ibid.

62. Ibid.; *Windber Journal,* 18 February 1904, 25 February 1904, 3 March 1904.

63. *Windber Journal,* 5 May 1904, 19 May 1904.

64. *Windber Era,* 14 July 1904, 10 August 1905, 11 July 1907.

Chapter 8: The Strike of 1906

1. *New York Times,* 2 April 1906, 1; 3 April 1906, 2; 31 March 1906, 2; *Somerset Herald,* 4 April 1906; *Johnstown Democrat,* 6 April 1906; *Somerset County Democrat,* 4 April 1906.

2. "Report of President Mitchell to the Sixteenth Annual Convention of the United Mine Workers of America," *UMWJ,* 19 January 1905, 2, 3; "No Wage Agreement," *UMWJ,* 15 February 1906, 1, 2, 7.

3. "The Joint Convention," *Coal Trade Journal,* 7 February 1906, 95; "Lust of Power Among the Bituminous Coal Miners," *Black Diamond,* 24 February 1906, 28–29. For evidence that many operators favored a suspension to reduce coal stocks and increase profits, see "A Suspension Is Necessary," *Black Diamond,* 20 January 1906, 26.

4. George E. Scott, Philadelphia, "[Report of the Association Bituminous Coal Operators Central Penna.]," 17 March 1906, in the William Bauchop Wilson Papers, Historical Society of Pennsylvania, Philadelphia; "The State of Trade," *Coal Trade Journal,* 11 April 1906, 253; "The Clearfield Conference," *Coal Trade Journal,* 11 April 1906, 267.

5. A. R. Hamilton, Pittsburgh, to John Mitchell, New York, 2 April 1906, in John Mitchell Papers, Series I, reel 11.

6. Ibid.; A. R. Hamilton, Pittsburgh, to John Mitchell, New York, telegram, 2 April 1906, in John Mitchell Papers, Series I, reel 11; George E. Scott, Clearfield, to H. A. Berwind, Philadelphia, 3 April 1906, in John Mitchell Papers, Series I, reel 11. For the impact of Berwind-White's independent actions on the position of the Central Pennsylvania operators at Clearfield, see "Situation in the Central Pennsylvania District," *Coal Trade Journal,* 4 April 1906, 242.

7. *New York Times,* 26 May 1906, 1; 27 May 1906, 1.

8. "Report of President to Sixteenth Annual Convention," *UMWJ,* 19 January 1905, 5; "Minutes of the Seventeenth Annual Convention," *UMWJ,* 8 February 1906, 3, 8; "No Wage Agreement," *UMWJ,* 15 February 1906, 1, 2, 7.

9. *New York Times,* 27 February 1906, 1; 28 February 1906, 1; 20 March 1906, 5; 24 March 1906, 6; 25 March 1906, 2; "Verbatim Report U.M.W. Special Convention," *UMWJ,* 5 April 1906, 1, 2, 3, 6, 7; "Observations of the Conference," *Coal Trade Journal,* 28 March 1906, 229.

10. IC *Reports,* 6:500.

11. "Situation at Windber Unchanged," *UMWJ,* 5 May 1906, 3; IC *Reports,* 6:477; *Somerset Herald,* 11 April 1906.

12. "The Mining News," *UMWJ,* 29 March 1906, 5.

13. Ibid.; A. R. Hamilton to John Mitchell, Indianapolis, 31 March 1906, in John Mitchell Papers, Series I, reel 11; "Resolutions and Proposed Constitutional Amendments, District No. 2, U.M.W. of A. Convention, Johnstown, Pa., October 21st, 1919," UMWA District 2, UMWA Records.

14. Windber Borough Council, Meeting, 23 March 1906.

15. *Somerset Herald*, 4 April 1906; "The Windber Miners," *UMWJ*, 19 April 1906, 2.

16. *Johnstown Democrat*, 30 March 1906; "Strike Notice Is Sent Out," *UMWJ*, 29 March 1906, 2; "Strike Is Called in the Central Pennsylvania Field," *Black Diamond*, 5 May 1906, 28–29.

17. *Somerset Herald*, 4 April 1906; *Somerset County Democrat*, 25 April 1906, 4 April 1906.

18. *National Labor Tribune*, 12 April 1906, 5.

19. "Talks in Many Tongues," *UMWJ*, 12 April 1906, 8.

20. Ibid.

21. IC *Reports*, 6:500; "The Windber Miners," *UMWJ*, 19 April 1906, 1; *Johnstown Democrat*, 6 April 1906.

22. Martin Smolko, "[Letter]," *Slovák v Amerike*, 17 April 1906, 3.

23. *Johnstown Democrat*, 6 April 1906; "The Windber Miners," *UMWJ*, 19 April 1906, 2; "Talks in Many Tongues," *UMWJ*, 12 April 1906, 8.

24. "Talks in Many Tongues," *UMWJ*, 12 April 1906, 8; A. R. Hamilton, Pittsburgh, to John Mitchell, New York, 2 April 1906, in John Mitchell Papers, Series I, reel 11; A. R. Hamilton to John Mitchell, 9 April 1906, in John Mitchell Papers, Series I, reel 11.

25. *Somerset Herald*, 4 April 1906; IC *Reports*, 6:500; *Johnstown Democrat*, 6 April 1906.

26. *Johnstown Democrat*, 13 April 1906.

27. Ibid.

28. *New York Times*, 31 March 1906, 2; "The Situation," *UMWJ*, 12 July 1906, 1; "New Scale Is Signed," *UMWJ*, 19 July 1906, 2.

29. "Mine Workers Program," *UMWJ*, 17 May 1906, 3; "Miners in Politics," *UMWJ*, 17 May 1906, 4; "Miners—Operators," *UMWJ*, 17 May 1906, 8; "Legal Opinion on the State Police," *UMWJ*, 17 May 1906, 8; "The State Constabulary," *UMWJ*, 24 May 1906, 1; "Now Is the Time," *UMWJ*, 24 May 1906, 4; "Call for a Convention," *UMWJ*, 24 May 1906, 1; "Miners Held Convention," *UMWJ*, 24 May 1906, 8; and "Bituminous Mining," *UMWJ*, 7 June 1906, 8.

30. *Somerset County Democrat*, 11 April 1906; *Somerset Herald*, 4 April 1906; *Johnstown Democrat*, 13 April 1906. A good example of how little even responsible American newspapers understood foreigners or the labor movement was contained in a county paper during this time. The *Somerset Herald* carried an article chastising ignorant foreigners in Windber. It claimed that these people did not seem to understand the lofty purpose of a fire department because they had called firemen "scabs" when they were putting out a fire on company property.

But it was the *Herald*, not the miners, which displayed ignorance in this case. Miners knew very well that the fire and police forces were almost interchangeable bodies. Physically they shared the same headquarters, were composed of the same American-born elite that gave its loyalty to the company, and were deputized with full police powers during moments of crisis, such as labor conflicts. They stood side by side with police and detectives throughout the strike and events of 16 April. The fire company was not a neutral institution in Windber. See *Somerset Herald*, 11 April 1906.

31. *New York Times*, 17 April 1906, 1; *Johnstown Democrat*, 20 April 1906; *Somerset County Democrat*, 18 April 1906; "[Editorial]," *UMWJ*, 26 April 1906, 4. My composite reconstruction of events is based on a comparison of various newspaper accounts but relies primarily on Michael Balogh, "Oprava viesty podanej v zaležitosci kervprelievania v Windber, Pa.," *Amerikansky Russky Viestnik*, 26 April 1906, 4; *Johnstown Democrat*, 20 April 1906; "Windber Massacre," *UMWJ*, 26 April 1906; H. Bousfield, "From Pennsylvania," *UMWJ*, 26 April 1906.

32. Captain John W. Borland, Greensburg, Pa., to Captain [Superintendent] John C. Groome, Harrisburg, Pa., "Monthly Report of Troop 'A', Pennsylvania State Police Force for the month ending April 30th, 1906," Records of the Pennsylvania State Police, Record Group 30, Pennsylvania State Archives, Pennsylvania Historical and Museum Commission, Harrisburg, Pa.

33. *New York Times*, 18 April 1906, 1.

34. *Johnstown Democrat*, 20 April 1906; *Somerset County Democrat*, 18 April 1906; *Somerset Herald*, 18 April 1906.

35. Borland to Groome, "Monthly Report," 30 April 1906. That Central Pennsylvania operators believed that the violence in Windber and the calling in of state troops, along with Governor Pennypacker's proclamation supporting the constabulary and deploring violence, strengthened their movement to destroy the union, see "The State of Trade," *Coal Trade Journal*, 9 May 1906, 334.

36. Borland to Groome, "Monthly Report," 30 April 1906.

37. Ibid.

38. *Johnstown Democrat*, 20 April 1906; *Somerset Herald*, 25 April 1906.

39. *National Labor Tribune*, 19 April 1906, 1; *Somerset Herald*, 16 May 1906; "Pennsylvania Notes," *Coal Trade Journal*, 23 May 1906, 382.

40. *Somerset Herald*, 30 May 1906; *Johnstown Democrat*, 25 May 1906.

41. "Editorial," *UMWJ*, 26 April 1906, 4; "Always the Same," *UMWJ*, 26 April 1906, 4.

42. Balogh, "Oprava viesty," 4.

43. Ibid.

44. The *Johnstown Democrat* article was reproduced in Bousfield, "From Pennsylvania," *UMWJ*, 26 April 1906, 5.

45. Some newspaper accounts of the events of April 16 carried false reports that miners had voted at their afternoon meeting to return to work. Other papers did not. But one leading operators' journal later justified the violence there by saying that the riot took place because drunk, knife-wielding foreigners resented the defeat of the strike. Union officials, priests, and others at the meeting denied that miners intended to return to work, and no one received any knife wounds. For the operators statement, see "An Affray at Windber," *Coal Trade Journal*, 18 April 1906, 280.

There is conflicting evidence about the role of Father Saas. A. R. Hamilton, an unscrupulous Berwind-White insider who supplied information to John Mitchell, habitually arranged for Saas, not the other two priests, to explain the ongoing crisis to visiting district and national union officials, and he wrote Mitchell that Saas had rightly told the men to go back to work on the day of the shootings. In the 1922 strike, Saas unquestionably did use his pulpit to preach to strikers that they should return to work. See A. R. Hamilton to John Mitchell, Indianapolis, 17 April 1906, in John Mitchell Papers, Series I, reel 11.

46. "Windber Priests Refute Statements Affecting Character of Miners," *UMWJ*, 26 April 1906, 1. See also "Windber Massacre," *UMWJ*, 21 April 1906.

47. *National Labor Tribune*, 19 April 1906, 1, 26 April 1906, 1; *Johnstown Democrat*, 4 May 1906; "Wholesale Arrests," *UMWJ*, 5 May 1906, 6; *Somerset Herald*, 9 May 1906.

48. *Somerset Herald*, 27 June 1906; Zahurak, interview, 7 October 1986; Leonardis, interview, 5 March 1984.

49. Captain John W. Borland, Greensburg, Pa., to Captain John C. Groome, Harrisburg, Pa., "Monthly Report of Troop 'A,' Pennsylvania State Police Force for the month ending May 31st, 1906," and Captain John W. Borland, Greensburg, Pa., to Captain John C. Groome, Harrisburg, Pa., "Monthly Report of Troop 'A', Pennsylvania State Police Force for the month ending June 30th, 1906," Records of the Pennsylvania State Police.

50. Ibid.

51. *National Labor Tribune*, 3 May 1906, 5; *Johnstown Democrat*, 4 May 1906; "The State 'Constabulary,'" *UMWJ*, 21 June 1906, 4; "Earning Their Salary," *UMWJ*, 28 June 1906, 2; "Wouldn't Entertain Cossack," *UMWJ*, 19 July 1906, 5; *Somerset County Democrat*, 15 August 1906; *Windber Era*, 19 December 1907.

52. "Bituminous Mining," *UMWJ*, 28 March 1907, 8; IC *Reports*, 6:500; *Johnstown Democrat*, 4 May 1906.

53. *Johnstown Democrat*, 4 May 1906; Novak, interview, 19 May 1985; Alcamo, *Windber Story*, 261; St. Cyril and Methodius Church, *Golden Jubilee*; St. Anthony's Church, *40th Anniversary*.

54. *National Labor Tribune*, 7 June 1906, 5; *Somerset Herald*, 6 June 1906; "Bituminous Mining Situation," *Coal Trade Journal*, 6 June 1906, 414; "The Windber Mines," *UMWJ*, 21 June 1906, 1; Pennsylvania Department of Mines, *Report of the Department of Mines of Pennsylvania*, part 2, *Bituminous 1905* (Harrisburg, Pa., 1906), 284; Pennsylvania Department of Mines, *Report of the Department of Mines of Pennsylvania*, part 2, *Bituminous 1906* (Harrisburg, Pa., 1907), 344. Along with the false reports of the strike's end were false reports that the state constabulary had left town. The state police force was reduced in June, but a sizable contingent remained into July, and the state police continued to maintain a presence in town and close connection with local officers for years after the strike's end. See *Somerset Herald*, 13 June 1906; *Windber Era*, 26 September 1907, 31 October 1907.

55. George E. Scott, Philadelphia, to Morrisdale Coal Co., Morrisdale, Pa., 9 June 1906, in William Bauchop Wilson Papers.

56. "The Situation," *UMWJ*, 28 June 1906, 1; "Bituminous Mining," *UMWJ*, 5 July 1906, 8; "The Situation," *UMWJ*, 12 July 1906, 1; "Bituminous Mining," *UMWJ*, 12 July 1906, 1; "New Scale Is Signed," *UMWJ*, 19 July 1906, 2. For operators' criticism that Central Pennsylvania operators had split ranks and hurt their cause, see "A Contrast Between Bituminous and Anthracite Operators," *Coal Trade Journal*, 11 July 1906, 508.

57. "New Scale Is Signed," *UMWJ*, 19 July 1906; "The Coal Strike Over," *UMWJ*, 26 July 1906, 3; Patrick Gilday, Morrisdale, Pa., to John Mitchell, Indianapolis, 1 April 1907, in John Mitchell Papers, Series I, reel 11; "The State of Trade," *Coal Trade Journal*, 1 August 1906, 550; Brophy, *A Miner's Life*, 85–86, 91, 99; "Vice-President Lewis' Report," *UMWJ*, 17 January 1907, 4–5; "President Mitchell's Report," *UMWJ*, 17 January 1907, 2; "Bituminous Mining," *UMWJ*, 28 March 1907, 8.

58. "Bituminous Mining," *UMWJ*, 28 March 1907, 8.

59. "Somerset Aroused," *UMWJ*, 7 June 1906, 3; *Somerset Herald*, 30 May 1906; *Somerset County Democrat*, 30 May 1906, 6 June 1906; "Somerset Miners," *UMWJ*, 6 September 1906, 5. For similar complaints by farmers, see *Johnstown Democrat*, 8 June 1906, 17 August 1906; *Somerset Herald*, 30 May 1906, 27 June 1906; "Farmers Ask State to Probe Coal Grabs," *UMWJ*, 2 August 1906, 6.

60. *Johnstown Democrat*, 21 September 1906, 25 January 1907; *Somerset County Democrat*, 26 September 1906; "Windber Rioters Pay $137 Each," *UMWJ*, 7 February 1907, 5.

61. "Secretary Wilson's Report," *UMWJ*, 17 January 1907, 5; "Bituminous Mining," *UMWJ*, 28 March 1907, 8.

62. Brophy, *A Miner's Life*, 85–86, 91, 98–99, 190–192.

63. P. J. Drain, Creekside, Pa., to John Mitchell, 28 May 1906, in John Mitchell Papers, Series I, reel 11; Jim Purcell to W. B. Wilson, Clearfield, Pa., 26 April 1906, in William Bauchop Wilson Papers; Brophy, *A Miner's Life*, 134–137. In 1918, when Brophy was president of District 2 and Hamilton was vice-president of the National Coal Association, the two men met each other for the first time at a conference in Washington. The coal industry representative offered to help Brophy rise within the national union if the union leader left Berwind-White mines alone, thereby "protecting" the company's enterprises from unionization. The district president indignantly rejected the deal.

64. The following account of the relationship between Hamilton and Mitchell is based on the correspondence between the two men. See their letters and telegrams, especially from April to July 1906, contained in John Mitchell Papers, Series I, reel 11.

65. A. R. Hamilton to H. N. Taylor, 5 April 1906, in John Mitchell Papers, Series I, reel 11; A. R. Hamilton, Pittsburgh, to John Mitchell, 6 April 1906, in John Mitchell Papers, Series I, reel 11; A. R. Hamilton to John Mitchell, 9 April 1906, in John Mitchell Papers, Series I, reel 11; A. R. Hamilton to John Mitchell, 18 May 1906, in John Mitchell Papers, Series I, reel 11; A. R. Hamilton to John Mitchell, 2 June 1906, in John Mitchell Papers, Series I, reel 11.

66. A. R. Hamilton to H. N. Taylor, 5 April 1906, in John Mitchell Papers, Series I, reel 11.

67. John Mitchell to H. N. Taylor, 4 April 1906, in John Mitchell Papers, Series I, reel 11; John Mitchell to H. N. Taylor, Chicago, 7 April 1906, in John Mitchell Papers, Series I, reel 11; John Mitchell to H. N. Taylor, Chicago, 9 April 1906, in John Mitchell Papers, Series I, reel 11; John Mitchell to W. D. Ryan, Springfield, Ill., 13 April 1906, in John Mitchell Papers, Series I, reel 11.

68. *New York Times*, 6 February 1906, 1.

69. Gutman, *Work, Culture, and Society*, 256.

70. *Windber Era*, 18 March 1909.

71. "Bituminous Mining," *UMWJ*, 28 March 1907, 8.

Chapter 9: Rising Expectations

1. "The Committee on Coal Production," *Coal Age* 11 (12 May 1917): 817–819; "Fuel Committee Meets in Washington," *Coal Age* 11 (12 May 1917): 835.

2. For a study of the relationship between the coal industry and the federal government from World War I through the New Deal, see James P. Johnson, *The Politics of Soft Coal: The Bituminous Industry from World War I Through the New Deal* (Urbana: University of Illinois Press, 1979).

3. *New York Times*, 22 January 1915, 1; 28 August 1917, 1.

4. *Windber Era*, 16 January 1913, 20 February 1913, 6 August 1914, 10 September 1914.

5. *Windber Era*, 20 August 1914, 4 March 1915, 1 April 1915.

6. *New York Times*, 16 November 1915, 3; 17 November 1915, 3; 24 November 1915, 1; 25 November 1915, 1; 26 November 1915, 1, 2, 3; 30 November 1915, 3; 2 December 1915, 1.

7. *Windber Era*, 4 March 1915, 1 April 1915, 14 December 1916, 18 January 1917.

8. *Windber Era*, 4 January 1917, 11 January 1917, 1 March 1917, 9 August 1917.

9. See Chapter 5.

10. See Chapter 5 and *Windber Era*, 1 November 1917.

11. *Windber Era*, 18 February 1915, 29 April 1915, 16 December 1916.

12. *Windber Era*, 23 November 1916; Joseph A. Borkowski, Central Council of Polish Organizations, Pittsburgh, Pa., to Mildred Beik, DeKalb, Illinois, 24 January 1984.

13. *Windber Era*, 7 January 1915, 14 January 1915, 11 February 1915, 8 July 1915.

14. *Windber Era*, 1 February 1917.

15. "Report of International President John P. White," *UMWJ*, 20 January 1916, 22, 39; "Official Circular, Program to Stimulate Organization," *UMWJ*, 25 January 1917, 4; "The Opportune Time: Now," *UMWJ*, 22 February 1917, 4.

16. [Resolution], 29 January 1916, UMWA, District 2, UMWA Records; "Further Convention Proceedings," *UMWJ*, 3 February 1916, 15.

17. K. C. Adams, "Agreement Reached for Bituminous Field," *UMWJ*, 9 March 1916, 11; "Anthracite Mine Workers Win Eight-Hour Day, Recognition of Districts 1, 7, and 9 and Wage Raise in Tentative Agreements Reached After Eight Weeks of Negotiations," *UMWJ*, 4 May 1916, 11; William Green, "Labor Day's Reflections," *UMWJ*, 31 August 1916, 6; "Miners of District 2

Win Recognition of Union," *UMWJ*, 20 April 1916, 11; "Scale Agreement for District 2," *UMWJ*, 20 April 1916, 11.

18. "Victory for Miners in District No. 2, Central Pennsylvania," *UMWJ*, 29 June 1916, 12; "From District 2," *UMWJ*, 10 August 1916, 9; "District 16 Reviving," *UMWJ*, 12 October 1916, 9; "Progress in Central Pennsylvania," *UMWJ*, 19 October 1916, 4; "Good Work in Meyersdale Field," *UMWJ*, 2 November 1916; "District 16 (Maryland) Rehabilitated," *UMWJ*, 9 November 1916, 10.

19. James Purcell, Clearfield, to John P. White, Indianapolis, 5 December 1916, UMWA, District 2, UMWA Records.

20. *New York Times*, 22 January 1915, 1; and "General Labor Review," *Coal Age* 11 (13 January 1917): 103.

21. M. Nelson, "Organization and Progress," *UMWJ*, 8 February 1917, 10.

22. Thomas Fisher, Philadelphia, to Francis S. Peabody, Washington, D.C., 30 July 1917, [copy of letter], UMWA, District 2, UMWA Records.

23. Ibid.

24. Ibid.

25. Ibid.; John Brophy, Clearfield, to W. B. Wilson, Washington, D.C., 4 June 1917, UMWA, District 2, UMWA Records; John Brophy, Clearfield, to George Koller, Beccaria, 30 June 1917, UMWA, District 2, UMWA Records.

26. Charles S. Ling, Johnstown, to Francis S. Peabody, Washington, D.C., 1 August 1917, [copy of letter], UMWA, District 2, UMWA Records.

27. Ibid.

28. P[at] H. Egan, Johnstown, to John Brophy, Clearfield, 5 July 1917, UMWA, District 2, UMWA Records.

29. Ibid.

30. John Brophy, Clearfield, to William Green, Indianapolis, 2 June 1917, UMWA, District 2, UMWA Records; Frank J. Hayes, Indianapolis, to John Brophy, Clearfield, 31 July 1917, UMWA, District 2, UMWA Records; John Brophy, Clearfield, to D. Irvine, Seattle, 12 June 1917, UMWA, District 2, UMWA Records.

31. John Brophy, Clearfield, to Robert Foster, Birmingham, Ala., 8 August 1917, UMWA, District 2, UMWA Records; "Coal and Coke News," *Coal Age* 11 (9 June 1917): 1015.

32. Edward J. Robinson, Windber, to John Brophy, Clearfield, 1 April 1918, District 2, UMWA Records. In citing this and subsequent letters in this chapter, I have made an occasional minor change in punctuation or capitalization to make the original text more readable.

33. Edward J. Robinson, Windber, to John Brophy, Clearfield, 15 April 1918, UMWA, District 2, UMWA Records; Edward J. Robinson, Windber, to James Feeley, Dunlo, 7 April 1918, UMWA, District 2, UMWA Records; and James Feeley, Dunlo, to John Brophy, Clearfield, 14 April 1918, UMWA, District 2, UMWA Records.

34. John Brophy, Clearfield, to Edward J. Robinson, Windber, 16 April 1918, UMWA, District 2, UMWA Records; [Handwritten notation on letter], Robinson to Brophy, 15 April 1918.

35. John L. Lewis, Indianapolis, to John Brophy, Clearfield, 15 April 1918, UMWA, District 2, UMWA Records.

36. Ibid.

37. Edward J. Robinson, Windber, to John Brophy, 13 May 1918, UMWA, District 2, UMWA Records.

38. Ibid.

39. Ibid.

40. Ibid.

41. John Brophy, Clearfield, to Edward J. Robinson, Windber, 21 May 1918, UMWA, District 2, UMWA Records.

42. Edward J. Robinson, Windber, to John Brophy, Clearfield, 18 June 1918, UMWA, District 2, UMWA Records.

43. Ibid.

44. "Resolution Passed at Special Convention Held at Portage, Pa., August 15, 1918," UMWA, District 2, UMWA Records; Frank J. Hayes, Indianapolis, to John Brophy, Clearfield, 3 September 1918, UMWA, District 2, UMWA Records; Frank J. Hayes, Indianapolis, to Norman Boden, Garrett, 1 September 1918, UMWA, District 2, UMWA Records.

45. "General Labor Review," *Coal Age* 14 (24 October 1918): 787.

46. See, for example, "President White on the Workers' War Duties," *UMWJ*, 31 August 1917, 5.

47. Edward J. Robinson, Windber, to John Brophy, Clearfield, 12 November 1918, UMWA, District 2, UMWA Records.

48. "General Labor Review," *Coal Age* 14 (19 December 1918): 1128; "The Epidemic Costs Very Heavily in Death Benefits," *UMWJ*, 15 January 1919, 10; Leonardis, interview, 8 March 1984; Windber Borough Council, Meetings, 18 November 1918, 12 December 1918, 3 January 1919.

49. Windber Citizens' Association, "An Outline of the Purposes and Principles of the Windber Citizens' Association of Windber, Pa.," 2, UMWA, District 2, UMWA Records.

50. Millott, "Extracts from Speech to Citizens of Windber, Pa.," 22 November 1918, Records of the Federal Mediation and Conciliation Service, Case File 33/2987 for Berwind-White Coal Co., Windber, Record Group 280, National Archives, Washington, D.C.

51. Ibid.

52. Ibid.

53. Ibid.

54. "Are Not Organized to Unionize the Mines" [Unidentified newspaper clipping], 11 December 1918, in Federal Mediation and Conciliation Service, Case File 33/2987.

55. J. W. Kelly, H. M. Habeeb, E. M. Berkhimer, Wm. H. Bolopur, Petition, "Prospective Citizens Club of Windber," Windber, to Hon. [William B.] Wilson, Washington, D.C., Federal Mediation and Conciliation Service, Case File 33/2987.

56. Ibid.

57. Edward J. Robinson, Windber, to John Brophy, Clearfield, 28 December 1918, UMWA, District 2, UMWA Records.

58. John Brophy, Clearfield, to Edward J. Robinson, Windber, 1 January 1919, UMWA, District 2, UMWA Records; Edward J. Robinson, Windber, to John Brophy, Clearfield, 12 January 1919, UMWA, District 2, UMWA Records.

59. Petition, Windber Citizens' Association, Windber, Pa., to The Honorable Secretary of Labor, Washington, D.C., 13 January 1919, Federal Mediation and Conciliation Service, Case File 33/2987.

60. Ibid.

61. Ibid.

62. J. W. Kelly, Windber, to Hon. William B. Wilson, 22 January 1919, Federal Mediation and Conciliation Service, Case File 33/2987.

63. *The Leader*, 29 January 1919. This article and the affidavits are in Federal Mediation and Conciliation Service, Case File 33/2987.

64. Frederick G. Davis, "Final Report Re Citizens' Association vs. Berwind White Coal Mining

Company, Windber, Pa.," 18 February 1919, 1–3, Federal Mediation and Conciliation Service, Case File 33/2987.

65. Ibid., 4.

66. Edward J. Robinson, Windber, to John Brophy, Clearfield, 28 January 1919, UMWA, District 2, UMWA Records; John Brophy, Clearfield, to W. B. Wilson, Washington, D.C., 6 February 1919, UMWA, District 2, UMWA Records; H. L. Kerwin, Washington, D.C. to John Brophy, Clearfield, 19 February 1919, UMWA, District 2, UMWA Records.

67. "Brophy-Committee from Windber," Memorandum, 9 February 1918, UMWA, District 2, UMWA Records.

68. Windber Citizens' Association, "Outline," 1.

69. Ibid., 6.

70. Ibid., 10.

71. Ibid., 12, 13.

72. J. H. Davis, Windber, to John Brophy, Clearfield, 8 February 1919, UMWA, District 2, UMWA Records.

73. Ibid.

74. David Cappett and Harry Smull, Holsopple, to the Members of Local Union 2722, 11 February 1919, UMWA, District 2, UMWA Records; Paul Miller, Michael Musilek, John McKay, Dunlo, to John Brophy, Clearfield, 14 February 1919, UMWA, District 2, UMWA Records; Charles Dorka, Lucerne Mines, to John Brophy, Clearfield, UMWA, District 2, UMWA Records; Thomas Keenan, Portage, to Richard Gilbert, Clearfield, 5 March 1919, UMWA, District 2, UMWA Records.

75. Alcamo, *Windber Story*, 134.

76. John Brophy, Pittsburgh, to Robert Harlin, Indianapolis, 22 March 1919, UMWA, District 2, UMWA Records.

77. Ibid.

78. Melvyn Dubofsky and Warren Van Tine, *John L. Lewis: A Biography* (New York: Quadrangle/New York Times Book Co., 1977), 36, 38–39, 63, 73.

79. Powers Hapgood, "Journal of My Trip in the Non-Union Coal Fields in Pennsylvania," [unpublished diary], 30–31, Powers Hapgood Papers, Manuscripts Department, Lilly Library, Indiana University, Bloomington, Ind.

80. "General Labor Review," *Coal Age* 15 (30 January 1919): 238; "Coal and Coke News," *Coal Age* 15 (26 June 1919): 1184.

81. "Changes in the Cost of Living July 1914–July 1919," *Coal Age*, 16 (28 August 1919): 370–372; Boyer and Morais, *Labor's Untold Story*, 204.

82. "Will Encourage Miners Who Quit the Mines for Other Industries to Return to Old Work in Coal Production," *UMWJ*, 1 September 1918, 2; "Government Urges Former Miners to Return to Work in the Mines," *UMWJ*, 15 October 1918, 2; "The Government's Duty," *UMWJ*, 1 May 1919, 6; "They Waited Too Long," *UMWJ*, 15 August 1919, 6.

83. "Policy Committee Goes on Record in Favor of the Six-Hour Day and Five Days a Week; a Substantial Wage Increase for All of the Mine Workers, and for Nationalization of the Coal Mines of the Country," *UMWJ*, 1 April 1919, 3–4; "Cleveland Convention Adjourns, Adopting Most Progressive Policy Ever Enunciated by the United Mine Workers in the History of the Organization," *UMWJ*, 1 October 1919, 3–5, 8–10.

84. "The Agreement in Full," *UMWJ*, 11 October 1917, 6; "General Labor Review," *Coal Age* 16 (16 October 1919): 657.

85. John Brophy, Clearfield, to John L. Lewis, Indianapolis, 16 December 1919, UMWA, District 2, UMWA Records; John L. Lewis, Indianapolis, to John Brophy, 19 December 1919,

UMWA, District 2, UMWA Records; Frank Weismuller and Paul K. Herbs, [Appeal For the Benefit of the Johnstown Men], to the Membership of the U.M.W. of A., District No. 2, 31 December 1919, UMWA, District 2, UMWA Records; *Johnstown Democrat*, 21 October 1919.

86. *Johnstown Democrat*, 25 October 1919.

87. "Resolutions and Proposed Constitutional Amendments, District No. 2, U.M.W. of A. Convention, Johnstown, Pa., October 21st, 1919," 7, UMWA, District 2, UMWA Records.

88. *Johnstown Democrat*, 3 November 1919.

89. Windber Borough Council, "An Ordinance Prohibiting the holding of any meeting, gathering or assemblage of persons in the Borough of Windber, unless a permit therefor shall have been issued by the Burgess, and providing penalties for the violation of the same," 2 December 1919, "Ordinances," Windber Borough Council, Windber, Pennsylvania.

90. *Johnstown Democrat*, 5 November 1919.

91. "Order for General Strike of Bituminous Miners Is Withdrawn and Canceled by the Direction of Judge A. B. Anderson, of the United States Court," *UMWJ*, 15 November 1919, 3–5.

92. "Miners Accept President Wilson's Proposition for Settlement of Bituminous Coal Strike and Mining Operations Generally Have Been Resumed," *UMWJ*, 15 December 1919, 3–4, 17; "Miners and Operators Seek End to Strike Negotiations," *Coal Age* 16 (27 November and 4 December 1919): 856; "Operators Protest Settlement," *Coal Age* 16 (25 December 1919): 935.

93. "Bituminous Coal Commission Makes Divided Report; President Wilson Affirms the Majority Report Which Ignores Vital Reforms Urged by the Mine Workers," *UMWJ*, 1 April 1920, 3; "Joint Wage Agreement for Bituminous Industry Signed at New York Conference, Based on Decision of the President's Bituminous Coal Commission," *UMWJ*, 15 April 1920, 3–4; Irving Bernstein, *The Lean Years: A History of the American Worker, 1920–1933* (Boston: Houghton Mifflin, 1960), 146–157; "President Lewis Sets at Rest All Talk About a Possible Reduction of Wages for the Coal Miners," *UMWJ*, 15 May 1921, 5.

94. Robinson to Brophy, 18 June 1918, UMWA, District 2, UMWA Records.

95. John Brophy, Clearfield, to John L. Lewis, Indianapolis, 3 May 1920, UMWA, District 2, UMWA Records; John Brophy, Clearfield, to Local Unions, Somerset County, 16 July 1920, UMWA, District 2, UMWA Records; John L. Lewis, Indianapolis, to John Brophy, Clearfield, 30 July 1920, UMWA, District 2, UMWA Records; John Brophy, Clearfield, to John L. Lewis, Indianapolis, 6 June 1921, UMWA, District 2, UMWA Records; John L. Lewis, Denver, to John Brophy, Clearfield, 11 June 1921, UMWA, District 2, UMWA Records.

96. United Mine Workers of America, District 2, *Why the Miners' Program?* (Clearfield, Pa.: UMWA District 2, 1921); Alan Singer, "John Brophy's 'Miners' Program': Workers' Education in UMWA District 2 during the 1920s," *Labor Studies Journal* 13 (Winter 1988): 50–64.

97. John Brophy, Clearfield, to Powers Hapgood, Clearfield, 14 July 1921, UMWA, District 2, UMWA Records; Powers Hapgood, *In Non-Union Mines: The Diary of a Coal Digger in Central Pennsylvania August–September, 1921* (New York: Bureau of Industrial Research, 1922), 48.

Chapter 10: The Strike of 1922

1. Dubofsky and Van Tine, *John L. Lewis*, 80, 147.

2. David Montgomery, *The Fall of the House of Labor: The Workplace, the State, and American Labor Activism, 1865–1925* (Cambridge: Cambridge University Press, 1987), 370, 388, 407–408. For more on the strike wave, see Montgomery's "The 'New Unionism' and the Transformation of Workers' Consciousness in America, 1909–22."

3. For the operators' reactions, see "Miners' Union Gets Hearing," *Coal Age* 21 (13 April 1922): 604; and "Sixth Week of the Coal Strike," *Coal Age* 21 (18 May 1922): 853. The national UMWA was also surprised. See "Labor Stands Firm," *UMWJ*, 15 April 1922, 7.

4. Heber Blankenhorn, *The Strike for Union* (1924; reprint, New York: Arno and New York Times, 1969). The importance of the strike comes through clearly in oral histories, life stories, and institutional records.

5. Bernstein, *Lean Years*.

6. Dubofsky and Van Tine, *John L. Lewis*, 200, 215–219, 222–223, 227, 236, 268–270. See also Gilbert J. Gall, "Heber Blankenhorn, the LaFollette Committee, and the Irony of Industrial Repression," *Labor History* 23 (Spring 1982): 246–253.

7. Greene, *Slavic Community on Strike*, xiii–xvi, 207–215.

8. Brody, *Steelworkers in America*, 96–111, 135–136; for a symposium on Brody's work, see "Symposium on David Brody, *Steelworkers in America: The Nonunion Era*, and the Beginnings of the 'New Labor History,'" *Labor History* 34 (Fall 1993): 457–514.

9. Clark Kerr and Abraham Siegel, "The Interindustry Propensity to Strike—An International Comparison," in *Industrial Conflict*, ed. A. Kornhauser et al. (New York: McGraw-Hill, 1954), 195, 203.

10. Göbel, "Becoming American," 173–198.

11. Blankenhorn, *Strike for Union*, 17, 44–45.

12. W[illia]m P. Parks, Beaverdale, to John Brophy, Clearfield, 30 March 1922, UMWA, District 2, UMWA Records.

13. *New York Herald*, 22 April 1922.

14. John Brophy, [Clearfield], to W[illia]m P. Parks, Beaverdale, 1 April 1922, UMWA, District 2, UMWA Records.

15. "Policy Committee Decides Against Any Separate Agreements with Bituminous Coal Operators," *UMWJ*, 1 April 1922, 3; Blankenhorn, *Strike for Union*, 42.

16. "Coal Production Equalled by Few Other Industries," *Coal Age*, 21 (19 January 1922): 77–78; "Questions for Coal Barons," *New Republic* 30 (24 May 1922): 361; "Coal on Verge of Upturn in Country's Business Cycle," *Coal Age* 21 (19 January 1922): 76; Paul Wooton, "Coal Consumption Exceeding Output; Danger of Stocks Dwindling to Unsafe Level in Event of Strike," *Coal Age* 20 (15 December 1921): 975; "Coal Industry Must Withstand Spotlight During 1922," *Coal Age* 21 (19 January 1922): 99–100; "The Strike Cloud's Silver Lining," *Coal Age* 21 (26 January 1922): 155; "About Wage Reductions," *UMWJ*, 1 April 1921, 7; "Non-Union Wages," *UMWJ*, 1 September 1921, 6; "Without a Spokesman," *UMWJ*, 15 September 1921, 6.

17. Brophy, "Reminiscences," 415–425; Brophy, *A Miner's Life*, 180–182.

18. Zahurak, interview, 7 October 1986.

19. Brophy, "Reminiscences," 415–425; Brophy, *A Miner's Life*, 180–182.

20. Arthur Taylor, Boswell, to John Brophy, 31 March 1922, UMWA, District 2, UMWA Records; "Affidavits on Coal and Iron Police," [copies], UMWA, District 2, UMWA Records; Blankenhorn, *Strike for Union*, 17, 54.

21. *Johnstown Democrat*, 24 May 1922.

22. Blankenhorn, *Strike for Union*, 14–18, 35, 50, 56–57, 83.

23. Ibid., 42–43.

24. Ibid., 18–20; *Johnstown Democrat*, 25 May 1922.

25. Zahurak, interview, 7 October 1986.

26. Blankenhorn, *Strike for Union*, 18–20.

27. Ibid.

28. Ibid., 20.

29. Ibid., 20, 50. See *New York Herald*, 22 April 1922.

30. Blankenhorn, *Strike for Union*, 20–21; James Gibson, Steve Foster, Joseph Green, Martin Madigan, [Resolution of Local 4207], Windber, to John Brophy, Clearfield, 4 July 1922, UMWA, District 2, UMWA Records; Andrew King, Stanley Ford, John Kunkle, John Bolcor, [Resolution of Local 5229], Windber, to John Brophy, Clearfield, 4 July 1922, UMWA, District 2, UMWA Records; Mike Vislocky, Andy Durcha, Joe Brazins, [Resolution of Local 5231], Windber, to John Brophy, Clearfield, 4 July 1922, UMWA, District 2, UMWA Records.

31. Blankenhorn, *Strike for Union*, 20, 53.

32. Berwind-White Coal Mining Company, "Time Books," Mine 30, April 1922; "Record of Starts," Mine 35, April 1922; "Employment Records," Berwind Corporation, Windber.

33. Blankenhorn, *Strike for Union*, 21.

34. "Central Pennsylvania," *Coal Age*, 20 April 1922, 676; and *Somerset Herald*, 10 May 1922.

35. *Johnstown Democrat*, 24 May 1922.

36. *New York Herald*, 22 April 1922; Hapgood, *In Non-Union Mines*, 6–12, 46–48.

37. Joseph Foster, "Why Are the Windber Miners Out?" in Blankenhorn, *Strike for Union*, 254–259.

38. John Brophy, [Statement on Civil Liberties], to Hon. John Hays Hammond, Chairman, and Members of the United States Coal Commission, Washington, D.C., 28 May 1923, in Powers Hapgood Papers.

39. Blankenhorn, *Strike for Union*, 66–70, 83.

40. Ibid.

41. Lynn G. Adams, Harrisburg, to [Coal Companies], 18 March 1922, UMWA, District 2, UMWA Records.

42. Thomas Fisher, Philadelphia, to Lynn G. Adams, Harrisburg, Pa., 29 March 1922, General Correspondence, Coal Strike, Records of the Pennsylvania State Police; John Lochrie, Windber, to Lynn G. Adams, Harrisburg, 25 March 1922, General Correspondence, Coal Strike, Records of the Pennsylvania State Police.

43. John Brophy, Cresson, to Hon. William C. Sproul, Harrisburg, 10 April 1922, UMWA, District 2, UMWA Records; *Johnstown Democrat*, 25 May 1922; Blankenhorn, *Strike for Union*, 71–72; John Brophy, Clearfield, to Hon. William C. Sproul, Harrisburg, 28 June 1922, UMWA, District 2, UMWA Records; Lynn G. Adams, Harrisburg, to John Brophy, Clearfield, 30 June 1922, UMWA, District 2, UMWA Records.

44. Windber Borough Council, Meeting, 6 April 1922; Blankenhorn, *Strike for Union*, 54.

45. Windber Borough Council, Meetings, 1 May 1922, 6 June 1922, 6 July 1922, 16 March 1923, 3 April 1923; Blankenhorn, *Strike for Union*, 97; Berwind-White Coal Mining Company, Time Records, Special Roll, Police, Guards, and Deputies, April–July 1922, Berwind Corporation, Windber.

46. Blankenhorn, *Strike for Union*, 118–120; *Johnstown Tribune*, 5 May 1922.

47. Berwind-White Coal Mining Company vs. John Brophy et al., "Bill of Complaint," in the Court of Common Pleas of Somerset County, Pennsylvania, 19 April 1922, 2, 5, 6, 7, UMWA, District 2, UMWA Records.

48. *Johnstown Democrat*, 20 April 1922.

49. *Penn-Central News*, 6 May 1922; Blankenhorn, *Strike for Union*, 36–38.

50. *Johnstown Ledger*, 23 May 1922; *Johnstown Democrat*, 23 May 1922, 24 May 1922; Blankenhorn, *Strike for Union*, 50–53; John Brophy et al., "Answer" [to Bill of Complaint], Berwind-White Coal Mining Company vs. John Brophy et al., in the Court of Common Pleas of Somerset County, [May 1922], 2, 9, UMWA, District 2, UMWA Records.

51. *Johnstown Democrat*, 23 May 1922, 24 May 1922; Blankenhorn, *Strike for Union*, 50–53.

52. *Johnstown Democrat*, 24 May 1922, 25 May 1922.

53. Blankenhorn, *Strike for Union*, 123; *Johnstown Democrat*, 26 May 1922.

54. *Johnstown Tribune*, 5 June 1922; *Johnstown Democrat*, 27 May 1922; *Johnstown Ledger*, 1 August 1922.

55. *Penn-Central News*, 6 May 1922; Blankenhorn, *Strike for Union*, 40, 97, 122–124.

56. Molnar and Molnar, interview, 23 February 1984; Ann Gary, interview by author, tape recording, Windber, 29 February 1984.

57. Leonardis, interview, 5 March 1984.

58. Zahurak, interview, 7 October 1986; Conjelko, interview, 13 March 1984.

59. Blankenhorn, *Strike for Union*, 50–66.

60. Zahurak, interview, 7 October 1986.

61. Ibid.

62. Windber Borough Council, Meeting, 1 May 1922.

63. Sväti Tomas, Odbor 304, Národný Slovenský Spolok, Windber, Pennsylvania, "Zápisnica," Meeting, 23 April 1922; Spolok Sväti Jána Krstitela, číslo 48, Pennsylvánska Slovenská Katolícka Jednota, Windber, Pennsylvania, "Zápisnica," Meetings, 7 May 1922; 2 July 1922; Spolok Sv. Jána Krstitela, číslo 292, Prvá Katolícka Slovenská Jednota, Windber, Pennsylvania, "Zápisnica," Meeting, 21 May 1922.

64. Kunrad Valent, Mine 40, to Michael Lenko, First Catholic Slovak Union, 20 June 1922, Box 86, File 868, First Catholic Slovak Union Papers, the Czech and Slovak American Collection, Immigration History Research Center, University of Minnesota, Minneapolis–St. Paul.

65. [Michael Lenko] to Kunrad Valent, Mine 40, 24 June 1922, Box 86, File 868, First Catholic Slovak Union Papers.

66. Sväti Tomas, Odbor 304, "Zápisnica," Meeting, 16 July 1922; Slovenský Politický a Poučný Klub, "Zápisnica," Meeting, 2 July 1922; Spolok Sv. Jána Krstitela, číslo 48, "Zápisnica," Meeting, 4 June 1922.

67. *Johnstown Democrat*, 7 June 1922.

68. Blankenhorn, *Strike for Union*, 131–137; "Affidavits on Coal and Iron Police," [copies], 6 October 1922, UMWA, District 2, UMWA Records.

69. *Johnstown Democrat*, 7 June 1922.

70. Paul Wooton, "Congress and Administration Favor 'Hands-Off' Policy Toward Threatened Coal Strike," *Coal Age* 21 (2 March 1922): 379; Paul Wooton, "Administration Continues Hands-Off Policy but Expects to Be Dragged into Strike," *Coal Age* 21 (20 April 1922): 669; "President Harding to Force Wage Settlement; Lewis and Secretary Davis Confer with President," *UMWJ*, 1 July 1922, 9; "Thirteenth Week of the Coal Strike," *Coal Age* 22 (6 July 1922): 25; Blankenhorn, *Strike for Union*, 139–141.

71. James Gibson, Steve Foster, Joseph Green, Martin Madigan, [Resolution of Local 4207], Windber, to John Brophy, Clearfield, 4 July 1922, UMWA, District 2, UMWA Records; Andrew King, Stanley Ford, John Kunkle, John Bolcor, [Resolution of Local 5229], Windber, to John Brophy, Clearfield, 4 July 1922, UMWA, District 2, UMWA Records; Mike Vislocky, Andy Durcha, Joe Brazins, [Resolution of Local 5231], Windber, to John Brophy, Clearfield, 4 July 1922, UMWA, District 2, UMWA Records.

72. Dubofsky and Van Tine, *John L. Lewis*, 84–86; Blankenhorn, *Strike for Union*, 146–149; "Harding Proposes Resumption of Mining at Old Wage Pending Fixing of Basic Scale by Commission," *Coal Age* 22 (13 July 1922): 63–64; "Policy Committee to Consider the Proposal of President Harding to End the Strike of Miners," *UMWJ*, 15 July 1922, 3–5; Paul Wooton,

"President Harding 'Invites' Operators to Return to Mine Properties and Resume Operations," *Coal Age* 22 (20 July 1922): 99; "Sproul Calls Troops to Maintain Order in Pennsylvania Bituminous Region," *Coal Age* 22 (27 July 1922): 143.

73. James Gibson, Stanley Ford, and Joseph Green, [Resolution of Local Unions No. 4207, 5229, 5231], Windber, to John Brophy, Clearfield, 24 July 1922, UMWA, District 2, UMWA Records.

74. Dubofsky and Van Tine, *John L. Lewis,* 86–88; Blankenhorn, *Strike for Union,* 150–153; "United Mine Workers of America Have Won Their Strike in the Bituminous Fields and Achieved the Greatest Victory Ever Won in Industrial Struggle," *UMWJ,* 15 August 1922, 3–4; "Greatest Strike in History," *UMWJ,* 15 August 1922, 6–7; "Lewis Abandons Four State Contract to End Coal Strike; Gets High Wages and Avoids Arbitration," *Coal Age,* 22 (17 August 1922): 257–259; *Penn-Central News,* 23 September 1922.

75. Dubofsky and Van Tine, *John L. Lewis,* 122; Powers Hapgood, Boswell, to mother and father, 16 August 1922, Box 2, "Correspondence," Powers Hapgood Papers.

76. Blankenhorn, *Strike for Union,* 156–159; Zahurak, interview, 7 October 1986.

77. Blankenhorn, *Strike for Union,* 156–159; John Brophy, Clearfield, to John L. Lewis, Indianapolis, 28 August 1922, UMWA, District 2, UMWA Records; John Brophy, Clearfield, to the Officers & Members of District No. 2, U.M.W. of A., 28 August 1922, UMWA, District 2, UMWA Records; Brophy, "Reminiscences," 440–447.

78. Brophy, "Reminiscences"; Blankenhorn, *Strike for Union,* 159–160.

79. Powers Hapgood, Johnstown, to mother, 14 September 1922, Box 2, "Correspondence," Powers Hapgood Papers.

80. *Penn-Central News,* 30 September 1922.

81. Blankenhorn, *Strike for Union,* 122–125, 161–162, 186; *New York Call,* 19 October 1922; "Photos of Berwind-White Tent Colonies," in Hapgood "Scrapbook 1921–1927."

82. *Penn-Central News,* 30 September 1922; L. R. Thomas, Connellsville, Pa., to H. L. Kerwin, Washington, 16 October 1922, and L. R. Thomas, Pittsburgh, to H. L. Kerwin, Washington, 30 November 1922, Federal Mediation and Conciliation Service, Case File 170/1795.

83. Geo[rge] Swetz, Windber, to National Secretary, First Catholic Slovak Union, 14 November 1922, Box 124, File 779, First Catholic Slovak Union Papers.

84. Zahurak, interview, 7 October 1986.

85. Ibid.

86. *The World,* 22 October 1922.

87. Ibid.

88. Blankenhorn, *Strike for Union,* 174–176; "Documents in District No. 2 Campaign in New York from Sept. 25, 1922 to Oct. 25, 1922," in Heber Blankenhorn Papers, Series III.

89. "Documents in District No. 2 Campaign"; Blankenhorn, *Strike for Union,* 176–178; *Johnstown Democrat,* 28 September 1922.

90. *Johnstown Democrat,* 28 September 1922; Blankenhorn, *Strike for Union,* 176–178.

91. Blankenhorn, *Strike for Union,* 176–178; *Johnstown Democrat,* 28 September 1922.

92. *Johnstown Democrat,* 28 September 1922; Blankenhorn, *Strike for Union,* 179.

93. Blankenhorn, *Strike for Union,* 179–180; *Johnstown Democrat,* 28 and 29 September 1922.

94. Blankenhorn, *Strike for Union,* 180–181; *Johnstown Democrat,* 28, 29, and 30 September 1922.

95. Blankenhorn, *Strike for Union,* 183–184; *New York Times,* 4 October 1922, 12.

96. Blankenhorn, *Strike for Union,* 185–201; *New York Times,* 14 October 1922, 13; 18 October 1922, 31; *New York Call,* 19 October 1922; *The World,* 15, 16, and 22 October 1922.

97. "News Items from Field and Trade," *Coal Age* 22 (9 November 1922): 780.

98. Blankenhorn, *Strike for Union*, 185–193; *New York Times*, 29 October 1922, 6; 30 October 1922, 19; 31 October 1922, 19; 1 November 1922, 21; 2 November 1922, 21; 3 November 1922, 19; 4 November 1922, 17.

99. Hirshfield, *Statement of Facts*, 8–9; *The World*, 15 October 1922.

100. *New York Times*, 27 December 1922, 29.

101. Hirshfield, *Statement of Facts*, 23; *New York Times*, 2 January 1923, 1.

102. Hirshfield, *Statement of Facts*, 31, 33.

103. Ibid., 30–35.

104. "Berwind-White Mine Workers, Worse Off than Slaves, Says Report of Investigators," *Coal Age* 23 (4 January 1923): 26; "Slave Conditions," *UMWJ*, 1 May 1925, 11.

105. [Harry B.] Dynes, Streator, Ill., to H. L. Kerwin, Washington [telegram], 18 October 1922, Federal Mediation and Conciliation Service, Case File 170/1795.

106. H. L. Kerwin, "Coal-Strike in Somerset-Connellsville-Cambria, etc." [Typewritten Memo, c. early December 1922], Federal Mediation and Conciliation Service, Case File 170/1795.

107. Edward J. Berwind, New York, to Hon. James J. Davis, Washington, 10 January 1923, Federal Mediation and Conciliation Service, Case File 170/1795.

108. Blankenhorn, *Strike for Union*, 205–206; Powers Hapgood, en route to Clearfield, to mother and father, 25 January 1923, Box 2, "Correspondence," Powers Hapgood Papers.

109. Hapgood to mother and father, 25 January 1923; Powers Hapgood, Boswell, to mother, 30 January 1923, Box 2, "Correspondence," Powers Hapgood Papers. The *UMWJ* reported that the decision to call off the Connellsville strike was a unanimous one and that the strike had been a "notable" victory. See "Notable Victory," *UMWJ*, 1 February 1923, 14; "Strike Declared Ended," *UMWJ*, 1 February 1923, 16.

110. "'*How to Run Coal*,'" *Coal Age* 23 (4 January 1923): 1; "Miners' Nationalization Committee Submits Plan for Public Ownership and Operation of Coal Industry," *Coal Age* 23 (4 January 1923): 23–24; "Ellis Searles Attacks Mine-Nationalization Plan as Work of 'Greenwich Village Coal Miners,'" *Coal Age* 23 (1 February 1923): 233–234; "Who Wants Nationalization?" *Coal Age* 23 (15 February 1923): 283–284; "Resigning from Nationalization Committee Brophy Charges Lewis with Evasion," *Coal Age* 23 (15 February 1923): 306; "President Brophy's Representative Suggests a 20 Per Cent Cut in the Wages of Mine Workers," *UMWJ*, 1 April 1923, 3–4; "That 20 Percent Reduction," *UMWJ*, 1 April 1923, 6.

111. John Brophy, "Circular from District 2, U.M.W.A., 26 March 1923, Box 2, "Correspondence," Powers Hapgood Papers; "Letter to Somerset County Operator Repudiates Charges Against Union," *UMWJ*, 1 April 1923, 16; "More Trouble-Making," *UMWJ*, 15 April 1923, 6; "Revolutionary Interests Have Huge Fund with Which to Finance Their Activities in America," *UMWJ*, 1 May 1923, 8–10; "Virtuous Indignation at Brophy," *Coal Age* 23 (12 April 1923): 587–588.

112. Blankenhorn, *Strike for Union*, 209; Richard Gilbert to the Members of the Local Unions, District No. 2, U.M.W. of A., 21 February 1923, in Hapgood "Scrapbook 1921–1927."

113. Blankenhorn, *Strike for Union*, 105, 194–195, 203–204, 261.

114. Ibid., 210–212; Powers Hapgood, Johnstown, to mother and father, 1 April 1923, Box 2, "Correspondence," Powers Hapgood Papers; "Somerset County Miners STRIKE" and "Strike Call," 1 April 1923, in Hapgood "Scrapbook 1921–1927."

115. UMWA, District 2, *Minutes of the Proceedings of Special Convention of the United Mine Workers of America, District No. 2, . . . DuBois, Pa., . . . 1923* (DuBois, Pa.: Gray Printing, 1923), 226–227, UMWA, District 2, UMWA Records; Powers Hapgood, Boswell, to mother, 23 April 1923, Box 2, "Correspondence," Powers Hapgood Papers.

116. Blankenhorn, *Strike for Union*, 214–216.

117. "Miners Decide to Carry on Strike in Somerset County," [unidentified newspaper clipping, June 1923], in Hapgood "Scrapbook 1921–1927"; Albert Ramsell, Powers Hapgood, U. S. G. Gallager, Charles Ghizzoni, Mike Mizu, [Resolution to Gifford Pinchot; Resolution to International Executive Board; and Resolution], 6 June 1923, in Hapgood "Scrapbook 1921–1927."

118. Powers Hapgood, Boswell, to mother and father, 27 July 1923, Box 2, "Correspondence," Powers Hapgood Papers; *Johnstown Ledger*, 15 August 1923; *Johnstown Democrat*, 15 August 1923.

119. "Resolution Adopted by Delegates of Somerset County Local Unions," 14 August 1923, in Hapgood "Scrapbook 1921–1927."

120. Zahurak, interview, 7 October 1986.

121. Blankenhorn, *Strike for Union*, 56–63. Production for Berwind-White mines in the Windber area totaled 2,612,897 tons in 1920, 2,610,237 tons in 1921, and 1,353,610 tons in 1922. See Pennsylvania, *Report of the Department of Mines of Pennsylvania*, part 2, *Bituminous 1919–1920* (Harrisburg, Pa., 1925), 330; Pennsylvania, *Report of the Department of Mines of Pennsylvania*, part 2, *Bituminous 1921–1922* (Harrisburg, Pa., 1925), 160, 329.

122. Dubofsky and Van Tine, *John L. Lewis*, 87–90; Montgomery, *Fall of the House of Labor*, 7, 407–409.

123. Blankenhorn, *Strike for Union*, 220, 229.

Chapter 11: The Long Depression and the New Deal

1. Higham, *Strangers in the Land*, chap. 10, 264–299; Bernstein, *Lean Years*.

2. *New York Times*, 29 September 1929, sec. 2, 22.

3. One of the best studies of the national Klan is David M. Chalmers, *Hooded Americanism: The History of the Ku Klux Klan*, 3rd ed. (Durham, N.C.: Duke University Press, 1987). The definitive work on the Pennsylvania Klan is Emerson Hunsberger Loucks, *The Ku Klux Klan in Pennsylvania: A Study in Nativism* (Harrisburg, Pa.: Telegraph Press, 1936). An excellent documentary that portrays the Klan in Central Pennsylvania coal fields is Jim Dougherty's *Struggle for an American Way of Life: Coal Miners and Operators in Central Pennsylvania, 1919–1933*, produced, directed, and written by Jim Dougherty, 56 min., Indiana University of Pennsylvania Folklife Documentation Center: Indiana, Pa., 1992, videocassette.

4. Loucks, *Ku Klux Klan*, 103–113, 134–147; Chalmers, *Hooded Americanism*, 110–112; and Hiram W. Evans, "The Klan of Tomorrow," *Imperial Night-Hawk*, 15 October 1924, 7, 2. Almost every issue of the Klan's national publication, *Imperial Night-Hawk*, 1923–1924, and its successor, *The Kourier*, contain articles that attack Catholicism, but see especially the 1928 *Kourier* issues for their attacks on Al Smith's presidential campaign.

5. *Imperial Night-Hawk*, 2 May 1923, 7; 20 June 1923, 2; 25 April 1923, 6.

6. Maier B. Fox, *United We Stand: The United Mine Workers of America, 1890–1990* (Washington, D.C.: UMWA, 1990), 260–261; Chalmers, *Hooded Americanism*, 54, 155–157, 167. For a critical account of UMWA racial policies in the Pittsburgh region in the 1920s, see Linda Nyden, "Black Miners in Western Pennsylvania, 1925–1931: The National Miners Union and the United Mine Workers of America," *Science and Society* 41 (Spring 1977): 69–101.

7. *Imperial Night-Hawk*, 25 April 1923, 5; 15 October 1924, 1–3, 6–8; 2 May 1923, 7.

8. Loucks, *Ku Klux Klan*, 50–54; Chalmers, *Hooded Americanism*, 238–239.

9. *Imperial Night-Hawk*, 26 December 1923, 4; Loucks, *Ku Klux Klan*, 25–27.

10. Zahurak, phone interview by author, 30 December 1988, Windber.

11. *Windber Era*, 26 February 1925, 28 February 1924, 27 August 1925, 3 June 1926, 21 February 1924.

12. Zahurak, interview, 30 December 1988; Loucks, *Ku Klux Klan*, 162–198.

13. Loucks, *Ku Klux Klan*, 50–57; Chalmers, *Hooded Americanism*, 237–240.

14. *Penn-Central News*, 20 September 1922; Blankenhorn, *Strike for Union*, 158–159, 162–168.

15. Blackenhorn, *Strike for Union*, 168–170; UMWA, District 2, *Annual Report of Richard Gilbert, Secretary Treasurer District No. 2, United Mine Workers of America, . . . 1923* (DuBois, Pa.: Gray, 1924), 9; UMWA, District 2, *Minutes . . . Convention of the United Mine Workers of America, District No. 2, . . . DuBois, Pa., . . . 1923*, 230–232.

16. Powers Hapgood, Johnstown, to mother, 14 September 1922, Box 2, Powers Hapgood Papers.

17. Dubofsky and Van Tine, *John L. Lewis*, 36, 38–39, 63, 73–74; Blankenhorn, *Strike for Union*, 12; John Walker, to Heber Blankenhorn, New York City, 1 November 1920, Box 11, Walker Collection, Illinois Historical Survey Library, University of Illinois, Urbana-Champaign; John Walker, to Heber Blankenhorn, New York City, 6 November 1920, Box 11, Illinois Historical Survey Library, University of Illinois, Urbana-Champaign.

Others have countered that Lewis was up against the odds in 1922. Certainly the national did have to contend with fierce operator opposition and an unsympathetic or hostile government. Unlike given districts, it also had responsibility for the welfare of the national organization. Perhaps Lewis made the best settlement possible at the time. Yet such opinions are no more provable than are the opposing assertions.

18. Montgomery, *Fall of the House of Labor*, 408–409; Dubofsky and Van Tine, *John L. Lewis*, 122–130; Alan Singer, "Communists and Coal Miners: Rank-and-File Organizing in the United Mine Workers During the 1920s," *Science and Society* 55 (Summer 1991): 132–157.

19. Singer, "Communists and Coal Miners," 139–141.

20. Ibid., 143–145; Alan Singer, "John Brophy's 'Miners' Program': Workers' Education in UMWA District 2 During the 1920's," *Labor Studies Journal* 13 (Winter 1988): 50–64.

21. For a general description of conditions in regional coal fields in the 1920s, see "The Long Depression of Southwestern Coal Miners," in *People, Poverty, and Politics: Pennsylvanians During the Great Depression*, ed. Thomas H. Coode, John F. Bauman, et al. (Lewisburg, Pa.: Bucknell University Press, 1981), 23–48.

22. *Windber Era*, 24 April 1924; Pennsylvania Department of Mines, annual reports, 1922–1940 (Harrisburg, Pa., 1921–1940).

23. *Windber Era*, 10 April 1924, 5 March 1925, 9 April 1925, 4 November 1926.

24. *Windber Era*, 9 July 1925.

25. *Windber Era*, 10 June 1926.

26. *Windber Era*, 3 February 1927.

27. Nyden, "Black Miners," 69–101; Bernstein, *Lean Years*, 358–390.

28. Powers Hapgood, Windber, to mother and father, 10 April 1927, Box 4, Powers Hapgood Papers.

29. Ibid.

30. "Miners of the Berwind-White Co.," [circular], Scrapbook 1925–1928, Powers Hapgood Papers.

31. Powers Hapgood, Cresson, to mother and father, 20 April 1927, Box 4, Powers Hapgood Papers.

32. Ibid.

33. Zahurak, interview, 7 October 1986.

34. Hapgood to mother and father, 20 April 1927.

35. Ibid.

36. Ibid., "Company's Dishonesty Exposed!" [circular], Scrapbook 1925–1928, Powers Hapgood Papers.

37. Powers Hapgood, en route Pittsburgh to Somerset, to mother, 25 April 1927, Powers Hapgood Papers.

38. Hapgood, to mother and father, 20 April 1927.

39. Hapgood, en route, to mother, 25 April 1927.

40. Ibid.

41. Dubofsky and Van Tine, *John L. Lewis,* 145–148; Irwin M. Marcus, James P. Dougherty, and Eileen M. Cooper, "Judge Jonathan Langham and the Use of the Labor Injunction in Indiana County, 1919–1931," *Pennsylvania Magazine of History and Biography* 118 (January–April 1994): 63–85.

42. Singer, "Communists and Coal Miners," 146–148; Nyden, "Black Miners," 85–87.

43. Dubofsky and Van Tine, *John L. Lewis,* 170–172; Nyden, "Black Miners," 69, 87–101.

44. Jeanette D. Pearl, Johnstown, to James Mark, Clearfield, 13 June 1927, 25 June 1927; 29 June 1927, and 11 July 1927, District 2 UMWA Records.

45. Ibid.

46. *Windber Era,* 14 July 1927; Jeanette D. Pearl, Johnstown, to James Mark, Clearfield, 14 July 1927, District 2 UMWA Records.

47. James Mark, Clearfield, to Jeanette D. Pearl, Johnstown, 16 July 1927, District 2 UMWA Records.

48. *Windber Era,* 1 November 1928, 8 November 1928; *Pennsylvania State Manual 1929* (Harrisburg, Pa., 1929), 653.

49. *Windber Era,* 6 November 1924, 9 November 1916; *Pennsylvania State Manual 1925–1926* (Harrisburg, Pa., 1926), 557. Windber traditionally elected Republican candidates for major races by large majorities. The one exception had occurred in November 1926, when factional divisions within the state's Republican party let a Democratic candidate for the U.S. Senate carry Windber and Somerset County in a nontypical election. See *Windber Era,* 4 November 1926. For a useful study of the Republican party in the 1920s, see Samuel John Astorino, "The Decline of the Republican Dynasty in Pennsylvania, 1929–1934" (Ph.D. diss., University of Pittsburgh, 1962).

50. *Pennsylvania State Manual 1929,* 653.

51. Leonardis, interview, 8 March 1984.

52. Leonardis, interview, 5 March 1984.

53. *Windber Era,* 6 May 1926.

54. *Pennsylvania State Manual 1929,* 73; Alcamo, *Windber Story,* 117.

55. *Industrial Worker,* 2 August 1932, 9 August 1932; Nyden, "Black Miners," 92–98; Dubofsky and Van Tine, *John L. Lewis,* 171–172.

56. *Pennsylvania State Manual 1933* (Harrisburg, Pa., 1933), 558.

57. Ibid.

58. *Somerset Daily Herald,* 16 May 1933.

59. *Somerset Daily Herald,* 22 May 1933.

60. Dubofsky and Van Tine, *John L. Lewis,* xv, 108, 136–139, 172–178.

61. Curtis Seltzer, *Fire in the Hole: Miners and Managers in the American Coal Industry* (Lexington: University Press of Kentucky, 1985), 42–55, 61, 83–84.

62. Irving Bernstein, *The Turbulent Years: A History of the American Worker, 1933–1941* (Boston: Houghton Mifflin, 1970); Dubofsky and Van Tine, *John L. Lewis,* 179–386. Miners interviewed invariably gave positive assessments of Lewis.

63. UMWA Local 6186, Windber, Minutes, Meetings, 23 October 1934, 7 August 1934; James P. Johnson, "Reorganizing the United Mine Workers of America in Pennsylvania During the New Deal," *Pennsylvania History* 37 (April 1970): 120–121.

64. By July, coal operators had noted the upsurge of spontaneous organizing prompted by section 7(a) of the NRA. See "United Mine Workers Presses Organization; Wage Increases Spread in East," *Coal Age* 38 (July 1933): 249.

65. James P. Johnson, "Reorganizing," 122–123; *New York Times*, 4 July 1933, 1; Alcamo, *Windber Story*, 178.

66. Novak, interview, 19 May 1985.

67. Zahurak, interview, 7 October 1986.

68. UMWA Local 6186 Minutes, 21 September 1933–30 December 1939.

69. Dubofsky and Van Tine, *John L. Lewis*, 190–201.

70. Zahurak, interview, 7 October 1986.

71. See Edwin B. Bonner, "The New Deal Comes to Pennsylvania: The Gubernatorial Election of 1934," *Pennsylvania History* 27 (January 1960): 44–68; W. Wayne Smith, "1934: The Critical Election in Indiana County," in Coode et al., *People, Poverty, and Politics*, 108–128; Richard C. Keller, *Pennsylvania's Little New Deal* (New York: Garland, 1982).

72. UMWA Local 6186 Minutes, 6 February 1934, 3 July 1934, 17 April 1934, 24 April 1934.

73. Ibid., 10 July 1934, 7 August 1934, 11 September 1934, 18 September 1934, 25 September 1934; "Plan Big Labor Day Event," *UMWJ*, 15 August 1934, 10; and "'Code Day' Big Success," *UMWJ*, 1 November 1934, 14.

74. UMWA Local 6186 Minutes, 16 October 1934, 23 October 1934, 5 November 1934, 13 November 1934, 20 November 1934, 27 November 1934. At a later date, an unspecified fine was reduced to $1.00 each. See ibid., 1 April 1935.

75. *Pennsylvania State Manual 1935–1936* (Harrisburg, Pa., 1936), 493, 592; UMWA Local 6186 Minutes, 20 November 1934, 3 January 1935; *Windber Era*, 3 January 1935, 24 January 1935, 31 January 1935.

76. UMWA Local 6186 Minutes, 21 September 1933–30 December 1939.

77. *New York Times*, 25 October 1935, 6; Keller, *Pennsylvania's Little New Deal*, 184–196; "Good Riddance," *UMWJ*, 1 July 1935, 8; "Kills Security Bills," *UMWJ*, 15 July 1935, 6; "New Child Labor Law," *UMWJ*, 15 August 1935, 9.

78. UMWA Local 6186 Minutes, 20 November 1934, 18 December 1934.

79. Ibid., 26 February 1935, 5 March 1935, 12 March 1935, 1 April 1935, 9 April 1935.

80. Novak, interview, 19 May 1985.

81. Ibid.; UMWA Local 6186 Minutes, 13 August 1935, 20 August 1935, 3 September 1935, 9 September 1935, 15 October 1935, 4 November 1935; *Windber Era*, 7 November 1935.

82. *Windber Era*, 2 January 1936, 6 March 1936; "Form Roosevelt Club," *UMWJ*, 1 April 1936, 22; "Club at Scalp Level," *UMWJ*, 1 May 1936, 15; UMWA Local 6186 Minutes, 3 March 1936.

83. *Windber Era*, 22 October 1936, 29 October 1936, "To Celebrate Labor Day," *UMWJ*, 15 August 1936, 19; "Big Day at Windber, Pa.," *UMWJ*, 1 September 1936, 19.

84. *Pennsylvania State Manual 1937* (Harrisburg, Pa., 1937), 322; *Windber Era*, 5 November 1936; Keller, *Pennsylvania's Little New Deal*, 229–242; "Governor Earle Pledges Full Protection of Labor's Rights in Pennsylvania," *UMWJ*, 15 September 1936, 6.

85. *New York Times*, 19 August 1936, 21; "Big Coal Producer Dead," *UMWJ*, 1 September 1936, 13.

86. "'Company Towns' Put Under Ban by Governor Earle of Pennsylvania," *UMWJ*, 15 January 1937, 20; Keller, *Pennsylvania's Little New Deal*, 6–7, 262–289.

87. Keller, *Pennsylvania's Little New Deal*; "Pennsylvania Legislature Makes Record for Liberal Laws," *UMWJ*, 15 June 1937, 6; "Pennsylvania Will Weed Out Labor Spies," *UMWJ*, 15 December 1937. For a study of labor and civil liberties at the federal level, see Jerold S. Auerbach, *Labor and Liberty: The LaFollette Committee and the New Deal* (Indianapolis, Ind.: Bobbs-Merrill, 1966).

88. Alcamo, *Windber Story*, 142.

Chapter 12: The Achievements and Limits of Worker Protest

1. For an excellent article on the complexities of Polish culture, see John J. Bukowczyk, "Polish Rural Culture and Immigrant Working Class Formation, 1880–1914," *Polish American Studies* 41 (Autumn 1984): 23–44.

Epilogue

1. For two recent regional studies of deindustrialization, see Dale A. Hathaway, *Can Workers Have a Voice? The Politics of Deindustrialization in Pittsburgh* (University Park: The Pennsylvania State University Press, 1993), and William Serrin, *Homestead: The Glory and Tragedy of an American Steel Town* (New York: Times Books, 1992).

2. Alcamo, *Windber Story*, 75.

3. Department of the Interior, Coal Mines Administration, *A Medical Survey of the Bituminous-Coal Industry* (Washington, D.C., 1947).

4. Alcamo, *Windber Story*, 128, 209; *History of Berwind*, 92–103.

5. See Doyle, *50th Anniversary*.

6. *Windber Journal*, 17 March 1903.

7. Alcamo, *Windber Story*, 171–174.

8. For examples, see Somerset County Deed Books, 468:299, 470:596, 920:282; Cambria County Deed Book, 657:289.

9. Department of Commerce, Bureau of the Census, *1990 Census of Population, General Population Characteristics: Pennsylvania* (Washington, D.C., 1992), 27.

10. Ibid.

11. *History of Berwind*, 1.

Selected Bibliography

Primary Sources, by Depository

Archives of Industrial Society, University of Pittsburgh Library, Pittsburgh, Pa.:
Heber Blankenhorn Papers
Berwind Corporation, Windber, Pa.:
Berwind-White Coal Mining Company Employment Records, Record of Starts, Time Records
The Catholic University of America, Department of Archives and Manuscripts, Washington, D.C.:
John Mitchell Papers. Microform Edition. Glen Rock, N.J.: Microfilming Corporation of America, 1974
Historical Collections and Labor Archives, University Libraries, Pennsylvania State University, State College, Pa.:
T. R. Johns Collection
Historical Society of Pennsylvania, Philadelphia, Pa.:
William Bauchop Wilson Papers
Illinois Historical Survey Library, University of Illinois, Urbana-Champaign, Ill.:
John H. Walker Papers
Immigration History Research Center, University of Minnesota, Minneapolis–St. Paul, Minn.:
Adam Podkrivacky Papers, the Czech and Slovak American Collection
First Catholic Slovak Union Papers, the Czech and Slovak American Collection
Foreign-Language Newspaper Collections
Miniszterelnökségi levéltár [Hungarian State Archives, Prime Minister's Records], 1909–1916, the Hungarian American Collection
National Slovak Society Papers, the Czech and Slovak American Collection
Indiana University Library, Indiana University of Pennsylvania, Indiana, Pa.:
United Mine Workers of America, District 2 Records, 1900–1940s
Lilly Library, Manuscripts Department, Indiana University, Bloomington, Ind.:
Powers Hapgood Papers
National Archives, Washington, D.C.:
Federal Mediation and Conciliation Records, RG 280, Case Files 33/423, 33/2987, 170/1795, 199/8563
U.S. Bureau of Census Manuscript Schedules, 12th Census 1900: Pennsylvania. Washington, D.C.: Bureau of Census, 1900. Microfilm edition, rolls 1387, 1388, 1389, 1486, 1487. Washington, D.C.: National Archives, 1978
U.S. Bureau of Census Manuscript Schedules, 13th Census 1910: Pennsylvania. Washington, D.C.: Bureau of Census, 1910. Microfilm edition, rolls 1324, 1420, 1421. Washington, D.C.: National Archives, 1978

Pennsylvania State Archives, Pennsylvania Historical and Museum Commission, Harrisburg, Pa.:
Pennsylvania State Police Records, RG 30, especially Annual and Biennial Reports; Bulletins and Circular Letters; General and Special Orders; General Correspondence; Records of the Ku Klux Klan; Strike Reports; Troop Reports
Samuel Pennypacker Papers, MG 171, Reports 1906
State Police Memorabilia, MG 360, Philip Doddridge Album and Diary

Slovak Hall, Windber, Pa.:
Slovenská Národná Budova [Slovak National Hall Association], Windber, Pennsylvania. Zápisnica [Minute Books], 1914–1923
Slovenský Národniský Politický Klub [Slovak Political Club], Windber, Pa. Zápisnica [Minute Books], 1901–1902
Slovenský Neodvislý Politický Klub [Slovak Independent Political Club], Windber, Pa. Zápisnica [Minute Books], 1907–1909
Slovenský Poučný a Politický Klub [Slovak Educational and Political Club], Windber, Pa. Zápisnica [Minute Books], 1909–1934
Spolok Sväti Jána Krstitela, číslo 48, Pennsylvánska Slovenská Katolícka Jednota [St. John the Baptist Beneficial Society, Branch 48, Pennsylvania Roman and Greek Catholic Union], Windber, Pa. Membership and Dues Books, 1899–1920; Zápisnica [Minute Books], 1899–1924
Spolok Sväti Jána Krstitela, číslo 292, Prvá Katolícka Slovenská Jednota [St. John the Baptist Society, Branch 292, First Catholic Slovak Union], Windber, Pa. Membership and Dues Books, 1898–1920; Zápisnica [Minute Books], 1898–1924
Spolok Sväti Tomas, Odbor 304, Národný Slovenský Spolok [St. Thomas Society, Branch 304, National Slovak Society], Windber, Pa. Membership and Dues Books, 1898–1920; Zápisnica [Minute Books], 1898–1924

Somerset County Courthouse, Somerset, Pa.:
Somerset County Deed Books
Somerset County Miscellaneous Docket Books
United States Naturalization Records, Somerset County, 1898–1904, and 1929–1945

SS. Cyril and Methodius Church, Windber, Pa.:
Baptismal Records, 1906–1940
Burial Records, 1906–1919
Marriage Records, 1906–1940

St. John Cantius Roman Catholic Church, Windber, Pa.:
Baptismal Records, 1898–1904
Marriage Records, 1899–1912

St. Mary's Byzantine Catholic Church, Windber, Pa.:
Baptismal Records, 1900–1904
Marriage Records, 1900–1930

St. Mary's Hungarian Roman Catholic Church, Windber, Pa.:
Baptismal Records, 1916–1928

United Mine Workers of America, Local 6186, Windber, Pa.:
Minute Books, 1933–1975

Windber Borough Building, Windber, Pa.:
Windber Borough Council Minute Books, 1900–1950
Windber Borough Ordinances

Public Documents

Atlantic Reporter, 1895–1935. St. Paul, Minn.: West Publishing Co., 1896–1935.

Hirshfield, David, New York City, Committee on Labor Conditions at the Berwind-White Company's Coal Mines in Somerset and Other Counties, Pennsylvania. *Statement of Facts and Summary of Committee Appointed by Honorable John F. Hylan, Mayor of the City of New York, to Investigate the Labor Conditions of the Berwind-White Company's Coal Mines in Somerset and Other Counties, Pennsylvania.* New York: M. B. Brown, December 1922.

Pennsylvania Courts. *The District Reports of Cases Decided in All the Judicial Districts of the State of Pennsylvania During the Year 1896.* Philadelphia: n.p., 1896.

———. *The District Reports of Cases Decided in All the Judicial Districts of the State of Pennsylvania During the Year 1907.* Philadelphia: Howard W. Page, 1908.

Pennsylvania Department of Internal Affairs. *Annual Report of the Secretary of Internal Affairs of the Commonwealth of Pennsylvania for the Year Ending November 30, 1906.* Harrisburg: State Printer, 1907.

Pennsylvania Department of Internal Affairs. Bureau of Mines. *Report of the Bureau of Mines of Pennsylvania,* annual reports, 1890–1902. Harrisburg: State Printer, 1891–1903.

Pennsylvania Department of Labor and Industry. *Report to Governor Gifford Pinchot by the Commission on Special Policing in Industry.* Harrisburg: State Printer, 1934.

Pennsylvania Department of Mines. *Report of the Department of Mines of Pennsylvania,* annual reports, 1903–1944. Harrisburg: State Printer, 1904–1959.

Pennsylvania Department of State Police. *Annual Report of the Department of State Police of the Commonwealth of Pennsylvania,* annual reports, 1906–1919. Harrisburg: State Printer, 1907–1919.

———. *Biennial Report of the Department of the Pennsylvania State Police,* biennial reports, 1920–1936. Harrisburg: State Printer, 1920–1936.

Pennsylvania General Assembly. *Laws of the General Assembly of the Commonwealth of Pennsylvania (Excepting Appropriation Laws) Passed at the Session of 1911 on the One Hundred and Thirty-Fifth Year of Independence.* Harrisburg: C. E. Aughinbaugh, State Printer, 1911.

Pennsylvania Supreme Court. *Pennsylvania State Reports, 1895.* New York: Banks & Brothers, Law Publishers, 1896.

———. *Pennsylvania State Reports, 1901–1910.* New York: Banks Law Publishing Co., 1902–1911.

———. *Pennsylvania State Reports, 1912–1920.* Philadelphia: George T. Bisel Co., 1912–1920.

Reese, John F., and James D. Sisler. *Bituminous Coal Fields of Pennsylvania.* Part 3, *Coal Resources.* Harrisburg: Pennsylvania Geological Survey, Fourth Series, 1928.

Smull's Legislative Hand Book and Manual of the State of Pennsylvania, 1896–1920. Harrisburg: State Printer, 1897–1921.

U.S. Department of Commerce. Bureau of the Census. *Fifteenth Census of the United States: 1930*, vol. 1: *Population: Number and Distribution of Inhabitants*. Washington, D.C.: GPO, 1931.

———. *Fifteenth Census of the United States: 1930 Population*, vol. 3, part 2: *Reports by States, Showing the Composition and Characteristics of the Population for Counties, Cities, and Townships or Other Minor Civil Divisions*. Washington, D.C.: GPO, 1932.

———. *Fifteenth Census of the United States: 1930 Population*, vol. 6: *Families*. Washington, D.C.: GPO, 1933.

———. *Fourteenth Census of the United States Taken in the Year 1920*, vol. 1: *Population: Number and Distribution of Inhabitants*. Washington, D.C.: GPO, 1921.

———. *Fourteenth Census of the United States Taken in the Year 1920*, vol. 3: *Population: Composition and Characteristics of the Population by States*. Washington, DC: GPO, 1923.

———. *Historical Statistics of the United States Colonial Times to 1970*, bicentennial edition. Washington, D.C.: Bureau of the Census, 1975.

———. *A Report of the Seventeenth Decennial Census of the United States. Census of Population: 1950*, vol. 2, part 38: *Characteristics of the Population*. Washington, D.C.: GPO, 1952.

———. *Sixteenth Census of the United States: 1940 Housing*, vol. 2, part 4: *General Characteristics*. Washington, D.C.: GPO, 1943.

———. *Sixteenth Census of the United States: 1940 Population*, vol. 2, part 6: *Characteristics of the Population*. Washington, D.C.: GPO, 1943.

———. *Thirteenth Census of the United States Taken in the Year 1910*, vol. 1: *Population: General Report and Analysis*. Washington, D.C.: GPO, 1913.

———. *Thirteenth Census of the United States Taken in the Year 1910*, vol. 3: *Population Reports by States, with Statistics for Counties, Cities and Other Civil Divisions*. Washington, D.C.: GPO, 1913.

U.S. Department of Commerce. Census Office. *Twelfth Census of the United States, Taken in the Year 1900*, vols. 1 and 2: *Population*. Washington, D.C.: GPO, 1901–1902.

U.S. Department of Commerce and Labor. Bureau of the Census. *Special Reports. Religious Bodies: 1906*, part 1: *Summary and General Tables*. Washington, D.C.: GPO, 1910.

———. *Special Reports. Religious Bodies: 1906*, part 2: *Separate Denominations: History, Description and Statistics*. Washington, D.C.: GPO, 1910.

U.S. Department of Labor. Bureau of Naturalization. *Historical Sketch of Naturalization in the U.S.* Washington, D.C.: GPO, 1926.

U.S. Department of Labor. Women's Bureau. *Home Environment and Employment Opportunities of Women in Coal-Mine Workers' Families*. Washington, D.C.: GPO, 1925.

U.S. Department of the Interior. Census Office. *Compendium of the Eleventh Census: 1890*. 3 parts. Washington, D.C.: GPO, 1892–1897.

U.S. Department of the Interior. Coal Mines Administration. *A Medical Survey of the Bituminous-Coal Industry*. Washington, D.C.: GPO, 1947.

U.S. Immigration Commission. *Reports of the Immigration Commission*, vols. 1 and 2: *Abstracts of Reports of the Immigration Commission, with Conclusions and Recommendations of the Minority*. Washington, D.C.: GPO, 1911; reprint, with an Introduction by Oscar Handlin, New York: Arno and New York Times, 1970.

———. *Reports of the Immigration Commission*, vol. 4: *Emigration Conditions in Europe.* Washington, D.C.: GPO, 1911; reprint, New York: Arno and New York Times, 1970.

———. *Reports of the Immigration Commission*, vols. 6 and 7: *Immigrants in Industries*, part 1, *Bituminous Coal Industry.* Washington, D.C.: GPO, 1911; reprint, New York: Arno and New York Times, 1970.

———. *Reports of the Immigration Commission*, vol. 19: *Immigrants in Industries*, part 23, *Summary Report on Immigrants in Manufacturing and Mining.* Washington, D.C.: GPO, 1911; reprint, New York: Arno and New York Times, 1970.

———. *Reports of the Immigration Commission*, vol. 41: *Statements and Recommendations Submitted by Societies and Organizations Interested in the Subject of Immigration.* Washington, D.C.: GPO, 1911; reprint, New York: Arno and New York Times, 1970.

U.S. Industrial Relations Commission. *Final Report and Testimony Submitted to Congress by the U.S. Commission on Industrial Relations Created by the Act of August 23, 1912*, vol. 8. Washington, D.C.: GPO, 1916.

Oral Histories

Unless otherwise indicated, the following interview tapes are currently in the author's possession but will be deposited in a major university archive in the near future. Pseudonyms have been used when those interviewed requested anonymity.

Blankenhorn, Heber. "The Reminiscences of Heber Blankenhorn." Interview by Harlan B. Phillips, 1955–1956. Transcript, Columbia University Oral History Research Collection. New York: Columbia University, 1957. Sanford, N.C.: Microfilming Corporation of America, 1979. Microfilm.

Brophy, John. "The Reminiscences of John Brophy." Interview by Dean Albertson, 1954–1955. Transcript, Columbia University Oral History Collection. New York: Columbia University, 1972. Microfilm.

Conjelko, Steve. Interview by author, Windber, Pa., 13 March 1984.

Czajkowski, Rose Koot. Interview by author, Scalp Level, Pa., 3 March 1984.

Czajkowski, Rose Koot, and Sophia Koot Fluder Yarzumbeck. Hagevo, Pa., 3 March 1984.

Dembinsky, Victoria, Mary Chmaj, Verna Trunack, Stanley F. May, Henry Hirsh, Joseph Zaczek, and Mary Zaczek. Interview by the Rev. Stanley J. Zabrucki and the author, Windber, Pa., 5 October 1986.

Dutzman, Elizabeth Petroci. Interview by author, Windber, Pa., 25 February, 28 February, and 15 March 1984.

Gary, Ann Grace Rosa. Interview by author, Windber, Pa., 29 February and 14 March 1984.

Gerula, Pete. Interview by author, Windber, Pa., 26 February and 15 March 1984.

Greathouse, Ernest. Interview by author, Johnstown, Pa., 1 October 1986.

Hoffman, Chauncey, M.D. Interview by author, Geistown, Pa., 11 July 1989.

Horvath, John. Interview by author, Windber, Pa., 6 March 1984.

Jankosky, Katherine [pseud.]. Interview by author, Windber, Pa., 24 February 1984.

Jankosky, Katherine [pseud.] and Mike [pseud.]. Interview by author, Windber, Pa., 5 and 8 March 1984.

Jankosky, Mike [pseud.]. Interview by author, Windber, Pa., 24 February 1984.
Keleschenyi, Joseph, William Kosturko, and Mary Kosturko. Interview by author, Windber, Pa., 13 March 1984.
Kovalsky, Mary S. Interview by author, Windber, Pa., 5 March and 6 March 1984.
Kush, Thaddeus. Interview by author, Windber, Pa., 23 February and 29 February 1984.
Leonardis, Pearl Camille. Interview by author, Windber, Pa., 5 March and 8 March 1984.
Molnar, Andrew [pseud.], and Elizabeth Molnar [pseud.]. Interview by author, Windber, Pa., 23 February and 26 February 1984.
Mucker, Agnes. Interview by author, Windber, Pa., 14 March 1984.
Novak, Joseph J. Interview by author, Windber, Pa., 19 May and 21 May 1985.
Repko, Thressa, Helen Vargo, and Julia Orlovsky. Interview by author, Windber, Pa., 21 July 1985.
Rohalley, John Henry. Interview by author, Windber, Pa., 2 March 1984.
Santucci, Maria, and Ludwig Santucci. Interview by author, Windber, Pa., 3 June 1985.
Sekeres, Rosella. Interview by author, Windber, Pa., 9 August 1985.
Shuster, Steve. Interview by author, Windber, Pa., 4 March 1984.
Shuster, Susie, and Katherine Rodish. Interview by author, Windber, Pa., 12 March 1984.
Taryan, Vilma. Interview by author, Windber, Pa., 6 March 1984.
Toth, John [pseud.]. Interview by author, Windber, Pa., 30 September 1986.
Washko, Stephen R. Interview by author, Windber, Pa., 16 March and 29 September 1984.
Zahurak, Joseph. Interview by author, Windber, Pa., 7 October 1986 and phone interview, 30 December 1988.

Printed Source Materials

American Carpatho-Russian Orthodox Greek Catholic Diocese of U.S.A. *Silver Anniversary, 1938–1963 American Carpatho-Russian Orthodox Greek Catholic Diocese of U.S.A., Johnstown, Pennsylvania.* Johnstown, Pa.: American Carpatho-Russian Orthodox Greek Catholic Diocese of U.S.A., 1963.
Beers, Frederick W. *County Atlas of Somerset, Pennsylvania.* New York: F. W. Beers and Co., 1876.
Byzantine Slavonic Rite Catholic Diocese of Pittsburgh. *Byzantine Slavonic Rite Catholic Diocese of Pittsburgh Silver Jubilee, 1924–1949.* Pittsburgh, Pa.: Byzantine Slavonic Rite Catholic Diocese of Pittsburgh, 1949.
Fehr, J. L., comp. *Windber–Scalp Level and Vicinity, Eureka Mines, Berwind-White Coal Mining Co. Operators.* Windber: Windber Publishing Co., December 1900.
Holy Child Jesus Church, Windber, Pa. *Souvenir of the Silver Jubilee Celebration of Rev. Father Morgan A. McDermott, B.A., Sunday July 31, 1938, Church of the Holy Child Jesus, Windber, Pennsylvania.* Windber, Pa.: Holy Child Jesus Church, 1938.
National Slavonic Society of the United States of America. *Constitution and By-Laws of the National Slavonic Society of the United States of America adopted at the VII. Convention Held at Chicago, Ill. on the 22nd, 23rd, 24th, 25th and 26th of May 1899.* Pittsburgh: P. V. Rovnianek, 1899.

———. *Constitution and By-Laws of the Nat. Slov. Society of U.S.A. adopted at the XI. Regular Convention Held at Milwaukee, Wis.* N.p: National Slovak Society, 1914.

Patriotic Order Sons of America. State Camp of Pennsylvania. *Pennsylvania State Camp and Subordinate Camp Constitutions, Preceded by National Camp Constitution and General Laws of the Patriotic Order Sons of America, U.S.A.* Philadelphia: Patriotic Order Sons of America, 1894.

Pennsylvania Slovak Catholic Union. *Zápisnica XXII. Konvencie, Pennsylvánskej Slovenskej Rímsko a Grécko Katolícky Jednoty, odbyvanej vdnoch od 16. do 21. mája v Pittsburgh, Pa. 1927.* Wilkes-Barre: Bratstva, 1927.

———. *Zápisnica Dvadsiatejtretej Konvencie Pennsylvánskej Slovenskej Rímsko a Grécko Katolíckej Jednoty odbyvanej od 23. do 28. September 1929 vo Wilkes-Barre, Pa.* N.p.: Bratsvo, 1930.

———. *Zápisnica 26. Konvencie Pennsylvánskej Slovenskej Rímsko a Grécko Katolíckej Jednoty vydrziavanej v meste Trenton, New Jersey v Hoteli Hildebrandt od 12-ho Do 17-ho September 1938.* N.p.: n.p., 1938.

Peters, Rev. Urban J., comp. *Memorial Volume of the Dedication of the Cathedral of the Blessed Sacrament Altoona, Pa., July 16, 1930.* Altoona, Pa.: Altoona Diocese, 1930.

SS. Cyril and Methodius Catholic Church, Windber, Pa. *Golden Jubilee SS. Cyril and Methodius Parish, 1906–1956, Windber, Pennsylvania.* Windber, Pa.: SS. Cyril and Methodius Parish, 1956.

———. *75th Anniversary Celebration, 1906–1981, SS. Cyril & Methodius Catholic Church, Windber, Pa., Sunday, November 8, 1981.* Windber, Pa.: SS. Cyril and Methodius Church, 1981.

SS. Peter and Paul Orthodox Greek Catholic Church, Windber, Pa. *Souvenir Program Dedication of Iconostas and Decorated Interior of SS. Peter and Paul's Orthodox Greek Catholic Church, Windber, Pa., June 4, 1950.* Windber, Pa.: SS. Peter and Paul Orthodox Greek Catholic Church, 1950.

———. *Twenty-Fifth Anniversary of SS. Peter and Paul Orthodox Greek Catholic Church, Windber, Pennsylvania, 1936–1951.* Windber, Pa.: SS. Peter and Paul Orthodox Greek Catholic Church, 1951.

St. Anthony of Padua Parish, Windber, Pa. *Golden Jubilee, 1908–1958, St. Anthony of Padua Parish, Windber, Pennsylvania.* Windber, Pa.: St. Anthony of Padua Parish, 1958.

———. *40th Anniversary St. Anthony's Church, Windber, Pennsylvania, 1908–1948.* Windber, Pa.: St. Anthony of Padua Parish, 1948.

St. Mary's Hungarian Roman Catholic Church, Windber, Pa. *25th Silver Anniversary Souvenir of the Ordination to the Holy Priesthood of Rev. Stephen M. Hegedus, Pastor, Wednesday, June 1, 1955, And the 41st Anniversary of St. Mary Church, 705 Somerset Avenue, Windber, Pa., Sunday, June 5,1955.* Windber, Pa.: St. Mary's Hungarian Roman Catholic Church, 1955.

———. *60th Anniversary, 1914–1974, St. Mary's Hungarian Roman Catholic Church, Windber, Penna., August 18, 1974. Souvenir of Rev. Joseph Vadas, O.F.M., Farewell Dinner May 23, 1976.* Windber, Pa.: St. Mary's Hungarian Roman Catholic Church, 1974.

[Surkosky, Edward] *Anniversary Book of the Diamond Jubilee of St. Mary's Byzantine Church, Windber, Pennsylvania,1900–1975, October 12, 1975.* Windber, Pa.: St. Mary's Byzantine Church, 1975.

United Mine Workers of America. *Proceedings of the Twenty-Fifth Consecutive and Second Biennial Convention of the United Mine Workers of America Held in the City of Indianapolis, Indiana, January 18 to February 1, 1916 Inclusive,* 2 vols. Indianapolis: Bookwalter-Ball Printing Co., 1916.

United Mine Workers of America, District 2. *Annual Report of Richard Gilbert, Secretary Treasurer, District No. 2, United Mine Workers of America, January 1st 1923 to December 31st, 1923.* DuBois, Pa.: Gray Printing Co., 1924.

———. *The Government of Coal.* Clearfield, Pa.: Kurtz Bros., n.d.

———. *Minutes of the Proceedings of Special Convention of the United Mine Workers of America, District No. 2, Held at the Avenue Theatre, DuBois, Pa., Commencing on Tuesday, April 17th, 1923.* DuBois, Pa.: Gray Printing Co., 1923.

———. *Why the Miners' Program?* Clearfield, Pa.: UMWA District 2, 1921.

Young Men's Christian Association, Pennsylvania. *Yearbook of the Young Men's Christian Associations of North America,* annual yearbooks, 1905–1917. N.p.: n.d.

Newspapers and Journals

Amerikansky Russky Viestnik, 1906

Black Diamond, 1896–1906, 1922, 1936

Coal Age, 1911–1936

Coal Trade Bulletin

Coal Trade Journal, 1899–1906

The Imperial Night-Hawk, 1923–1924

Jednota, 1906

Johnstown Democrat, 1898–1912, 1919, 1922

Johnstown Ledger

Johnstown Tribune

Johnstown Tribune Democrat

The Kourier, 1924–1928

National Labor Tribune, 1906

New York Call

New York Herald

New York Times, 1895–1936

Penn-Central News, 1922

Philadelphia Times

"Scrapbook of Berwind-White advertisements in many different foreign-language newspapers," located at Windber Museum, Windber, Pa.

Slovák v Amerike, 1906

Somerset County Democrat, 1894–1912

Somerset Herald, 1906, 1922, 1933

United Mine Workers Journal, 1891–1940

Windber Daily Era, 1903

Windber Era, 1902–1905, 1907, 1909, 1912, 1914–1917, 1924–1928, 1935–1936, 1942

Windber Journal, 1903–1904

The World, 1922

Dissertations

Astorino, Samuel John. "The Decline of the Republican Dynasty in Pennsylvania, 1929–1934." Ph.D. diss., University of Pittsburgh, 1962.

Fox, Harry Donald, Jr. "Thomas T. Haggerty and the Formative Years of the United Mine Workers of America." Ph.D. diss., West Virginia University, 1975.

Kalassay, Louis A. "The Educational and Religious History of the Hungarian Reformed Church in the United States." Ph.D. diss., University of Pittsburgh, 1939.

Singer, Alan. "Which Side Are You On? Ideological Conflict in the United Mine Workers of America, 1919–1928." Ph.D. diss., Rutgers University, 1982.

Stolarik, Marian Mark. "Immigration and Urbanization: The Slovak Experience, 1870–1918." Ph.D. diss., University of Minnesota, 1974.

Williams, Bruce T. "Underground Bituminous Coal Miners of Cambria County, Pennsylvania." Ph.D. diss., University of Pittsburgh, 1975.

Winograd, Leonard. "The Horse Died at Windber, A History of Johnstown Jewry." D.H.L. diss., Hebrew Union College–Jewish Institute of Religion, 1966.

Secondary Works

Abramson, Harold J. *Ethnic Diversity in Catholic America.* New York: John Wiley and Sons, 1973.

Alcamo, Frank Paul. *The Windber Story: A 20th Century Model Pennsylvania Coal Town.* N.p.: n.p., 1983.

Andrews, J. Cutler. "The Gilded Age in Pennsylvania." *Pennsylvania History* 34 (1967): 1–24.

Auerbach, Jerold S. *Labor and Liberty: The LaFollette Committee and the New Deal.* Indianapolis: Bobbs-Merrill Co., 1966.

Balch, Emily Greene. *Our Slavic Fellow Citizens.* New York: Charities Publication Committee, 1910; reprint, New York: Arno Press and New York Times, 1969.

Beers, Paul B. *Pennsylvania Politics Today and Yesterday: The Tolerable Accommodation.* University Park: The Pennsylvania State University Press, 1980.

Beik, Mildred A. "The Competition for Ethnic Community Leadership in a Pennsylvania Bituminous Coal Town, 1890s-1930s." In *Sozialgeschichte des Berghaus im 19. und 20. Jahrhundert [Towards a Social History of Mining in the 19th and 20th Centuries]*, ed. Klaus Tenfelde, 223–241. Munich: C. H. Beck, 1992.

———. "The UMWA and New Immigrant Miners in Pennsylvania Bituminous: The Case of Windber." In *A Model of Industrial Solidarity? The United Mine Workers of America, 1890–1990*, ed. John H. M. Laslett. University Park: The Pennsylvania State University Press, 1996.

Bell, Thomas. *Out of This Furnace.* Little, Brown and Company, 1941; reprint, Pittsburgh: University of Pittsburgh Press, 1976.

Bernstein, Irving. *The Lean Years: A History of the American Worker, 1920–1933.* Boston: Houghton Mifflin, 1960.

———. *The Turbulent Years: A History of the American Worker, 1933–1941.* Boston: Houghton Mifflin, 1970.

Blankenhorn, Heber. *The Strike for Union.* New York: H. W. Wilson Company, 1924; reprint, New York: Arno and New York Times, 1969.
Blatz, Perry. *Democratic Miners: Work and Labor Relations in the Anthracite Industry, 1875–1925.* Albany: State University of New York Press, 1994.
Bodnar, John. *Immigration and Industrialization: Ethnicity in an American Mill Town, 1870–1940.* Pittsburgh: University of Pittsburgh, 1977.
———. "Immigration and Modernization: The Case of Slavic Peasants in Industrial America." *Journal of Social History* 10 (1976): 44–71.
———. "Immigration, Kinship, and the Rise of Working-Class Realism in Industrial America." *Journal of Social History* 14 (Fall 1980): 45–65.
———. *The Transplanted: A History of Immigrants in Urban America.* Bloomington: Indiana University Press, 1985.
———. *Workers' World: Kinship, Community and Protest in an Industrial Society, 1900–1940.* Baltimore: Johns Hopkins University Press, 1982.
Bonner, Edwin B. "The New Deal Comes to Pennsylvania: The Gubernatorial Election of 1934." *Pennsylvania History* 27 (January 1960): 44–68.
Boyer, Richard O., and Herbert M. Morais. *Labor's Untold Story.* New York: United Electrical, Radio, and Machine Workers of America, 1955.
Brody, David. "Labour Relations in American Coal Mining: An Industry Perspective." In *Workers, Owners, and Politics in Coal Mining,* ed. Gerald D. Feldman and Klaus Tenfelde, 74–117. Munich: Berg Publishers, 1990.
———. *Steelworkers in America: The Nonunion Era.* New York: Harper and Row, 1960.
Brophy, John. *A Miner's Life.* Edited by John O. P. Hall. Madison: University of Wisconsin Press, 1964.
Bukowczyk, John J. *And My Children Did Not Know Me: A History of the Polish-Americans.* Bloomington: Indiana University Press, 1987.
———. "Polish Rural Culture and Immigrant Working Class Formation, 1880–1914." *Polish American Studies* 41 (Autumn 1984): 23–44.
Byington, Margaret. *Homestead: The Households of a Mill Town.* N.p.: The Russell Sage Foundation, 1910; reprint with new intro., Samuel P. Hays. Pittsburgh: University of Pittsburgh, 1974.
Catholic Encyclopedia, 1909 ed. S.v. "Altoona, Diocese of," by Morgan M. Sheedy.
———. S.v. "Greek Catholics in America," by Andrew J. Shipman.
———. S.v. "Hungarian Catholics in America," by Andrew J. Shipman.
———. S.v. "Slavs in America," by Andrew J. Shipman.
Chalmers, David M. *Hooded Americanism: The History of the Ku Klux Klan.* 3rd ed. Durham, N.C.: Duke University Press, 1987.
Coleman, McAlister. *Men and Coal.* Foreword by John Chamberlain. New York: Farrar and Rinehart, 1943.
Coode, Thomas H., John F. Bauman, et al., eds. *People, Poverty, and Politics: Pennsylvanians During the Great Depression.* Lewisburg, Pa.: Bucknell University Press, 1981.
Corbin, David Alan. *Life, Work, and Rebellion in the Coal Fields: The Southern West Virginia Miners, 1880–1922.* Urbana: University of Illinois Press, 1981.
Dix, Keith. *What's a Coal Miner to Do? The Mechanization of Coal Mining.* Pittsburgh: University of Pittsburgh Press, 1988.
Dougherty, James. *The Struggle for an American Way of Life: Coal Miners and Operators in Central Pennsylvania, 1919–1933.* Produced, directed, and written by Jim

Dougherty. 56 min. Indiana University of Pennsylvania Folklife Documentation Center: Indiana, Pa., 1992. Videocassette.

Doyle, Fred C., ed. *50th Anniversary Windber, Pa.* Windber: N.p., 1947.

Dubofsky, Melvyn, and Warren Van Tine. *John L. Lewis: A Biography.* New York: Quadrangle/New York Times Book Co., 1977.

Dyrud, Keith P., Michael Novak, and Rudolph J. Vecoli, eds. *The Other Catholics.* New York: Arno Press, 1978.

Eavenson, Howard N. *The First Century and a Quarter of American Coal Industry.* Pittsburgh: privately printed, Koppers Building, 1942.

Evans, Chris. *History of United Mine Workers of America.* 2 vols. Indianapolis: n.p., 1918–1921.

Federal Council of the Churches of Christ in America. Commission on the Church and Social Service. *The Coal Controversy.* N.p.: Federal Council of the Churches of Christ, 1922.

Fenton, Edwin. *Immigrants and Unions, a Case Study: Italians and American Labor, 1870–1920.* New York: Arno Press, 1975.

Fox, Maier B. *United We Stand: The United Mine Workers of America, 1890–1990.* Washington, D.C.: United Mine Workers of America, 1990.

Fritz, Wilbert G., and Theodore A. Veenstra. *Regional Shifts in the Bituminous Coal Industry with Special Reference to Pennsylvania.* Pittsburgh: University of Pittsburgh Bureau of Business Research, 1935.

Gabaccia, Donna. "*The Transplanted*: Women and Family in Immigrant America." *Social Science History* 12 (Fall 1988): 243–253.

Galush, William J. "Faith and Fatherland: Dimensions of Polish-American Ethnoreligion, 1875–1975." In *Immigrants and Religion in Urban America,* ed. Randall M. Miller and Thomas D. Marzik, 84–102. Philadelphia: Temple University Press, 1977.

Glück, Elsie. *John Mitchell Miner: Labor's Bargain with the Gilded Age.* New York: John Day Company, 1929.

Göbel, Thomas. "Becoming American." *Labor History* 29 (Spring 1988): 173–198.

Gompers, Samuel. "Russianized West Virginia—Corporate Perversions of American Concepts of Liberty and Human Justice—Organized Labor to the Rescue." *American Federationist* (1913): 825–835.

Goodrich, Carter. *The Miner's Freedom: A Study of the Working Life in a Changing Industry.* Boston: Marshall Jones Company, 1925.

Gordon, David M., Richard Edwards, and Michael Reich. *Segmented Work, Divided Workers: The Historical Transformation of Labor in the United States.* Cambridge: Cambridge University Press, 1982.

Graebner, William. *Coal-Mining Safety in the Progressive Period: The Political Economy of Reform.* Lexington: University Press of Kentucky, 1976.

Greater Pennsylvania Council. *The Decline of the Bituminous Coal Industry in Pennsylvania.* Harrisburg: Greater Pennsylvania Council, 1932.

Green, Archie. *Only a Miner: Studies in Recorded Coal-Mining Songs.* Urbana: University of Illinois Press, 1972.

Greene, Victor R. *American Immigrant Leaders, 1800–1910: Marginality and Identity.* Baltimore: Johns Hopkins University Press, 1987.

———. "'Becoming American': The Role of Ethnic Leaders—Swedes, Poles, Italians, and Jews." In *The Ethnic Frontier: Essays in the History of Group Survival in Chicago and the Midwest*, ed. Melvin G. Holli and Peter d'A. Jones, 143–175. Grand Rapids, Mich.: William B. Eerdmans Publishing Company, 1977.

———. *For God and Country: The Rise of Polish and Lithuanian Ethnic Consciousness in America.* Madison: State Historical Society of Wisconsin, 1975.

———. *The Slavic Community on Strike: Immigrant Labor in Pennsylvania Anthracite.* Notre Dame: University of Notre Dame, 1968.

Gutman, Herbert. *Work, Culture, and Society in Industrializing America: Essays in American Working-Class and Social History.* New York: Vintage Books, 1966.

Guyer, John P. *Pennsylvania's Cossacks and the State Police.* N.p.: n.p., 1924.

Hamilton, Walton H., and Helen R. Wright. *The Case of Bituminous Coal.* New York: Macmillan Co., 1925.

Hapgood, Powers. *In Non-Union Mines: The Diary of a Coal Digger in Central Pennsylvania August–September, 1921.* New York: Bureau of Industrial Research, 1922.

Harvard Encyclopedia of American Ethnic Groups. Cambridge, Mass.: Belknap Press of Harvard University Press, 1980. S.v. "American Identity and Americanization," by Philip Gleason, 31–58.

———. S.v., "Carpatho-Rusyns," by Paul Robert Magocsi, 200–210.

———. S.v., "Hungarians," by Paula Benkart, 462–471.

———. S.v., "Italians," by Humbert S. Nelli, 545–560.

———. S.v., "Poles," by Victor Greene, 787–803.

———. S.v., "Russians," by Paul Robert Magocsi, 885–894.

———. S.v., "Slovaks," by Marian Mark Stolarik, 926–934.

Hathaway, Dale A. *Can Workers Have a Voice? The Politics of Deindustrialization in Pittsburgh.* University Park: The Pennsylvania State University Press, 1993.

Higham, John, ed. *Ethnic Leadership in America.* Baltimore: Johns Hopkins University Press, 1978.

———. "Integration vs. Pluralism: Another American Dilemma." *The Center Magazine* 7 (July–August 1974): 67–73.

———. *Strangers in the Land: Patterns of American Nativism 1860–1925.* New Brunswick, N.J.: Rutgers University Press, 1955.

Hinrichs, A. F. *The United Mine Workers of America and the Non-Union Coal Fields.* New York: Columbia University Press, 1923.

The History of Berwind (Philadelphia: Berwind Group, 1993).

Hoerder, Dirk. *Labor Migration in the Atlantic Economies: The European and North American Working Classes During the Period of Industrialization.* Westport, Conn.: Greenwood Press, 1985.

Hourwich, Isaac A. *Immigration and Labor: The Economic Aspects of European Immigration to the United States.* New York: G. P. Putnam's Sons, 1912.

Johnson, James P. "Reorganizing the United Mine Workers of America in Pennsylvania During the New Deal." *Pennsylvania History* 37 (April 1970): 117–132.

Keller, Richard C. *Pennsylvania's Little New Deal.* New York: Garland Publishing, 1982.

[Klein, H. M. J.]. *The Pennsylvania Young Men's Christian Association: A History, 1854–1950.* Kennett Square, Pa.: Kennett News and Advertiser Press, 1950.

Klein, Philip S., and Ari Hoogenboom. *A History of Pennsylvania.* 2nd ed. University Park: The Pennsylvania State University Press, 1973.

Korman, Gerd. *Industrialization, Immigrants, and Americanizers: The View from Milwaukee, 1866–1921.* Madison: State Historical Society of Wisconsin, 1967.

Lambie, Joseph T. *From Mine to Market: The History of Coal Transportation on the Norfolk and Western Railway*. New York: New York University Press, 1954.

Lane, Winthrop D. *Civil War in West Virginia*. New York: B. W. Huebsch, 1921; reprint ed., New York: Arno and New York Times, 1969.

Laslett, John H. M., ed. *A Model of Industrial Solidarity? The United Mine Workers of America, 1890–1990*. University Park: The Pennsylvania State University Press, 1996.

Lewin, Kurt. "The Consequences of Authoritarian and Democratic Leadership." In *Studies in Leadership: Leadership and Democratic Action*, ed. Alvin W. Gouldner, 409–417. New York: Harper and Brothers, 1950.

———. *Resolving Social Conflicts: Selected Papers on Group Dynamics*. Edited by Gertrud Weiss Lewin. Foreword by Gordon W. Allport. New York: Harper and Brothers, 1948.

Lewis, John L. *The Miners' Fight for American Standards*. Indianapolis, Ind.: Bell Publishing Co., 1925.

Loucks, Emerson Hunsberger. *The Ku Klux Klan in Pennsylvania: A Study in Nativism*. Harrisburg: Telegraph Press, 1936.

Lukács, George. *History and Class Consciousness: Studies in Marxist Dialectics*. Trans. Rodney Livingstone. Cambridge, Mass.: MIT Press, 1971.

Magocsi, Paul Robert. *Our People: Carpatho-Rusyns and Their Descendants in North America*. Preface by Oscar Handlin. Toronto: Multicultural History Society of Ontario, 1984.

Majumdar, Shyamal K., and E. Willard Miller, eds. *Pennsylvania Coal: Resources, Technology, and Utilization*. N.p.: Pennsylvania Academy of Science, 1983.

Marchbin, Andrew A. "Hungarian Activities in Western Pennsylvania." *Western Pennsylvania Historical Magazine* 23 (1940): 163–174.

Marcus, Irwin M., James P. Dougherty, and Eileen M. Cooper. "Judge Jonathan Langham and the Use of the Labor Injunction in Indiana County, 1919–1931." *The Pennsylvania Magazine of History and Biography* 118 (January–April 1994): 63–85.

Meyer, Stephen. "Adapting the Immigrant to the Line: Americanization in the Ford Factory, 1914–1921." *Journal of Social History* 14 (Fall 1980): 67–82.

Meyerhuber, Carl I., Jr. *Less than Forever: The Rise and Decline of Union Solidarity in Western Pennsylvania, 1914–1948*. London: Associated University Press, 1987.

Miller, George. *A Pennsylvania Album: Picture Postcards, 1900–1930*. University Park: The Pennsylvania State University Press, 1979.

Montgomery, David. *The Fall of the House of Labor: The Workplace, the State, and American Labor Activism, 1865–1925*. Cambridge: Cambridge University Press, 1987.

———. "The 'New Unionism' and the Transformation of Workers' Consciousness in America, 1909–1922." *Journal of Social History* 7 (Summer 1974): 509–529.

———. "Strikes in Nineteenth-Century America." *Social Science History* 4 (Winter 1980): 81–104.

———. *Workers' Control in America; Studies in the History of Work, Technology, and Labor Struggles*. Cambridge: Cambridge University Press, 1979.

Morawska, Ewa. *For Bread with Butter: The Life-Worlds of East Central Europeans in Johnstown, Pennsylvania, 1890–1940*. Cambridge: Cambridge University Press, 1985.

———. " 'For Bread with Butter': Life-Worlds of Peasant-Immigrants from East Central Europe, 1880–1914." *Journal of Social History* 17 (Spring 1984): 387–404.

———. "The Internal Status Hierarchy in the East European Immigrant Communities of Johnstown, Pa., 1890–1930's." *Journal of Social History* 16 (Fall 1982): 75–107.

Myrdal, Gunnar. *An American Dilemma: The Negro Problem and Modern Democracy*. New York: Harper and Row, 1944.

———. "The Case Against Romantic Ethnicity." *The Center Magazine* 7 (July–August 1974): 26–30.

———. "Mass Passivity in America." *The Center Magazine* 7 (March–April 1974): 72–75.

Nash, Michael. *Conflict and Accommodation: Coal Miners, Steel Workers, and Socialism, 1890–1920*. Westport, Conn.: Greenwood Press, 1982.

"Naši pioneri." *Národný Kalendár* 24 (1927): 140–148.

Novak, Michael. *The Rise of the Unmeltable Ethnics: Politics and Culture in the Seventies*. New York: Macmillan Company, 1972.

Nyden, Linda. "Black Miners in Western Pennsylvania, 1925–1931: The National Miners Union and the United Mine Workers of America." *Science and Society* 41 (Spring 1977): 69–101.

O'Grady, Joseph P., ed. *The Immigrants' Influence on Wilson's Peace Policies*. Lexington: University Press of Kentucky, 1967.

Olson, James S. *Catholic Immigrants in America*. Chicago: Nelson-Hall, 1987.

Pankuch, John H. "Fifty Years of Progress." *Pamätnica N.S.S. 1890–1940, Národný Kalendár* 47 (1940): 86–96.

Paučo, Joseph, ed. *60 Years of the Slovak League of America*. Middletown, Pa.: Slovak League of America, 1967.

Pritchard, Paul W. "William B. Wilson, Master Workman." *Pennsylvania History* 12 (April 1945): 81–109.

Puskás, Julianna. *From Hungary to the United States, 1880–1914*. Budapest: Akadémiai Kiadó, 1982.

———. *Kivándorló Magyarok az Egyesült Államokban, 1880–1940*. Budapest: Akadémiai Kiadó, 1982.

Roberts, Peter. "The Y.M.C.A. Among Immigrants." *The Survey*, 15 February 1913, 697–700.

Rochester, Anna. *Labor and Coal*. New York: International Publishers, 1931.

Roy, Andrew. *A History of the Coal Miners of the United States from the Development of the Mines to the Close of the Anthracite Strike of 1902, Including a Brief Sketch of Early British Miners*. Columbus, Ohio: J. L. Trauger Printing Company, 1905; reprint, Westport, Conn.: Greenwood Press, 1970.

Sargent, A. Tappan. "Enduring Coal Enterprise." *Black Diamond*, 28 March 1936, 21–33.

Schwieder, Dorothy. *Black Diamonds: Life and Work in Iowa's Coal Mining Communities, 1895–1925*. Ames: Iowa State University Press, 1983.

Seltzer, Curtis. *Fire in the Hole: Miners and Managers in the American Coal Industry*. Lexington: University Press of Kentucky, 1985.

Serrin, William. *Homestead: The Glory and Tragedy of an American Steel Town*. New York: Times Books, 1992.

Shalloo, J. P. *Private Police with Special Reference to Pennsylvania*. Philadelphia: American Academy of Political and Social Science, 1933.

Singer, Alan. "Communists and Coal Miners: Rank-and-File Organizing in the United

Mine Workers of America During the 1920s." *Science and Society* 55 (Summer 1991): 132–157.
———. "John Brophy's 'Miners' Program': Workers' Education in UMWA District 2 During the 1920's." *Labor Studies Journal* 13 (Winter 1988), 50–64.
Smith, Arthur D. Howden. *Men Who Run America: A Study of the Capitalistic System and Its Trends Based on Thirty Case Histories.* Indianapolis: Bobbs-Merrill Co., 1935.
Stevens, Sylvester K. *Pennsylvania: Titan of Industry.* New York: Lewis Historical Publishing Co., 1948.
Stolarik, Marian Mark. "From Field to Factory: The Historiography of Slovak Immigration to the United States." *International Migration Review* 10 (Spring 1976): 81–102.
———. *The Role of American Slovaks in the Creation of Czecho-Slovakia, 1914–1918.* Cleveland, Ohio: Slovak Institute, 1968.
Tams, W. P., Jr. *The Smokeless Coal Fields of West Virginia.* Morgantown: West Virginia University Library, 1963.
Thomas, William I., and Florian Znaniecki. *The Polish Peasant in America,* 2 vols. New York: Alfred A. Knopf, 1927.
Thompson, Edward P. *The Making of the English Working Class.* New York: Vintage Books, 1963.
Trotter, Joe William, Jr. *Coal, Class, and Color: Blacks in West Virginia, 1915–1932.* Urbana: University of Illinois Press, 1990.
Vecoli, Rudolph. "Cult and Occult in Italian-American Culture: The Persistence of a Religious Heritage." In *Immigrants and Religion in Urban America,* ed. Randall M. Miller and Thomas D. Marzik, 25–47. Philadelphia: Temple University Press, 1977.
———. "Prelates and Peasants: Italian Immigrants and the Catholic Church." *Journal of Social History* 2 (Spring 1969): 217–268.
Warne, Frank Julian. *The Slav Invasion and the Mine Workers: A Study in Immigration.* Philadelphia: J. B. Lippincott Co., 1904.
Williams, Bruce T., and Michael D. Yates. *Upward Struggle: A Bicentennial Tribute to Labor in Cambria and Somerset Counties.* Johnstown, Pa.: Johnstown Regional Central Labor Council, 1976.

Index

About the Author

Mildred Allen Beik is the daughter of a Pennsylvania coal miner whose family emigrated from Hungary early in the twentieth century. This native of Mine No. 40, Windber, earned a Ph.D. in history at Northern Illinois University in 1989 and is currently an independent scholar living in Atlanta. She has taught at a variety of universities, most recently at the Georgia Institute of Technology, and has been a consultant and participant in various public history workshops, including occasional Oral History and Visual Ethnography Institutes sponsored by Indiana University of Pennsylvania. To honor the struggles of immigrant and American miners for freedom and social justice, she has nominated Windber for designation by the Newberry Library and National Park Service as a National Labor Landmark.